Advances in the Preparation and Characterization Techniques for Developing Coating Materials and Applications

Advances in the Preparation and Characterization Techniques for Developing Coating Materials and Applications

Guest Editors

Sheng-Rui Jian
Phuoc Huu Le

MDPI

Basel • Beijing • Wuhan • Barcelona • Belgrade • Novi Sad • Cluj • Manchester

Guest Editors

Sheng-Rui Jian
Department of Materials
Science and Engineering
I-Shou University
Kaohsiung
Taiwan

Phuoc Huu Le
Department of Physics
and Biophysics
Can Tho University of
Medicine and Pharmacy
Can Tho
Vietnam

Editorial Office
MDPI AG
Grosspeteranlage 5
4052 Basel, Switzerland

This is a reprint of the Special Issue, published open access by the journal *Coatings* (ISSN 2079-6412), freely accessible at: https://www.mdpi.com/journal/coatings/special_issues/Coating_Materials_Application.

For citation purposes, cite each article independently as indicated on the article page online and as indicated below:

Lastname, A.A.; Lastname, B.B. Article Title. *Journal Name* **Year**, *Volume Number*, Page Range.

ISBN 978-3-7258-4315-2 (Hbk)
ISBN 978-3-7258-4316-9 (PDF)
https://doi.org/10.3390/books978-3-7258-4316-9

Contents

About the Editors

Sheng-Rui Jian

Sheng-Rui Jian (Prof. Dr.) is a Distinguished Professor at the Department of Materials Science and Engineering, I-Shou University, Kaohsiung, Taiwan. He received his Ph.D. degree in Electrophysics from National Chiao Tung University, Hsinchu, Taiwan, in 2006. His research focuses on the nanoscale mechanical deformation behaviors and mechanisms of semiconductor materials and thin films using microscopic techniques and molecular dynamics simulations. He has published more than 180 journal papers and was awarded the Word's Top 2% Scientists (Career Impact) 2021 and 2022. Jian can be contacted by email at srjian@isu.edu.tw and srjian@gmail.com.

Phuoc Huu Le

Phuoc Huu Le (Assoc. Prof. Dr.) is an Associate Professor at the Department of Physics and Biophysics, Faculty of Basic Sciences, Can Tho University of Medicine and Pharmacy, Can Tho City, Vietnam. He received his Ph.D. degree in Materials Science and Engineering from National Chiao Tung University, Taiwan, in July 2014. His research interests have been concentrated on thin films and characterizations (i.e., thermoelectric, topological insulator, transparent conducting oxide, complex oxide; nanomechanical, magnetotransport, ultrafast dynamics, and other functional properties of thin-film materials), and nanomaterials and applications (i.e., Nanophotocatalysts for degradation of emerging pollutants/contaminants; nanomaterials for photochemical water splitting, energy conversion and storage, and electrochemical sensors). Dr. Le has published more than 50 journal papers and 8 book chapters (Google Scholar: >1240 citations, h-index: 20).

Preface

The field of coating materials has witnessed significant advancements, driven by the need for improved performance, durability, and functionality across various applications. This reprint, derived from the Special Issue "Advances in the Preparation and Characterization Techniques for Developing Coating Materials and Applications" in *Coatings*, presents a curated collection of research articles that reflect the current state-of-the-art in coating technologies.

The chapters are organized to guide the reader through fundamental growth and characterization techniques, insights into corrosion behavior and protective coatings, the impact of radiation on structural integrity, applications in biomedical and environmental contexts, and the development of advanced functional coatings. Each contribution offers a unique perspective, collectively providing a comprehensive overview of the challenges and innovations in the field.

We extend our sincere gratitude to all authors for their valuable contributions and to the reviewers for their critical evaluations, which have ensured the quality and relevance of the content. We hope this compilation serves as a valuable resource for researchers, engineers, and practitioners engaged in the development and application of coating materials.

Sheng-Rui Jian and Phuoc Huu Le
Guest Editors

coatings

MDPI

Article

Experimental and Simulation Studies on Thermal Shock of Multilayer Thermal Barrier Coatings with an Intermediate Transition Layer at 1500 °C

Pengpeng Liu [1,†], Shilong Yang [1,†], Kaibin Li [1,*], Weize Wang [1,2,*], Yangguang Liu [1] and Ting Yang [1]

1 Key Laboratory of Pressure System and Safety, Ministry of Education, East China University of Science and Technology, Shanghai 200237, China; y84220050@mail.ecust.edu.cn (P.L.); y84220052@mail.ecust.edu.cn (S.Y.); y10210065@mail.ecust.edu.cn (Y.L.); y10200060@mail.ecjust.edu.cn (T.Y.)

2 Shanghai Institute of Aircraft Mechanics and Control, Shanghai 200092, China

* Correspondence: y10210064@mail.ecust.edu.cn (K.L.); wangwz@ecust.edu.cn (W.W.)

† These authors contributed equally to this work.

Abstract: Strain tolerance is a crucial factor affecting the thermal life of coatings, and a higher strain tolerance can effectively alleviate the thermal stresses on coatings during thermal shock. To improve the strain tolerance, the coating structure was optimized by introducing an intermediate transition layer in this study. The intermediate transition layer material was prepared using a 1:1 volume ratio mixture of 6–8 wt. % Yttria-stabilized zirconia (YSZ) and NiCrAlY powders in the experiments. The coating structure consisted of an Al_2O_3-$GdAlO_3$ (AGAP) anti-erosion layer, a YSZ layer, an intermediate transition layer, and a bonding layer from top to bottom. After thermal shock experiments at 1500 °C, the coatings with the addition of the intermediate transition layer exhibited different failure modes, with the crack location shifting from between the YSZ and the bonding layer to within the intermediate transition layer, compared to the coatings without the intermediate transition layer. Finite element simulation analysis showed that the intermediate transition layer effectively increased the strain tolerance of the coating and significantly reduced the thermal stress. Furthermore, incorporating an embedded micron agglomerated particle-based (EMAP) thermal barrier coating structure into the intermediate transition layer effectively alleviated thermal stresses and enhanced the coating's thermal insulation performance.

Keywords: thermal barrier coatings; multilayer; thermal shock; finite element analysis

1. Introduction

Thermal barrier coatings (TBCs) are widely used in the hot-end components of aircraft engines to provide adequate thermal protection and insulation [1–4]. TBCs can increase the operating temperature of engines, resulting in significant improvements in engine efficiency and performance. Typical TBC systems consist of a super-alloy substrate, a metallic bond coat (BC) layer (MCrAlY, M = Ni or/and Co), and a ceramic topcoat (TC) layer [5,6]. In general, TBCs are deposited by atmospheric plasma spraying (APS) and have lamellar microstructures. Among them, MCrAlY (M = Ni or/and Co) is widely used for the metal bonding layer material; its primary function is to relieve the interfacial stresses and to improve the compatibility of the thermal expansion between the TC layer and the substrate. The TC layer, on the other hand, is mainly responsible for thermal insulation and possesses high heat resistance and low thermal conductivity [7,8]. In addition, considering the harsh environment in which the high-temperature alloys are exposed, the material for the TC layer also needs to have good corrosion resistance to ensure long-term stability. Among the known materials, 6–8 (wt. %) Yttria-stabilized zirconia (YSZ) is the most widely used ceramic layer material [9]. This is due to it exhibiting high toughness, heat resistance, strength, a coefficient of thermal expansion, and low cost.

However, the working temperature of YSZ coatings should not exceed 1200 °C for long periods of time. Otherwise, sintering will occur in the coating, leading to microstructural changes such as grain growth and reduced porosity. For YSZ, high-temperature sintering can trigger phase transitions, particularly from the stable cubic phase to the monoclinic phase. This phase change is often accompanied by volume expansion, which increases internal stresses and can result in crack formation and delamination. These microstructural changes significantly degrade the thermal performance and structural integrity of the coating, thereby reducing its thermal lifespan [10,11]. With the rapid development of advanced gas turbine technology, the requirements for the engine operating temperature continue to increase, and the working environment faced by the thermal barrier coatings becomes more and more severe. Therefore, it is essential to find ways to increase the service temperature of thermal barrier coatings [12–14].

Plasma-sprayed coatings are formed by the continuous stacking of solidified flakes, and they show a layered structure because the interfaces of the flakes are not fully bonded, leading to a large number of two-dimensional inter-plate pores between the flakes [15–20]. Many practices and research results show that the cracking failure of thermal barrier coatings prepared by APS technology is mainly related to the following three aspects [21–26]: (1) The characteristic laminated microstructure of APS coatings generates a large number of parallel microcracks inside the coatings. These microcracks reduce the bonding strength between the coating and the metal substrate as well as the cracking resistance of the coating itself, which can easily lead to coating spalling under thermal shock conditions [21,22]. (2) Under thermal shock conditions, thermal stresses are generated inside the coating due to the difference in thermal expansion coefficients of the TC layer and the metal substrate. Meanwhile, microcracks inside the coating will continue to propagate and expand under the effect of thermal shock stress, finally leading to the spalling failure of the coating [23,24]. Furthermore, the sintering of the coating in a high-temperature environment leads to an increase in its stiffness and a decrease in its strain tolerance, which also increases the thermal stresses due to thermal expansion mismatch. (3) Thermal growth oxides (TGOs) formed by oxidation of thermal barrier coatings at high temperatures are also one of the essential causes of coating spallation failure [25,26].

APS-prepared coatings with excellent fracture toughness and strain tolerance can significantly improve the thermal shock life of the coatings [27]. However, due to the loose and porous layered microstructure of APS thermal barrier coatings, the fracture toughness and strain tolerance of the coatings usually show an inverted relationship, making it difficult to achieve the advantages of high strain tolerance and high fracture toughness at the same time [28,29]. For example, in our previous studies [30,31], it was found that the Al_2O_3-$GdAlO_3$-YSZ (AGAP-YSZ) coating system demonstrates superior fracture toughness, while the AGAP-$Gd_2Zr_2O_7$-YSZ (AGAP-GZ-YSZ) coating system exhibits higher strain tolerance. Furthermore, the failure modes of the AGAP-YSZ coating system and the AGAP-GZ-YSZ coating system were completely different under the thermal shock experimental conditions at 1500 °C. The failure of the AGAP-YSZ coating system was mainly due to the excessive thermal stress, which led to interfacial cracking. In contrast, no significant cracking was observed inside the YSZ layer. In contrast, the AGAP-GZ-YSZ system, although with lower internal thermal stress, is prone to crack sprouting and expansion within the GZ layer during thermal shock due to the lower fracture toughness of the GZ layer, which leads to the premature failure of the coating. Overall, the AGAP-YSZ coating system exhibits good fracture toughness. However, cracking occurs due to excessive interfacial thermal stress during thermal shock, while the AGAP-GZ-YSZ system, although with lower thermal stress, leads to premature coating failure due to insufficient fracture toughness of the GZ layer. To achieve a coating system with higher thermal shock life, a coating with good fracture toughness and low thermal stress is needed. To provide excellent fracture toughness of AGAP-YSZ coatings, this study suggests the introduction of an intermediate transition layer in the AGAP-YSZ coating system. Through the intermediate transition layer, it may improve the strain tolerance of the coating system, relieve the thermal stresses inside the coating, and further improve the thermal shock life of the coating. For simplicity, the AGAP-YSZ coating system is denoted as the AGAP-YSZ-BC Layer (AYB) coating, while the AGAP-YSZ-Intermediate Layer coating system is referred to as the AGAP-YSZ-Intermediate Layer-BC Layer (AYIB) coating in this paper. Additionally, a novel embedded micron agglomerated particle-based (EMAP) thermal barrier coating, known as the EMAP coating, with high thermal shock resistance was developed by our group in previous research [32]. The novel structured coating is prepared by controllably embedding a second-phase powder particle with a specific pore structure into the coating matrix formed by conventional spraying during the plasma spraying process to prepare a composite thermal barrier coating with a dense conventional coating matrix and a dispersion of structurally specific secondary phase particles therein. The second-phase particles with specific pore structures are called porous embedded particle clusters (PEPC) [32,33]. The advantage of EMAP coatings over conventional coatings is that the dense coating matrix ensures high fracture toughness of the coating as a whole. The coating's strain tolerance and thermal insulation capacity can be improved by modulating the pore structure of the coating through the second-phase powder particles.

In the experiments conducted in this study, three different spray coating processes for the intermediate transition layer were designed to investigate their impact on the performance of traditional thermal barrier coatings. These processes were labeled as Group A, Group B, and Group C. In these three groups, the spraying parameters of the AGAP layer, YSZ layer, and BC layer were kept consistent. Among them, the spraying parameters of the intermediate transition layer for the coatings in Groups A and C were consistent with those of the BC and YSZ layers, respectively. In contrast, the spraying process parameters of the intermediate transition layer for the coatings in Group B were taken to be the middle value between Groups A and C. For the finite element simulations, the thermal stresses of AYB and AYIB coating systems during the thermal shock test were performed. The structure of the AYIB coating system was further optimized in the finite element simulation based on the EMAP structure strategy, resulting in the formation of the AGAP-YSZ-Intermediate Layer with EMAP coating-BC Layer (AYI(E)B) coating system. The thermal stress and thermal insulating properties at high temperatures of the AYIB and AYI(E)B coating systems were simulated and analyzed. These results may provide some strategies for designing TBC systems with excellent thermal shock resistance.

2. Experiments and Numerical Models

2.1. Coating Preparation

A disc of nickel-based alloy (GH3230) with a diameter of 25.4 mm and a thickness of 3 mm was chosen as the metal substrate for this experiment. Prior to the coating process, the metal substrate was ultrasonically cleaned with ethanol and roughened through sandblasting. To ensure optimal adhesion of the BC layer, the metal substrate was preheated before each coating application. An Al_2O_3/Gd_2O_3 granulated powder was utilized to prepare the top ceramic layer, namely AGAP. The YSZ layer is made of YSZ powder (45–60 μm). The intermediate transition layer is composed of a 1:1 volumetric ratio mixture of NiCrAlY powder (45–106 μm, Beijing, Sunsparying New Material Co., Ltd., Beijing, China) and YSZ powder. The BC layer is made of NiCrAlY powder. All coatings were prepared using the F4 atmospheric plasma spraying equipment (Oerlikon Metco, Zurich, Switzerland). The parameters of the coating process for each layer are shown in Table 1.

Table 1. Plasma spraying parameters.

Parameter	BC	YSZ	Intermediate Layer of A	Intermediate Layer of B	Intermediate Layer of C	AGAP
Current, A	550	600	550	575	600	650
Voltage, V	68.5	67.2	68.5	67.9	67.2	67
Primary gas flow rate, Ar, L/min	50	50	50	45	50	40

Table 1. *Cont.*

Parameter	BC	YSZ	Intermediate Layer of A	Intermediate Layer of B	Intermediate Layer of C	AGAP
Carrier gas flow rate, H_2, L/min	8	9	8	8.5	9	9
Spray distance, mm	120	90	120	105	90	120
Travers speed of gun, mm/s	900	500	900	700	500	900
Powder feeding rate, %	10	20	10	15	20	20
Thickness, μm	120	300	80	80	80	100

2.2. Thermal Shock Test

The thermal shock resistance of the coatings was examined using a gas torch flame thermal shock test. During the experiment, the coating side of the sample was heated by a high-temperature flame generated by a propane-oxygen lance (Shaanxi Dewei Automation Co., Ltd., Xi'an, China). In contrast, the substrate side was cooled using compressed air, which created a large temperature gradient inside the coating [34–37]. All specimens undergo thermal shock testing under identical conditions. The procedure is as follows: On the thermal shock test apparatus, the sample surface is heated by a high-temperature flame, while the backside is cooled using compressed air, typically maintaining a backside temperature of around 1050 °C. The sample surface is heated from room temperature to 1500 °C within 120 s and then held at this temperature for 1000 s. Finally, the sample surface is rapidly quenched with deionized water to bring the temperature down to room temperature. A non-contact infrared thermometer with a range value of 18–1650 °C was used to monitor the samples' coating surface and substrate surface temperatures. Thermal cycle life, as well as forms of failure, were observed experimentally. Usually, the failure of a coating is defined as the number of thermal cycles in which more than 10 percent of the surface of the coating buckles and flakes [38].

2.3. Characterization

In order to observe the sprayed microstructure of the samples as well as the failure mode of the samples after the thermal shock experiments, the samples were cut using a wire cutter (STX-202A, Shenyang Kejing Automation Equipment Co., Ltd., Shenyang, China). The cut samples were analyzed and characterized by scanning electron microscopy (SEM, HITACHI S-3400 N, Hitachi Scientific Instruments Co., Ltd., Tokyo, Japan) after curing with epoxy resin and surface polishing. The SEM images of the samples were post-processed using ImageJ software (version 1.8.0) to obtain the porosity of the coated samples.

2.4. Finite Element Analysis

Thermal barrier coatings generate thermal stresses inside the coating during thermal shock due to the different coefficients of thermal expansion of each material layer. ABAQUS 2022 built a finite element model to calculate the magnitude of thermal stresses inside the coating and their locations to better investigate the failure mode of the sample. The established geometrical model and the schematic of each component of the TC layer in the model are shown in Figure 1. A sequential coupling method is used in the finite element calculation of thermal stress analysis, i.e., heat transfer analysis first obtains the temperature field change inside the coating system during thermal cycling. Then, the temperature field is used to input load conditions during stress analysis to obtain the stress field change inside the coating during thermal cycling [36,37]. The temperature loads were set based on the flame thermal shock test parameters. In the warming phase, the sample surface was increased from room temperature to 1500 °C after 120 s, maintained at this temperature for 1000 s, and finally cooled quickly to room temperature by deionized water. During the thermal shock experiment, convective heat transfer was the dominant mode of heat exchange between the sample and the surrounding environment. The convective heat transfer coefficients were 110 W/m^2·°C during the heating phase and 1000 W/m^2·°C during the cooling phase. Given that convective heat transfer predominates within this temperature range and the contribution of radiative heat transfer is negligible, the impact of thermal radiation can be disregarded in the heat transfer analysis. The model is meshed using generalized plane strain and bilinear quadrilateral elements. When setting the boundary conditions, the left boundary of the model is set as a symmetric boundary, and the other boundaries are accessible [30–33]. The material physical parameters of the AGAP layer, the YSZ layer, the BC layer, and the substrate are set according to the literature [33]. The material of the intermediate transition layer can calculate its thermodynamic parameters using the practical medium theory [39]. The specific calculation results are shown in Table 2.

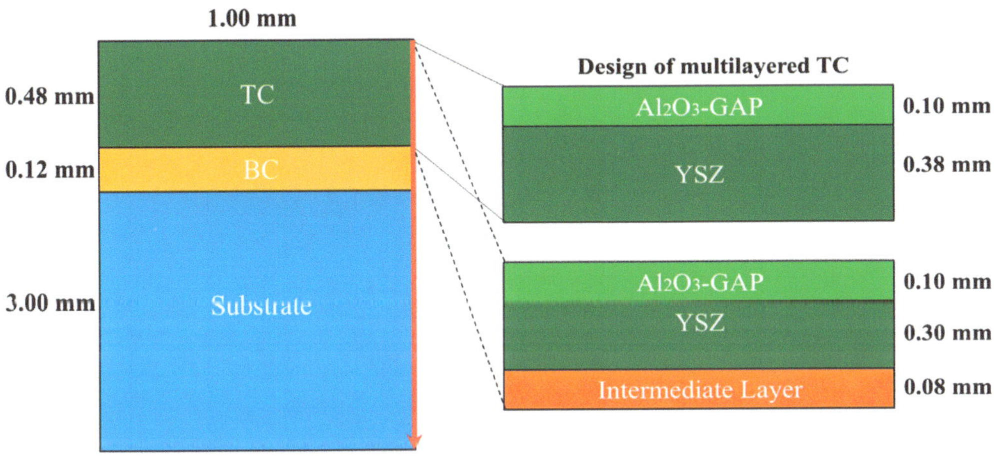

Figure 1. Schematic of the Geometrical Model for Finite Element Simulation (the red arrow represents the path to the data extraction location).

Table 2. Thermodynamic parameters of the intermediate transition layer.

Temperature (°C)	Elastic Modulus (10^9 Pa)	Coefficient of Thermal Expansion (10^{-6} °C^{-1})	Density (10^3 kg/m^3)	Thermal Conductivity (W/m·°C)	Thermal Capacity (J/kg·°C)	Poisson Rate
20	120	12.7	6.69	3.5	425	0.2
200		13.2		4.345		0.2
400		13.5		5.34		0.21
600		14.1		6.575		0.21
800		15.2		7.83		0.22
1000		16.2		8.67		0.23
1100		16.7		9.06		0.23

3. Result and Discussion

3.1. Experiment Characterization

3.1.1. Microscopic Morphology of Sprayed Coatings

The sprayed state micromorphology of three sets of AYIB coating process samples is shown in Figure 2. In Figure 2, we can observe the layered interfaces of the multilayer coatings, indicating that good bonding is maintained between the layers. Due to the characteristics of atmospheric plasma spraying, we can see the presence of pores, unmelted particles, and interlayer microcracks inside the coating. These properties give the thermal barrier coating an excellent thermal insulation capability. The average porosity of the AGAP, YSZ, and intermediate layers is 6.21%, 11.94%, and 14.44% in Group A. The average porosity of the AGAP, YSZ, and intermediate layers is 6.21%, 11.94%, and 14.44% in Group

B. The average porosity of the AGAP, YSZ, and intermediate layers is 6.21%, 11.94%, and 14.44% in Group B. The average porosities of the AGAP layer, YSZ layer, and intermediate layer in Group B samples were 6.68%, 11.95%, and 13.21%. The average porosities of the AGAP layer, YSZ layer, and intermediate layer in Group C samples were 6.44%, 11.26%, and 11.81%.

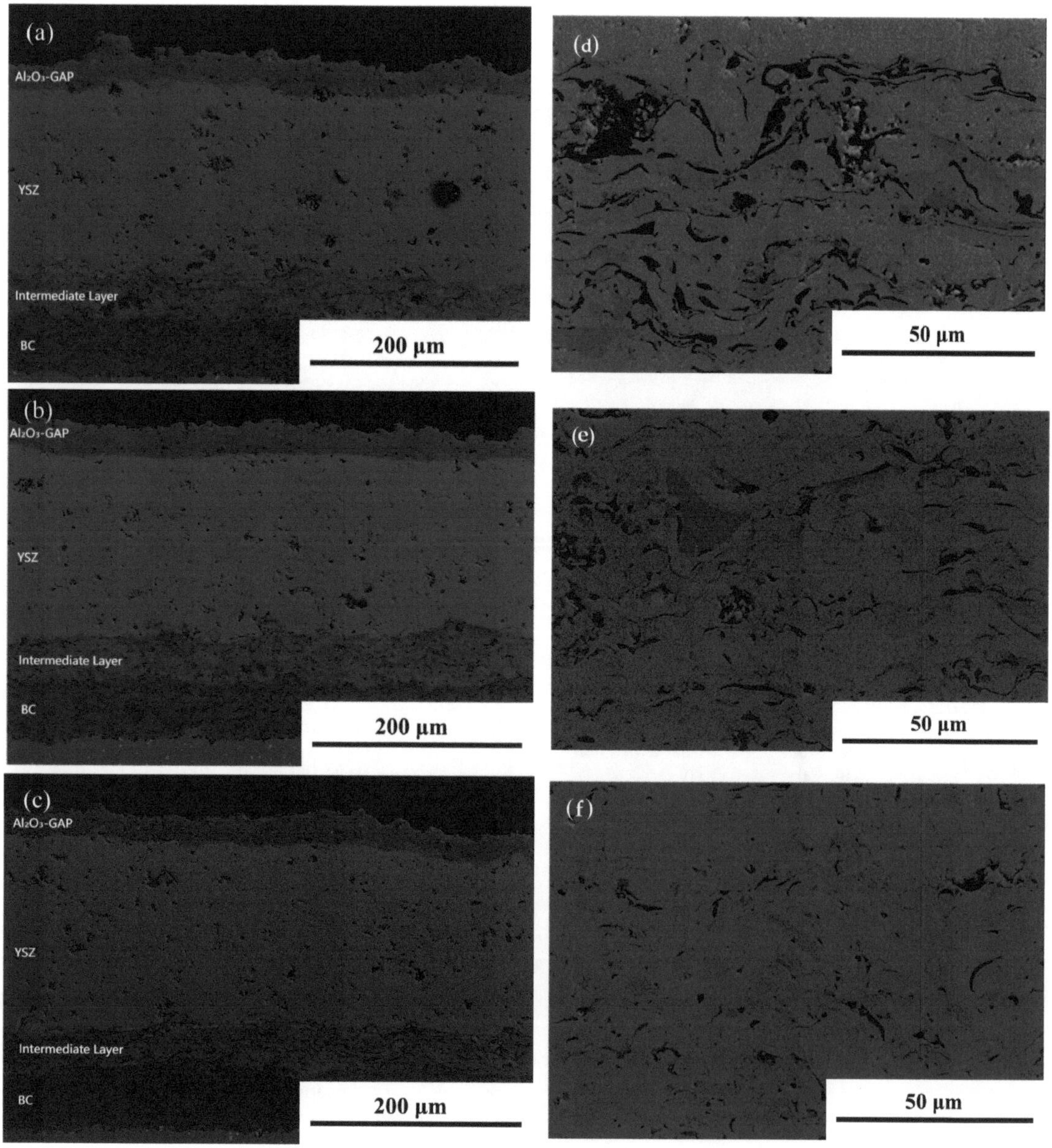

Figure 2. Cross-sectional morphologies of the as-sprayed (**a**,**d**) Group A, (**b**,**e**) Group B, and (**c**,**f**) Group C.

In the three groups of experiments, the spraying power, the speed of moving the gun, and the spraying distance of the intermediate transition layer were different in each group.

In the regulation of the plasma spraying process parameters, it has been shown [27,28] that as the spraying power increases, the melting of the powder in the plasma jet is more adequate, and the flight speed increases, resulting in a smaller number and size of interlayer pores in the coating layer. It has been shown [9] that the deposition efficiency of the coating increases as the gun shift velocity decreases. This is because a lower gun transfer speed allows the sprayed material to remain on the substrate surface for a more extended period, which enhances the adhesion and densification of the coating and also helps improve the microstructure of the coating. In addition, the spraying distance also has a direct effect on the microstructure of the coating. R.A. Abbas et al. [40] showed that the pore size radius increased continuously, and the pore density increased when the spraying distance was increased from 70 mm to 130 mm. This suggests that more considerable spraying distances may lead to cooling and diffusion of the sprayed powder in flight, which results in the formation of more and larger pores and affects the densification and properties of the coating. Based on the aforementioned findings, among the three sample groups, the intermediate transition layer of the Group C coating exhibited the highest spraying power, the lowest gun shift speed, and the shortest spraying distance. Consequently, when compared to Group A and Group B, the intermediate transition layer of the Group C coating demonstrated superior performance in thermal shock tests. In addition, the AGAP layer has a dense microstructure, which is an excellent improvement for the erosion resistance of the coating. The microstructure and the phase transition process of the AGAP layer in the present experimental samples are consistent with our previous studies [30,31].

3.1.2. Thermal Shock Resistance

The service ambient temperature of AYB coating systems is usually 1100 °C, but in this experiment, the three thermal barrier coatings were exposed to an extreme thermal environment of 1500 °C. Therefore, instead of using the number of previous thermal shock cycles as a comparison criterion for coating life, the length of thermal shock at the highest temperature is used for comparison in this paper. Generally, in conventional thermal shock tests, the initial cracks originate from the edges of the samples and gradually spread to the neighboring regions due to the edge effect [41,42]. The experimental results showed that the samples underwent 1980 s at 1500 °C; the samples in Group A and Group B spalled, while the samples in Group C remained intact. The samples in Group C lasted for 2400 s in the thermal shock test before spalling. The failure micro-morphology of the three groups of thermal barrier coating samples after the thermal shock test is shown in Figure 3. By observing the micro-morphology of the three groups of coatings after thermal shock, we found that the failure location of the AYIB coating system changed in the high-temperature environment compared with the AYB coating system. All three AYIB coatings produced transverse cracks inside the intermediate transition layer, which expanded and eventually led to coating failure. In our previous studies [30,31], on the other hand, the failure mode of the AYB coatings was mainly characterized by cracks sprouting and expanding between the YSZ layer and the BC layer, leading to coating failure, with no apparent cracks inside the YSZ layer. The observed difference in thermal shock life, with Group C demonstrating superior performance relative to Groups A and B, can be primarily attributed to the higher spraying power applied to the intermediate transition layer in Group C. This resulted in a significant enhancement of the fracture toughness of the intermediate transition layer in Group C, in contrast to the lower fracture toughness observed in Groups A and B. Consequently, the intermediate transition layer in Group C exhibits a greater ability to resist crack initiation and propagation under thermal shock conditions, thereby improving its overall thermal shock resistance. The exposure time of the coatings in the thermal environment in this experiment was too short, and the effect of TGO on the thermal shock behavior of the coatings was ignored.

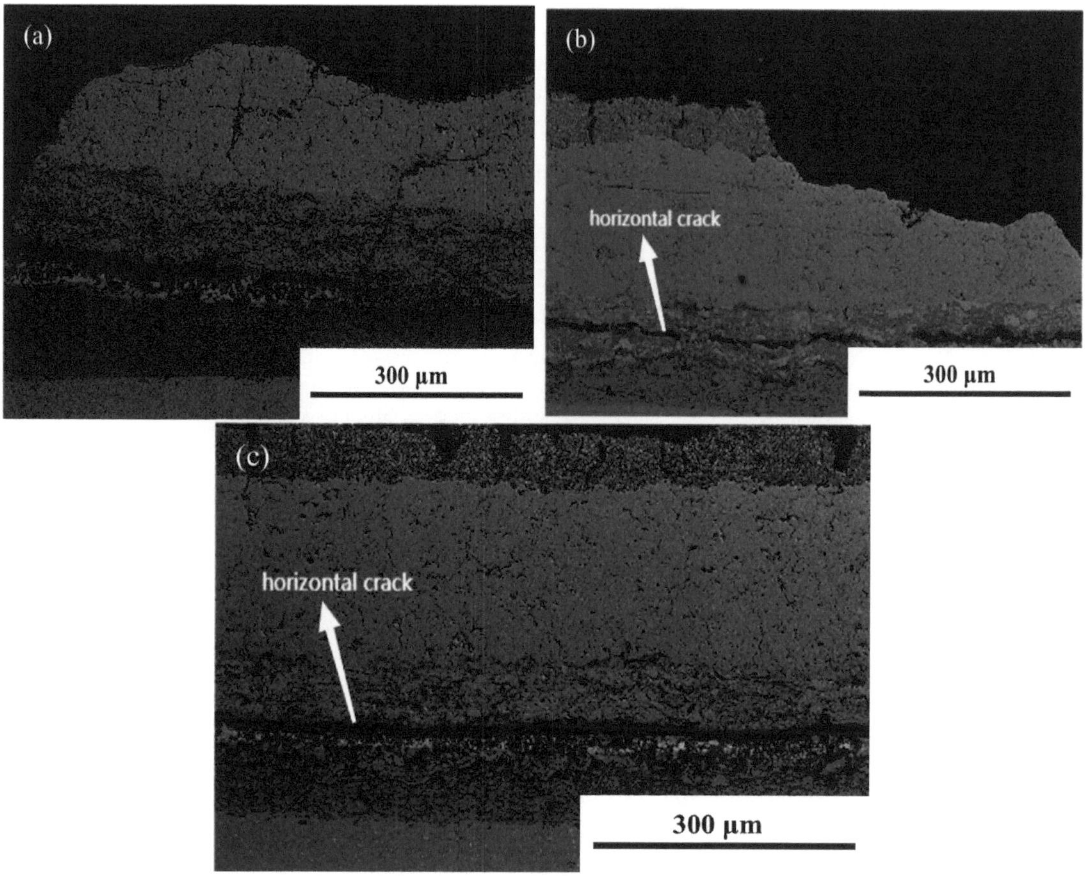

Figure 3. Cross-sectional morphologies of coatings after thermal shock test (**a**) Group A, (**b**) Group B, (**c**) Group C.

3.2. Simulation of Thermal Stress and Temperature Distributions of AYIB Coatings

The results of the thermal shock experiments show that adding an intermediate transition layer to the AYB coating system significantly changes the location of the coating crack initiation. In order to further investigate the differences in the thermal insulation performance and thermal stress distribution inside the coatings of AYB and AYIB coatings, numerical simulations of the temperature and stress fields of the two coatings were carried out in this paper using finite element software. In addition, the effects of different thicknesses of the intermediate transition layer on the thermal insulation performance of the substrate and the distribution of thermal stresses inside the coating are simulated.

3.2.1. Temperature Distributions of AYIB Coatings

In this study, the temperature field of the AYIB coating system was first simulated using ABAQUS software (version 2022). During the simulation process, the thickness of the intermediate transition layer was varied, and the intermediate transition layers of 60 μm, 80 μm, and 100 μm were selected, respectively. The overall thickness of the TC layer was kept at 480 μm, so the corresponding thicknesses of the YSZ layer were set as 320 μm, 300 μm, and 280 μm, respectively. In addition, the thickness of the AGAP layer was set to 100 μm, and the thickness of the BC layer was set to 120 μm. Given that the focus of this part of the study is on the effects of different coating structures on the temperature and stress distribution during thermal shock at 1500 °C, the interfacial structures (e.g., roughness and oxides), as well as the complex feature structures (e.g., pores and cracks) within the APS coatings, were neglected during the simulation. In order to clarify the relationship between the specific value of the temperature distribution and the depth, the temperature data from the top of the coating to the top layer of the substrate and the bottom of the substrate along the right boundary path are plotted according to the position and direction of the arrows as shown in Figure 1. The plotting results are shown in Figure 4. For the AYB coating system, the temperature of the substrate surface did not reach 1275 °C in the high-temperature environment of 1500 °C. After the introduction of the intermediate transition layers of 60 μm, 80 μm, and 100 μm, the temperatures of the substrate surface were 1283 °C, 1287 °C, and 1294 °C, respectively. The simulation results show that the thermal insulation performance of the AYIB coating system decreases compared to that of the AYB coating system after introducing the intermediate transition layer. This is because the overall thickness of the coating, the AGAP layer, and the BC layer always remain constant, and the thickness of the YSZ layer decreases accordingly with the thickness of the intermediate transition layer. According to the compelling medium principle, the thermal conductivity of the intermediate transition layer is higher than that of the YSZ layer, so the overall thermal insulation performance of the coating decreases as the thickness of the transition layer increases.

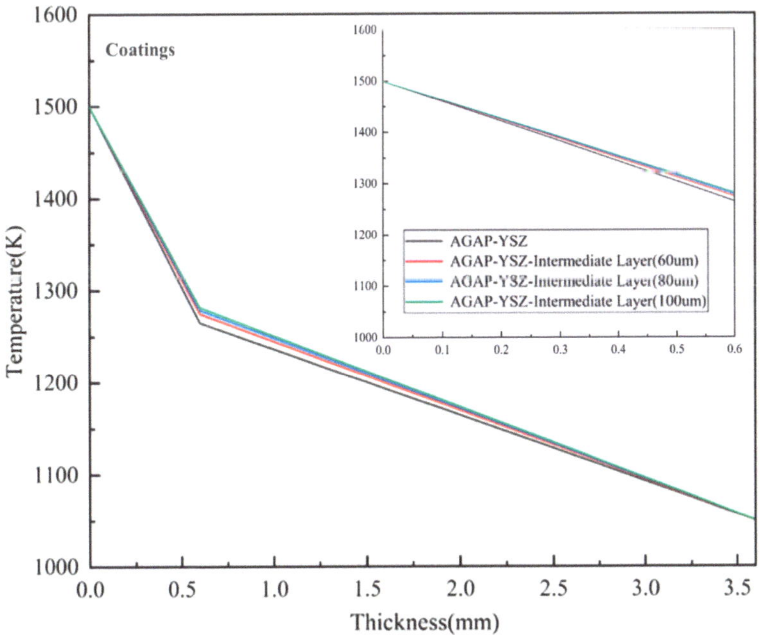

Figure 4. Temperature data plots of AYIB coatings with different intermediate transition layer thicknesses along the right boundary path at 1500 °C.

3.2.2. Thermal Stress Distribution of AYIB Coatings

In the flame thermal shock experiments, the main reason for the cracking of the coating interface is the thermal expansion mismatch between different materials due to the difference in thermal expansion coefficients, which generates thermal stresses. In the finite element simulation, this thermal stress manifests as tensile stress in the S22 direction. Figure 5 illustrates the variation in stress data along the right boundary path (Figure 1) for the AYIB coating system with different thicknesses of intermediate transition layers. As shown in Figure 5, the maximum thermal stress of the AYB coating system occurs between the YSZ layer and the BC layer under thermal shock conditions at 1500 °C, with a simulated stress value of 542.8 MPa. The location of the maximum thermal stress is the location of crack initiation and development, and the simulated results of AYB are consistent with the failure results after the actual thermal shock test. After the introduction of the intermediate transition layers of 80 μm and 100 μm, the maximum location of the maximum thermal stress of the AYIB coating system is shifted to the interface between the BC layer and the substrate compared with that of the AYB coating system. The simulated stress values are 457.1 MPa and 478.1 MPa, respectively, representing reductions of 15.79% and 11.92%. For the interior of the coating, the maximum thermal stress is within the intermediate transition layer, which is in line with the actual thermal shock assessment test results. However, when the intermediate transition layer with a thickness of 60 μm is introduced, the location of the maximum thermal stress of the AYIB coating system is within the BC layer, and the reduction in the maximum thermal stress is not noticeable compared with that of the AYB coating system. From the analysis results, increasing the thickness of the intermediate transition layer can effectively reduce the thermal stress within the coating and improve the strain tolerance of the coating.

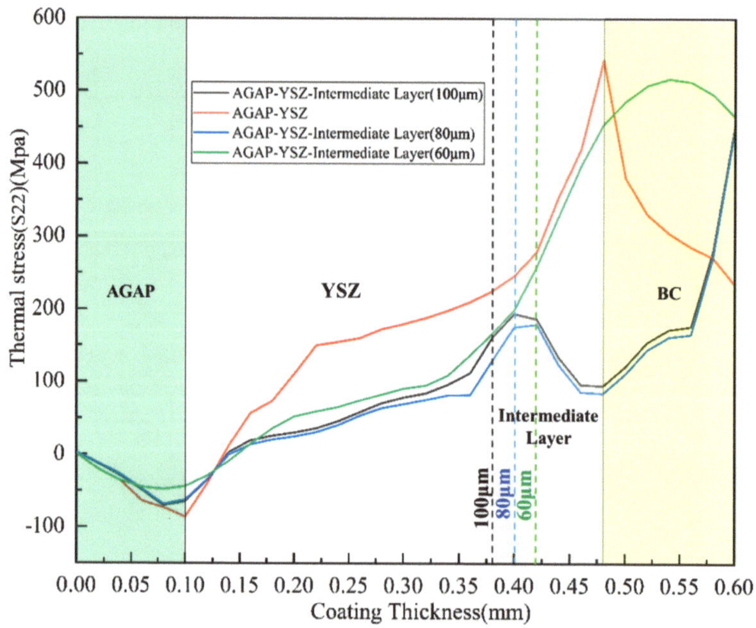

Figure 5. Distribution of coating internal stresses along the right boundary path at 1500 °C for AYIB coatings with different intermediate transition layer thicknesses.

The main reason for the different stress distributions of the two groups of coatings with different structures is due to the change in Young's modulus of the TBC layer after introducing the intermediate transition layer, which is also a significant influence on thermal stress. According to a related study [43], the stresses in the coating and the substrate as a whole can usually be divided into three components:

$$\sigma_c = \sigma_a + \sigma_g + \sigma_t \tag{1}$$

σ_a is aging stress, σ_g is growth stress, and σ_t is thermal expansion mismatch (CTE) stress coefficient. Aging stresses are internal stresses caused by changes in the organizational structure of a matrix material after it has undergone an aging treatment. During the aging process, the internal crystal structure of the material changes, which may lead to local changes in stress distribution, which are aging stresses. High-temperature deposition and rapid cooling usually involve preparing coatings or thin films. The different coefficients of thermal expansion between the deposited material and the substrate during the temperature change may lead to compressive or tensile stresses within the deposited layer. These stresses are growth stresses, and they may affect the coating's adhesion, stability, and durability. CTE mismatch stresses are due to the mismatch between the coating material and the substrate in the coefficient of thermal expansion. When the coating material's and substrate's thermal expansion coefficients are different or mismatched, the two will generate internal stresses due to the different thermal expansion rates during temperature changes [43]. These stresses can be positive or negative, depending on the specific properties of the material and the temperature history during the coating preparation process, where the CTE mismatch stresses are calculated as follows [44]:

$$\sigma_t = \frac{(\alpha_c - \alpha_{sub}) \triangle TE_C}{(1 - \upsilon_C)} \tag{2}$$

α_c and α_{sub} are the coefficients of thermal expansion of the coating and the substrate, respectively; υ_C is Poisson's coating ratio, and $\triangle T$ is the difference between the thermal shock test and initial temperatures. Therefore, according to Equation (2), when determining the thermal shock test temperature, a coating material with a coefficient of thermal expansion that better matches the substrate's, a lower Young's modulus, and a smaller Poisson's ratio can be selected to alleviate thermal stresses caused by thermal expansion mismatch.

3.3. Simulation of Thermal Stress and Temperature Distributions of AYI(E)B Coatings

Through the above experimental and simulation results, we found that the intermediate transition layer in the AYB coating system can effectively alleviate the thermal stress generated by the coating under thermal shock. However, due to the high thermal conductivity of the intermediate transition layer, the thermal insulation performance of the AYIB coating is reduced. Although the thermal stress caused by the thermal expansion mismatch was mitigated by structural modulation, the thermal insulation effect of the coating on the substrate was affected. Therefore, the structure of the intermediate transition layer needs to be further optimized to achieve a balance between high strain tolerance and thermal insulation performance of the AYIB coating system.

In previous studies [45,46], we proposed a novel deposition method for porous thermal barrier coatings by improving the conventional APS technique, which successfully prepared porous EMAP coatings using co-sprayed micrometer-sized agglomerated particles as the pore source. The PEPC structure is a typical feature of EMAP coatings and is the main factor influencing the porosity of the coatings. Previous studies have shown that EMAP coatings have significant advantages in strain tolerance, thermal insulation capability, and sintering resistance. Therefore, in this paper, we have chosen to introduce the EMAP structure into the intermediate transition layer to improve the coating's thermal insulation and strain tolerance. For ease of description, this novel coating system is referred to as AYI(E)B coating in this paper.

During the simulations in this section, we choose to blend three different contents (volume share) of PEPC, 3%, 6%, and 9%, inside the intermediate transition layer to analyze the effect of different contents of PEPC on the thermal insulation performance and strain tolerance of the AYI(E)B coating system. The geometric modeling of the finite element model is shown in Figure 6. Previous studies [17,18] have shown through microscopic observations that PEPC particles have an approximately ellipsoidal shape, resembling flattened spheres. In a 2D plane, they can be simplified as standard ellipses with a perimeter of 50 μm and a length-to-width ratio of 1:2. In addition, the material parameters of PEPC were kept consistent with previous studies [32,33], where the thermal conductivity of PEPC was set to 0.025 W/m-K and the modulus of elasticity to 140 kPa. The thickness of the intermediate transition layer was set to 80 µm. The rest of the coating thicknesses and material parameters were consistent.

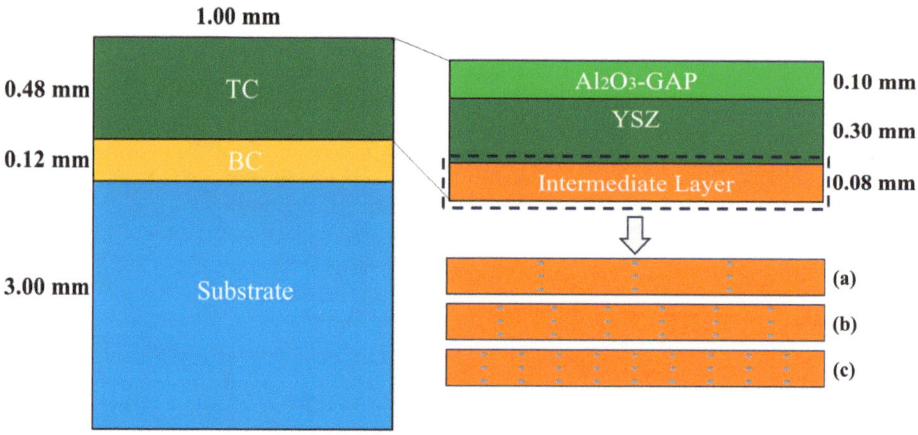

Figure 6. Finite element model of AYI(E)B with different PEPC contents: (**a**) 3%; (**b**) 6%; (**c**) 9%.

3.3.1. Temperature Distributions of AYI(E)B Coatings

Figure 7 illustrates the temperature change profiles of AYI(E)B coatings with different PEPC contents along the right boundary path (Figure 1) under thermal shock at 1500 °C. When the PEPC content in the intermediate transition layer is 3%, 6%, and 9%, the temperatures reaching the substrate surface are 1203 °C, 1193 °C, and 1183 °C, respectively. The simulation results show that the temperature of the substrate surface decreases gradually with the increase in the PEPC content in the intermediate transition layer. This indicates that the thermal insulation ability of the coating is positively correlated with the content of PEPC, and the higher the content of PEPC, the better the thermal insulation performance of the coating. Compared with the AYIB coating system, the thermal insulation performance of the AYI(E)B coating system was greatly improved.

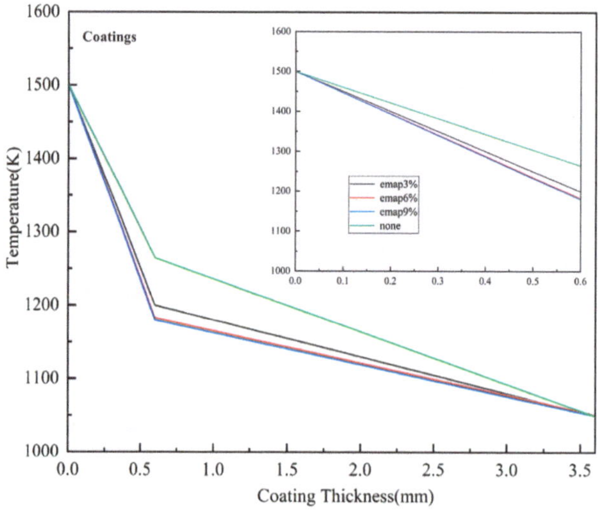

Figure 7. Temperature data plots of AYI(E)B coatings with different intermediate transition layer thicknesses along the right boundary path at 1500 °C.

3.3.2. Thermal Stress Distribution of AYI(E)B Coatings

Figure 8 illustrates the stress distribution cloud of AYI(E)B coatings with different PEPC contents under thermal shock at 1500 °C and the stress variation curves along the right boundary path (Figure 1). The simulation results show that the location of the maximum thermal stress in the AYI(E)B system is at the interface between the BC layer and the substrate after the introduction of the EMAP structure in the intermediate transition layer. For the inside of the coating, the maximum thermal stress values within the coating for the AYI(E)B coating system were 194.7 MPa, 175.5 MPa, and 171.9 MPa when 3%, 6%, and 9% of PEPC were introduced, respectively, and the locations were all inside the intermediate transition layer. In addition, the maximum intra-coating thermal stresses of the AYI(E)B coating system appeared inside the intermediate transition layer, with values of 150.39 MPa, 146.85 MPa, and 99.01 MPa, respectively. Compared with the AYIB coating system, the overall S22 stresses of the AYI(E)B coating system were reduced, and the strain tolerances were improved, further mitigating the thermal stress of the coating system. Compared to the traditional APS structure, the EMAP structure coating demonstrates superior resistance to sintering shrinkage and exhibits a lower degree of sintering hardening. This allows it to maintain a lower elastic modulus, even after prolonged exposure to high temperatures [32]. While the substrate stiffness of conventional structural coatings increases dramatically during sintering at high temperatures, the effective modulus of elasticity of EMAP coatings with embedded particles is reduced by approximately 75% compared to conventional coatings without particles, which mitigates thermal stresses due to thermal expansion mismatches in high-temperature environments. This property significantly improves the strain tolerance of the coating at high temperatures and increases the thermal shock life of the coating.

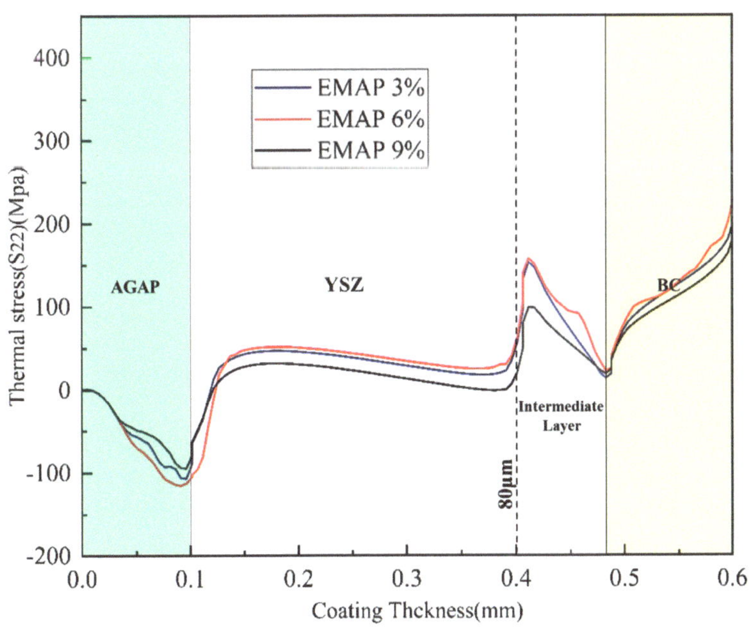

Figure 8. Stress distribution along the right boundary path at 1500 °C for coatings with different PEPC contents.

In summary, the AYI(E)B coating system has less thermal stress and better thermal insulation capability than the AYIB coating system. On the one hand, the EMAP intermediate transition layer adopts a porous structure. It contains many second-phase particles, which not only enhances the anti-sintering property of the coating but also improves its thermal insulation effect, thus effectively reducing the thermal stress of the coating at high temperatures. On the other hand, incorporating second-phase particles reduced Young's modulus of the intermediate transition layer of EMAP; according to the analysis of Equation (2), the reduction in Young's modulus helps reduce the stresses induced by CTE mismatch in the coating system, further enhancing the coating lifetime. Therefore, as shown in Figure 9, we propose a novel design scheme for the AYI(E)B multilayer thermal barrier coating.

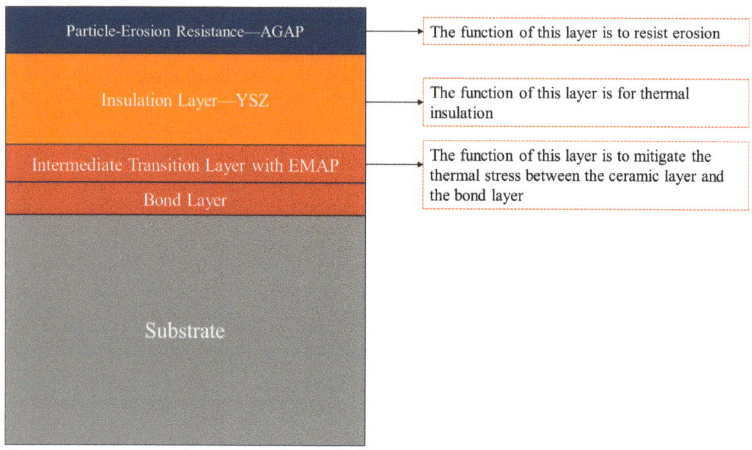

Figure 9. AYI(E)B multilayer thermal barrier coating design solution.

4. Conclusions

This study presents a novel thermal barrier coating system incorporating an intermediate transition layer. Three different AGAP/YSZ/intermediate layer/bond coat systems were fabricated and subjected to a 1500 °C flame thermal shock test. At the same time, finite element simulations of flame thermal shock at 1500 °C were carried out for the AGAP/YSZ/bonded coating system and AGAP/YSZ/intermediate layer/bonded coating system. Although the introduction of an intermediate transition layer increases the complexity of the spraying process, as well as the preparation time and cost of the coating, both experimental and simulation analyses indicate that this layer effectively alleviates thermal stresses during thermal shock, thereby significantly enhancing the strain tolerance of the coating. On this basis, this study further modulates the structure of the AGAP/YSZ/intermediate/bonding layer coating system and designs a new coating system for the AGAP/YSZ/EMAP intermediate/bonding layer. The finite element simulation results show that introducing the EMAP structure relieves the thermal stress during thermal shock and increases the strain tolerance of the coating but also improves its thermal insulation performance. The main conclusions of this study are summarized as follows:

1. By analyzing the results of the flame thermal shock experiments on the AYIB coating system at 1500 °C, the coating's failure mode changes due to the introduction of the intermediate transition layer, and the location of crack generation moves from the interface between the BC layer and the ceramic layer to the interior of the intermediate transition layer.
2. The thermal insulation performance of the coating decreases slightly due to the introduction of the intermediate transition layer. For the AYB system, the temperature of the substrate surface is 1275 °C in a high-temperature environment of 1500 °C. After the introduction of the intermediate transition layers of 60 μm, 80 μm, and 100 μm, the temperatures of the substrate surface are 1283 °C, 1287 °C, and 1294 °C, respectively.

3. Compared with the AYB coating system without an intermediate transition layer, the thermal mismatch stresses caused by thermal expansion mismatch in the high-temperature thermal shock experiments are significantly alleviated by adding different thicknesses of the intermediate transition layer in the AYIB coating system. When the thickness of the intermediate transition layer is 80 μm and 100 μm, the maximum thermal stresses are 457.1 MPa and 478.1 MPa, respectively, 15.78% and 12.12% are reduced.
4. The coatings' strain tolerance and thermal insulation properties were significantly improved by introducing the EMAP structure in the intermediate transition layer. When the contents of second-phase embedded particles were 3%, 6%, and 9%, respectively, the maximum thermal stresses inside the coatings were 150.39 MPa, 146.85 MPa, and 99.01 MPa. The temperatures on the substrate surface were 1203 °C, 1193 °C, and 1183 °C, respectively. The increase in the second-phase particle content effectively relieves the thermal stress inside the coating and enhances the thermal insulation performance of the coating system.

Author Contributions: Conceptualization, P.L., S.Y. and W.W.; methodology, P.L. and S.Y.; software, S.Y.; validation, P.L., S.Y., W.W., K.L., Y.L. and T.Y.; formal analysis, P.L. and S.Y.; investigation, P.L., S.Y. and K.L.; resources, W.W.; data curation, P.L. and S.Y.; writing—original draft preparation, P.L.; writing—review and editing, W.W.; visualization, P.L.; supervision, W.W.; project administration, W.W.; funding acquisition, W.W. All authors have read and agreed to the published version of the manuscript.

Funding: This research was funded by the Science Center for Gas Turbine Project (P2021-A-IV-002-002), Civil Space 'the 14th Five-Year Plan' Pre-research Project (D050101), National High Technology Research and Development Program of China (2023YFB3711200), National Natural Science Foundation of China (52175136, 52130511), Shanghai Gaofeng Project for University Academic Program Development.

Institutional Review Board Statement: Not applicable.

Informed Consent Statement: Not applicable.

Data Availability Statement: The raw/processed data required to reproduce these findings cannot be shared at this time as the data also forms part of an ongoing study.

Acknowledgments: The author thanks the contribution of W.W. for reviewing and discussing this work; S.Y. for providing experimental support for SEM microstructural analysis; Y.L. and T.Y. for discussing experimental results; and the Research Center of Analysis and Test of East China University of Science and Technology for the help on the characterization.

Conflicts of Interest: The authors declare no conflicts of interest.

References

1. Guo, H.; Guo, Y.; Xue, Z.; Gong, S.; Xu, H. Overview of thermal barrier coatings for advanced gas turbine engine. In *Thermal Barrier Coatings*; Woodhead Publishing: Cambridge, UK, 2023; pp. 1–20.
2. Cao, X.Q.; Vassen, R.; Stoever, D. Ceramic materials for thermal barrier coatings. *J. Eur. Ceram. Soc.* **2004**, *24*, 1–10. [CrossRef]
3. Clarke, D.R.; Oechsner, M.; Padture, N.P. Thermal-barrier coatings for more efficient gas-turbine engines. *MRS Bull.* **2012**, *37*, 891–898. [CrossRef]
4. Padture, N.P.; Gell, M.; Jordan, E.H. Thermal Barrier Coatings for Gas-Turbine Engine Applications. *Science* **2002**, *296*, 280–284. [CrossRef] [PubMed]
5. Feuerstein, A.; Knapp, J.; Taylor, T.; Ashary, A.; Bolcavage, A.; Hitchman, N. Technical and Economical Aspects of Current Thermal Barrier Coating Systems for Gas Turbine Engines by Thermal Spray and EBPVD: A Review. *J. Therm. Spray Technol.* **2008**, *17*, 199–213. [CrossRef]
6. Wright, P.K.; Evans, A.G. Mechanisms governing the performance of thermal barrier coatings. *Curr. Opin. Solid State Mater. Sci.* **1999**, *4*, 255–265. [CrossRef]
7. Dong, H.; Liang, X.; Wang, Z.; Luo, N.; Zhang, J.; Zhan, Y.; Zhang, X. Enhancing the performances of EB-PVD TBCs via overlayer Al-modification. *Surf. Coat. Technol.* **2023**, *473*, 130001. [CrossRef]

8. Fang, H.J.; Wang, W.Z.; Yang, Z.N.; Yang, T.; Wang, Y.H.; Huang, J.B.; Ye, D.D. Phase stability, thermal shock behavior and CMAS corrosion resistance of Yb_2O_3-Y_2O_3 co-stabilized zirconia thermal barrier coatings prepared by atmospheric plasma spraying. *Surf. Coat. Technol.* **2021**, *427*, 127864. [CrossRef]
9. Kulkarni, A.; Vaidya, A.; Goland, A.; Sampath, S.; Herman, H. Processing effects on porosity-property correlations in plasma sprayed yttria-stabilized zirconia coatings. *Mater. Sci. Eng. A* **2003**, *359*, 100–111. [CrossRef]
10. Li, G.R.; Yang, G.J.; Li, C.X.; Li, C.J. Sintering characteristics of plasma-sprayed TBCs: Experimental analysis and an overall modelling. *Ceram. Int.* **2018**, *44*, 2982–2990. [CrossRef]
11. Trice, R.W.; Su, Y.J.; Mawdsley, J.R.; Faber, K.T.; De Arellano-López, A.R.; Wang, H.; Porter, W.D. Effect of heat treatment on phase stability, microstructure, and thermal conductivity of plasma-sprayed YSZ. *J. Mater. Sci.* **2002**, *37*, 2359–2365. [CrossRef]
12. Lashmi, P.; Ananthapadmanabhan, P.; Unnikrishnan, G.; Aruna, S. Present status and future prospects of plasma sprayed multilayered thermal barrier coating systems. *J. Eur. Ceram. Soc.* **2020**, *40*, 2731–2745. [CrossRef]
13. Vaßen, R.; Jarligo, M.O.; Steinke, T.; Mack, D.E.; Stöver, D. Overview on advanced thermal barrier coatings. *Surf. Coat. Technol.* **2010**, *205*, 938–942. [CrossRef]
14. Fang, H.J.; Wang, W.Z.; Huang, J.B.; Li, Y.J.; Ye, D.D. Corrosion behavior and thermos-physical properties of a promising Yb_2O_3 and Y_2O_3 co-stabilized ZrO_2 ceramic for thermal barrier coatings subject to calcium-magnesium-aluminum-silicate (CMAS) deposition: Experiments and first-principles calculation. *Corros. Sci.* **2021**, *182*, 109230. [CrossRef]
15. McPherson, R. A review of microstructure and properties of plasma sprayed ceramic coatings. *Surf. Coat. Technol.* **1989**, *40*, 173–181. [CrossRef]
16. Li, C.; Wang, W. Quantitative characterization of lamellar microstructure of plasma-sprayed ceramic coatings through visualization of void distribution. *Mater. Eng. A* **2004**, *386*, 10–19. [CrossRef]
17. Allen, A.; Ilavsky, J.; Long, G.; Wallace, J.; Berndt, C.; Herman, H. Microstructural characterization of yttria-stabilized zirconia plasma-sprayed deposits using multiple small-angle neutron scattering. *Acta Mater.* **2001**, *49*, 1661–1675. [CrossRef]
18. Antou, G.; Montavon, G.; Hlawka, F.; Cornet, A.; Coddet, C.; Machi, F. Evaluation of modifications induced on pore network and structure of partially stabilized zirconia manufactured by hybrid plasma spray process. *Surf. Coat. Technol.* **2004**, *180*, 627–632. [CrossRef]
19. Fauchais, P.; Vardelle, M.; Goutier, S. Latest Researches Advances of Plasma Spraying: From Splat to Coating Formation. *J. Therm. Spray Technol.* **2016**, *25*, 1534–1553. [CrossRef]
20. Kulkarni, A.A.; Goland, A.; Herman, H.; Allen, A.J.; Ilavsky, J.; Long, G.G.; Johnson, C.A.; Ruud, J.A. Microstructure–Property Correlations in Industrial Thermal Barrier Coatings. *J. Am. Ceram. Soc.* **2004**, *87*, 1294–1300. [CrossRef]
21. Fleck, N.; Cocks, A.; Lampenscherf, S. Thermal shock resistance of air plasma sprayed thermal barrier coatings. *J. Eur. Ceram. Soc.* **2014**, *34*, 2687–2694. [CrossRef]
22. Heveran, C.M.; Xu, J.; Sarin, V.K.; Basu, S.N. Simulation of stresses in TBC–EBC coating systems for ceramic components in gas turbines. *Surf. Coat. Technol.* **2013**, *235*, 354–360. [CrossRef]
23. Levi, C.G. Emerging materials and processes for thermal barrier systems. *Curr. Opin. Solid State Mater. Sci.* **2004**, *8*, 77–91. [CrossRef]
24. Ozgurluk, Y.; Gulec, A.; Ozkan, D.; Binal, G.; Karaoglanli, A.C. Structural characteristics, oxidation performance and failure mechanism of thermal barrier coatings fabricated by atmospheric plasma spraying and detonation gun spraying. *Eng. Fail. Anal.* **2023**, *152*, 107499. [CrossRef]
25. Scrivani, A.; Rizzi, G.; Berndt, C.C. Enhanced thick thermal barrier coatings that exhibit varying porosity. *Mater. Sci. Eng. A* **2008**, *476*, 1–7. [CrossRef]
26. Tao, S.; Yang, J.; Shao, F.; Zhao, H.; Zhong, X.; Zhuang, Y.; Sheng, J.; Ni, J.; Li, Q.; Tao, S. Atmospheric plasma sprayed thick thermal barrier coatings: Microstructure, thermal shock behaviors and failure mechanism. *Eng. Fail. Anal.* **2021**, *131*, 105819. [CrossRef]
27. Liu, K.; Tang, J.; Bai, Y.; Yang, Q.; Wang, Y.; Kang, Y.; Zhao, L.; Zhang, P.; Han, Z. Particle in-flight behavior and its influence on the microstructure and mechanical property of plasma sprayed $La_2Ce_2O_7$ thermal barrier coatings. *Mater. Sci. Eng. A* **2015**, *625*, 177–185. [CrossRef]
28. Liu, L.; Wang, S.; Zhang, B.; Jiang, G.; Liu, H.; Yang, J.; Wang, J.; Liu, W. Present status and prospects of nanostructured thermal barrier coatings and their performance improvement strategies: A review. *J. Manuf. Process.* **2023**, *97*, 12–34. [CrossRef]
29. Pakseresht, A.; Sharifianjazi, F.; Esmaeilkhanian, A.; Bazli, L.; Nafchi, M.R.; Bazli, M.; Kirubaharan, K. Failure mechanisms and structure tailoring of YSZ and new candidates for thermal barrier coatings: A systematic review. *Mater. Des.* **2022**, *222*, 111044. [CrossRef]
30. Yang, Z.; Wang, W.; Deng, S.; Fang, H.; Yang, T.; Wang, L. Thermal Shock Behavior and Particle Erosion Resistance of Toughened GZ Coatings Prepared by Atmospheric Plasma Spraying. *Coatings* **2021**, *11*, 1477. [CrossRef]
31. Yang, Z.; Yang, K.; Wang, W.; Yang, T.; Fang, H.; Qiang, L.; Zhang, X.; Zhang, C. Investigation of Thermal Shock Behavior of Multilayer Thermal Barrier Coatings with Superior Erosion Resistance Prepared by Atmospheric Plasma Spraying. *Coatings* **2022**, *12*, 804. [CrossRef]
32. Yang, T.; Wang, W.; Huang, J.; Wang, L.; Yang, Z.; Fang, H.; Ye, D. Thermal Shock Resistance and Bonding Strength of Novel-Structured Thermal Barrier Coatings with Different Microstructure. *J. Therm. Spray Technol.* **2022**, *31*, 1540–1555. [CrossRef]

33. Yang, T.; Wang, W.; Tang, Z.; Liu, Y.; Li, K. Structural optimization for porous thermal barrier coating and analysis of thermomechanical properties by experimental and computational investigation. *Surf. Coat. Technol.* **2023**, *458*, 129347. [CrossRef]
34. Yedida, V.S.; Vasudev, H. A review on the development of thermal barrier coatings by using thermal spray techniques. *Mater. Today Proc.* **2022**, *50*, 1458–1464. [CrossRef]
35. Song, J.-B.; Wang, L.-S.; Dong, H.; Yao, J.-T. Long lifespan thermal barrier coatings overview: Materials, manufacturing, failure mechanisms, and multiscale structural design. *Ceram. Int.* **2022**, *49*, 1–23. [CrossRef]
36. Huang, J.; Wang, W.; Lu, X.; Liu, S.; Li, C. Influence of Lamellar Interface Morphology on Cracking Resistance of Plasma-Sprayed YSZ Coatings. *Coatings* **2018**, *8*, 187. [CrossRef]
37. Kim, D.-J.; Shin, I.-H.; Koo, J.-M.; Seok, C.-S.; Lee, T.-W. Failure mechanisms of coin-type plasma-sprayed thermal barrier coatings with thermal fatigue. *Surf. Coat. Technol.* **2010**, *205*, S451–S458. [CrossRef]
38. Yang, T.; Wang, W.; Tang, Z.; Liu, Y.; Li, K.; Liu, W.; Zhang, W.; Ye, D. CMAS infiltration behavior of atmospheric plasma-sprayed thermal barrier coating with tailored pore structures. *Ceram. Int.* **2023**, *50*, 7218–7229. [CrossRef]
39. Cai, Z.; Zhang, Z.; Liu, Y.; Zhao, X.; Wang, W. Numerical Study on Effect of Non-uniform CMAS Penetration on TGO Growth and Interface Stress Behavior of APS TBCs. *Chin. J. Mech. Eng.* **2021**, *34*, 128. [CrossRef]
40. Abbas, R.A.; Ajeel, S.A.; Bash, M.A.A.; Kadhim, M.J. Effect of plasma spray distance on the features and hardness reliability of YSZ thermal barrier coating. *Mater. Today Proc.* **2021**, *42*, 2553–2560. [CrossRef]
41. Jamali, H.; Mozafarinia, R.; Razavi, R.S.; Ahmadi-Pidani, R. Comparison of thermal shock resistances of plasma-sprayed nanostructured and conventional yttria stabilized zirconia thermal barrier coatings. *Ceram. Int.* **2012**, *38*, 6705–6712. [CrossRef]
42. Wu, J.; Guo, H.-B.; Zhou, L.; Wang, L.; Gong, S.-K. Microstructure and Thermal Properties of Plasma Sprayed Thermal Barrier Coatings from Nanostructured YSZ. *J. Therm. Spray Technol.* **2010**, *19*, 1186–1194. [CrossRef]
43. Li, K.; Wang, W.; Yang, S.; Liu, Y.; Yang, T. First-principles calculations and thermal-mechanical experimental studies on middle-entropy rare-earth disilicates. *Ceram. Int.* **2024**, *50*, 22290–22305. [CrossRef]
44. Li, K.; Wang, W.; Yang, T.; Liu, Y. Simulation of 1500 °C Thermal Shock for Novel Structural Thermal/Environmental Barrier Coatings System. *Coatings* **2023**, *13*, 96. [CrossRef]
45. Huang, J.B.; Wang, W.Z.; Li, Y.J.; Fang, H.J.; Ye, D.D.; Zhang, X.C.; Tu, S.T. Improve durability of plasma-sprayed thermal barrier coatings by decreasing sintering-induced stiffening in ceramic coatings. *J. Eur. Ceram. Soc.* **2020**, *40*, 1433–1442. [CrossRef]
46. Huang, J.B.; Wang, W.Z.; Li, Y.J.; Fang, H.J.; Ye, D.D.; Zhang, X.C.; Tu, S.T. A novel strategy to control the microstructure of plasma-sprayed YSZ thermal barrier coatings. *Surf. Coat. Technol.* **2020**, *402*, 126304. [CrossRef]

coatings

MDPI

Article

Improving Optical and Electrical Characteristics of GaN Films via 3D Island to 2D Growth Mode Transition Using Molecular Beam Epitaxy

Thi Thu Mai [1], Jin-Ji Dai [1], Wu-Ching Chou [1,*], Hua-Chiang Wen [1], Le Trung Hieu [2] and Huy Hoang Luc [3,*]

[1] Department of Electrophysics, National Yang Ming Chiao Tung University, Hsinchu 30010, Taiwan; maithucs@gmail.com (T.T.M.); jinjidai@gmail.com (J.-J.D.); a091316104@gmail.com (H.-C.W.)
[2] Wide Bandgap Compound Semiconductor Research Center, Industry Academia Innovation School, National Yang Ming Chiao Tung University, Hsinchu 30010, Taiwan; hieult@nycu.edu.tw
[3] Faculty of Physics, Hanoi National University of Education, 136 Xuan Thuy, Cau Giay, Hanoi 123105, Vietnam
* Correspondence: wcchou957@nycu.edu.tw (W.-C.C.); hoanglhsp@hnue.edu.vn (H.H.L.); Tel.: +886-3-5712121-56129 (W.-C.C.); +84-903200023 (H.H.L.)

Abstract: Molecular beam epitaxy (MBE) is demonstrated as an excellent growth technique for growing a low-defect GaN channel layer, which is crucial for controlling vertical leakage current and improving breakdown voltage (BV) in GaN-based high-electron mobility transistors (HEMTs). The 3D islands to 2D growth mode transition approach was induced by modulating substrate growth temperature (Tsub), displaying an overall improvement in film quality. A comprehensive investigation was conducted into the effects of Tsub on surface morphologies, crystal quality, and the optical and electrical properties of GaN films. Optimal results were achieved with a strain-relaxed GaN film grown at 690 °C, exhibiting significantly improved surface characteristics (root-mean-square roughness, Rq = 0.3 nm) and impressively reduced edge dislocations. However, the film with the smoothest surface roughness, attributed to the effect of the Ga-rich condition, possessed a high surface pit density, negatively affecting optical and electrical properties. A reduction in defect-related yellow emission further confirmed the enhanced crystalline quality of MBE GaN films. The optimized GaN film demonstrated outstanding electrical properties with a BV of ~1450 V, surpassing that of MOCVD GaN (~1180 V). This research significantly contributes to the advancement of MBE GaN-based high electron mobility transistor (HEMT) applications by ensuring outstanding reliability.

Keywords: 3D-to-2D growth; MBE; GaN HEMT; vertical leakage; breakdown voltage; dislocations

1. Introduction

GaN-on-Si-based high-electron mobility transistors (HEMTs) possess unique advantages for applications demanding high power and frequency due to their outstanding characteristics, such as high breakdown voltage (BV), excellent thermal stability, lower on-state resistance, high electron saturation velocity, and low cost [1,2]. However, crucial technological challenges remain due to the mismatches in lattice parameters and the coefficient of thermal expansion (CTE) between GaN and Si [3]. These issues limit the device's ultimate performance, resulting from the involvement of current collapse phenomena, high density of threading dislocations (TDD) (10^9–10^{10} cm^{-2}), and excessive residual strains [4,5]. The current collapse phenomenon arises from charge traps in the undoped (UID) GaN layer, which subsequently accumulated near the interface of two-dimensional electron gas (2DEG) at the AlGaN/GaN heterostructure, and from the GaN buffer layer, which deleteriously causes power losses [6]. These traps are frequently identified in MOCVD-grown GaN films due to the incorporation of undesirable impurity-related defects (such as carbon, oxygen, or silicon) during growth [7]. Moreover, dislocations pose a substantial risk, contributing to an increase in leakage current and sheet resistance of 2DEG and also reducing BV and free carrier mobility [8,9]. Hence, contemporary efforts to enhance HEMT devices focus on minimizing interfacial impurity accumulation and reducing dislocations to minimize power losses. Additionally, carefully managing the GaN structure thickness also requires particular attention for mitigating micro-cracks and wafer curvature while improving the breakdown field. Several approaches have been proposed to address these issues, including implementing thick strain-relief stacks (SRSs) and interlayers. These approaches encompass the mitigating stress buffer layers [10], super-lattice buffer layers [11], step-graded AlN/AlGaN [12], low-temperature LT-AlN and SiNx interlayers [13,14], 3D-to-2D growth mode transition [14,15], and epitaxial lateral overgrowth (ELOG) [16,17]. However, while addressing certain issues, the thick SRS can introduce a notable drawback of severe cracking and wafer bowing arising from the over-thickening effects of highly strained epitaxial films [18]. The minimum thickness for high-quality GaN HEMT structures with a TDD of ~10^9 cm^{-2} is typically around 6 μm [12]. Techniques involving interlayers, or ELOG, can generate nanomasks to facilitate the transition from 3D to 2D growth mode but may negatively affect device performance, including increasing vertical leakage current [19], cracking issues, and interrupted dislocation propagation in regions lacking mask contribution [17]. In contrast, the molecular beam epitaxy (PA-MBE) technique has shown remarkable breakthroughs in improving device performance. This technique produces high-pure epitaxial films at ultra-high vacuum, low growth temperatures, and optimal stoichiometric growth conditions [20–23]. Specifically, MBE-grown GaN homoepitaxial films in Ga-rich conditions have resulted in excellent film quality [20,21]. Zhang et al. [20] found that higher nitrogen flows led to a better crystal quality of homoepitaxial GaN films but also resulted in an undesirable rough surface. Surface roughness critically impacts carrier mobility, parasitic resistance, and the presence of negative deep traps [24,25]. Smooth surface morphologies are attainable when films are grown in the Ga droplet and intermediate regimes. However, the devices still exhibited inferior electron transport density and luminescence properties owing to the Ga droplet effect [26,27]. Controlling substrate growth temperature (Tsub) allows for the efficient desorption of excessive Ga atoms on the surface while effectively incorporating them into GaN films, which is facilitated by supplying adequate thermal energy for surface activation during coalescence [21,28]. However, excessive substrate thermal energy can lead to surface over-desorption, resulting in rough surface morphology and deteriorating the film quality. Conversely, insufficient desorption temperature may cause Ga droplet issues [27]. Thus far, there has been limited study of employing MBE for promoting 3D-to-2D growth mode transitions. Furthermore, a comprehensive study of the electrical characteristics of MBE-grown GaN films under growth mode transition is essential.

In this study, high-quality UID GaN layers on MOCVD GaN/Si (111) templates were obtained using MBE growth. This accomplishment involved implementing a 3D to 2D growth mode transition by adjusting Tsub under Ga-rich conditions. The effects of the growth mode transition on the surface morphologies, defect reduction, crystal quality, and optical and electrical characteristics of the MBE-grown GaN films compared with MOCVD GaN templates have been investigated. The findings of this work provide valuable insights for future research on the manufacturing of highly reliable GaN-based devices.

2. Materials and Methods

Figure 1a illustrates the schematic structure of MBE GaN film on MOCVD GaN/Si (111) templates. The templates were deposited by MOCVD on 6-inch Si (111) wafers, consisting of a 200 nm AlN nucleation layer followed by a 280 nm $Al_{0.25}Ga_{0.75}N$ layer. The 800 nm C-doped GaN insulating buffer layer was then grown to reduce defect-induced residual electrons. The final deposited layer was a 300 nm UID GaN layer. The 2 × 2 cm^2 GaN/Si templates were cut and cleaned with acetone, isopropyl alcohol, and deionized water (DI). The cleaned templates were then transferred into the growing chamber, which was followed by an in situ thermal cleaning process under an ultra-high vacuum (4.0×10^{-11} Torr) at 150 °C for 30 min in order to eliminate surface organic contaminants and native oxides. Subsequently, the MBE UID GaN films were processed at various Tsub of 670 °C (Sample A), 680 °C (Sample B), 690 °C (Sample C), and 700 °C (Sample D). Sample X refers to the MOCVD GaN/Si template. During the growth process, active nitrogen species were maintained at a flow rate of 0.6 sccm using an SVT radio-frequency plasma source operating at a constant RF-plasma power of 300 W. A standard gallium (Ga) dual-filament effusion cell was utilized to deliver thermal evaporation of the Ga precursor with an imposing flux of 6.70×10^{-7} Torr. Real-time monitoring of epitaxial film quality along the $(11\bar{2}0)$ azimuth was conducted via reflection high-energy electron diffraction (RHEED) powered by a 12 keV electron gun with a set filament current of 1.5 μA. The thickness of the MBE GaN films was approximately 300 nm ± 10 nm, which was measured using a JSM7001F scanning electron microscope (SEM), as shown in Figure 1b. The sample surface morphologies were studied using atomic force microscopy (AFM) and SEM measurements. X-ray diffraction (XRD) examination was carried out employing a Bruker D2 PHASER using CuKα X-ray source radiation (λ = 0.15406 nm). The scanning range of 2 theta (2θ) spanned from 10 to 80° with a scan rate of 0.03° s^{-1}, a slit width of 0.1 mm, and an exposure period of 0.2 s. High-resolution XRD of Bede D1 using CuKα radiation provided information on dislocation densities in GaN that were evaluated from the full width at half maximum (FWHM) of the ω-rocking curves of the (002) and (102) reflections. Optical characteristics were obtained utilizing Raman and photoluminescence (PL) examinations. The PL measurement was carried out using a 325 nm He-Cd laser excitation source. The room-temperature Raman measurement was performed using an Argon ion laser with a wavelength of 514 nm and a power of 50 mW. Room-temperature vertical leakage–voltage (I-V) characterization was conducted through the floating substrate with a top metal of 74×74 μm^2 Ti/Al/Ni/Au stack, which was followed by the formation of ohmic contact using rapid thermal annealing at 800 °C for 60 s in a nitrogen atmosphere.

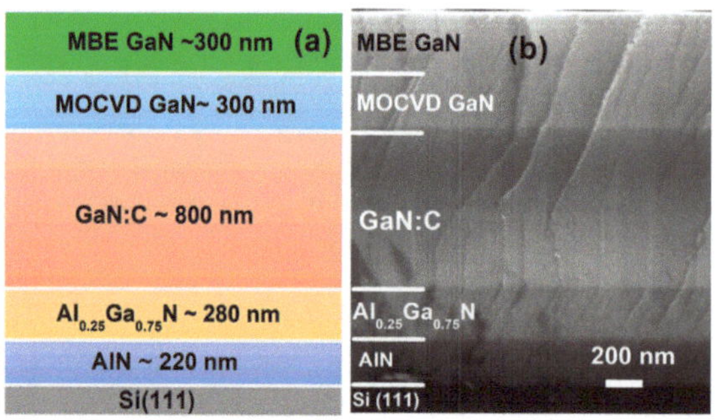

Figure 1. (**a**) Schematic diagram of MBE GaN film grown on MOCVD GaN/Si templates. (**b**) Cross-sectional SEM image of the GaN film.

3. Results and Discussion

3.1. Surface Reconstruction and Morphology Analysis

Figure 2 displays the captured RHEED patterns illustrating surface reconstruction of the GaN film at varying Tsub. Throughout the Tsub range up to 660 °C, the 2 × 2 streaky RHEED patterns of GaN templates were observed [Figure 2a inset]. Subsequently, as Tsub increased from 670 to 700 °C, the RHEED patterns evolved into spotty features with slight increases in size and intensities (Figure 2a–d). This transformation indicates the formation of the predominant three-dimensional (3D) islands on the surface of the GaN templates [29]. The initially spotty RHEED patterns then underwent an immediate transition into 1 × 1 streaky RHEED reconstruction shortly after the initiation of homoepitaxial growth. The rapid recovery of RHEED patterns indicates a transition from 3D islands to 2D growth modes, suggesting the achievement of smooth surface morphologies in MBE GaN [30]. These consistent streaky RHEED patterns persisted during the growth of samples A, B, and C (Figure 2e–g). Typically, the 1 × 1 streaky RHEED features are associated with

GaN films grown under the Ga-stable condition, which are characterized by Ga-rich and intermediate regimes where excessive Ga atoms on the surface manifest as Ga metallic droplet regions or wet regions [22,31]. It is evident that the RHEED characteristics are significantly affected by modulating Tsub. Notably, Sample A (Tsub = 670 °C) exhibited dimmed RHEED patterns, indicating the GaN film went through the Ga droplet regime. With the increasing Tsub, the streaky RHEED patterns of samples B and C became brighter, signifying well-desorbed excessive Ga atoms on the film surface. Conversely, persistence in spotted RHEED patterns with heightened intensities was observed during the growth of sample D (Tsub = 700 °C), indicating the adherence of the grown film to a 3D growth mode. The spotty RHEED features are acknowledged for creating a high surface roughness. This observation highlights the crucial role of modulating Tsub in facilitating the transition from 3D islands to 2D growth mode, thereby contributing to the preservation of smooth surface morphologies. The transition from streaky to spotty RHEED patterns has been reported as a function of substrate temperatures and the Ga/N ratio [31]. Hence, we infer that maintaining a growth temperature below 700 °C is essential to prevent over-desorption and provide a smooth film surface. For simplicity of the analysis and to emphasize the improved properties of GaN films, we primarily studied samples A, B, and C and compared them with sample X.

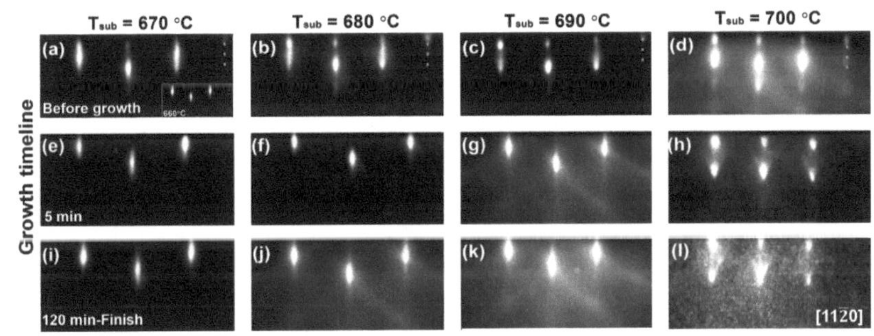

Figure 2. In situ RHEED patterns evolution of GaN films recorded at (**a–d**) different substrate temperatures of 670 °C, 680 °C, 690 °C, and 700 °C respectively; (**e–h**) after 5 minutes of deposition; and (**i–l**) immediately after growth completions. Insets in Figure 2a show RHEED patterns of GaN at a substrate temperature of 660°C.

The variation of surface roughness and pit densities (PD) of MBE GaN films with Tsub was explored using SEM and AFM examinations, as displayed in Figure 3. For reference, the AFM image of the template is also provided in Figure 3d. The relation between root mean square (Rq) and average PDs estimated relatively from SEM images is given in Figure 3h. The results revealed that the 3D-to-2D growth mode transition greatly improved the surface morphologies of MBE films, resulting in a decreased Rq relative to sample X (Rq = 0.50 nm). Sample A (Tsub = 670 °C) displayed an Rq of 0.18 nm, which was attributed to the lower surface desorption energy of Ga atoms during coalescence, enabling comparatively smooth surface morphology due to Ga droplets. However, the existence of excessive Ga atoms can be accompanied by the formation of nitrogen vacancies, thereby inducing a high surface pit density of ~2.12×10^9 cm^{-2}. As Tsub is raised to 680 °C and 690 °C, enhanced substrate thermal energy facilitates a more effective surface desorption of Ga atoms. This phenomenon resulted in a substantial decrease in surface pit densities. For sample B, the density dropped to 1.11×10^9 cm^{-2} with a corresponding Rq value of 0.28 nm. Notably, for sample C, the reduction was even more pronounced with the density dropping to 5.58×10^8 cm^{-2} and an associated Rq value of 0.30 nm. The results suggested that the transition from 3D-to-2D growth mode serves as a driving factor in decreasing surface pit densities. Improved surface morphology aids in mitigating the attenuation of free carrier density at the interface of AlGaN/GaN HEMT due to the scattering effect.

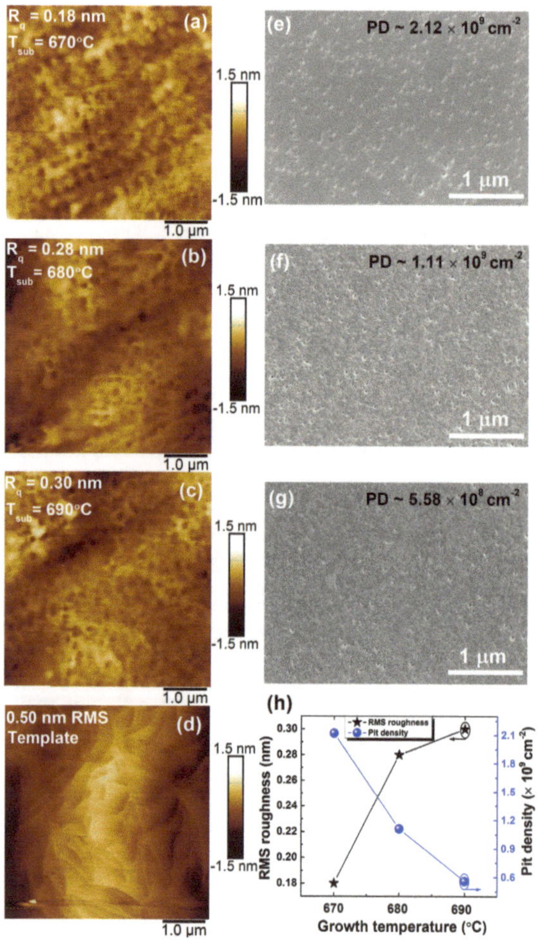

Figure 3. The 5 µm × 5 µm AFM images of the MBE GaN films grown under varied substrate temperatures of (**a**) 670, (**b**) 680, (**c**) 690 °C, and (**d**) MOCVD GaN template. The related SEM images displayed in (**e–h**) show the relation between RMS roughness and average surface pit densities plotted against substrate growth temperatures.

3.2. Structural Properties

Figure 4a shows the 2θ XRD patterns of the GaN film grown at 690 °C (sample C). It can be observed that the dominant diffraction peaks corresponding to (100), (002), (102), (003), (103), (004), and (202) planes of the GaN wurtzite structure according to the PDF 01-074-7289 card were detected in all samples (supporting information; see Figure S1). The K_β and K_α Si (111), Si (222), and AlN (002), (004) diffraction peaks arise from the template [32], while the broad and unidentified peak below 20° can be assigned to the X-ray source. The buildup of undesirable impurities in MBE films is anticipated to affect strain generation, leading to modified lattice parameters. Commonly, the relative lattice constants (a and c) were calculated from the peak positions of the common planes (002) and (102) using Bragg's equation [3]:

$$2d_{hkl}\sin\theta_{hkl} = n\lambda \quad (1)$$

$$\frac{1}{d_{hkl}^2} = \frac{4}{3}\left\{\frac{h^2+hk+k^2}{a^2}\right\} + \frac{l^2}{c^2} \quad (2)$$

where d_{hkl} is the interplanar spacing, which was calculated using Equation (2). The Miller indices (hkl) refer to the respective diffraction planes. θ_{hkl} is the reflection angle. The wavelength λ of the X-ray source is 0.15406 nm. The calculated lattice constants are given in Table 1. These lattice parameters are plotted and compared to those of the fully relaxed GaN values (a_o = 0.3189 nm and c_o = 0.5185 nm) [33], which are illustrated in Figure 4b. It is seen that the computed lattice parameters "a" for all homoepitaxial GaN were found to be slightly greater than the corresponding strain-free GaN value, revealing that GaN films suffered an in-plane tensile strain. This tensile strain typically arises from impurities, defects, and mismatches in the lattice parameters as well as variations in CTE when growing GaN on Si. The results showed that the lattice parameter "a" of the MBE films exhibited a decrease compared to that of the templates, and this reduction became more prominent with increasing Tsub. MOCVD-grown GaN films commonly exhibit a high concentration of free electrons, which is related to the incorporation of undesirable impurities and extended point defects during growth. Such factors probably cause the expansion of the "a" lattice parameter via the size effect [33,34]. Transitioning growth modes are suggested to effectively inhibit the accumulation of undesirable impurities into the growing film, resulting in better MBE GaN crystal quality. Additionally, the presence of Ga droplets has also made it possible to increase thermal expansion coefficients, which could be responsible for expanding the lattice parameter [35]. Indeed, the decrease in lattice parameter "a" in MBE GaN films with rising Tsub demonstrated efficient Ga atom desorption from the surface, which was facilitated by adequate surface desorption energy at a precisely controlled growth temperature. Therefore, the finding indicates that the reduction in lattice parameters can be attributed to either preventing undesirable impurity incorporation and limited extended defects or to the effect of growth temperatures.

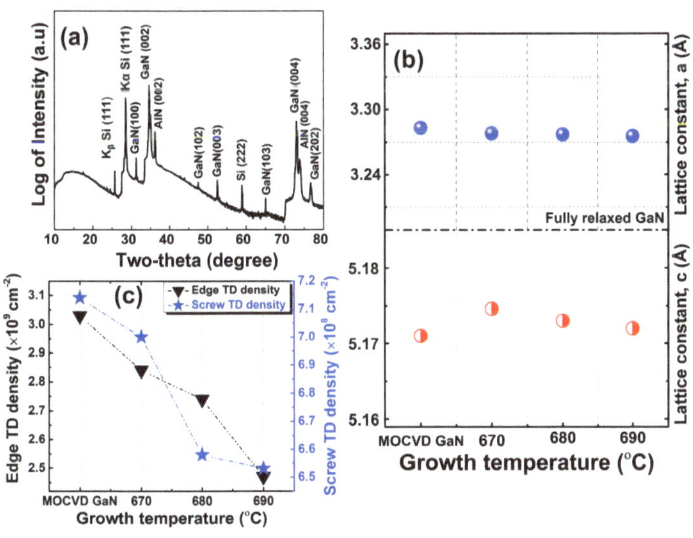

Figure 4. (**a**) The 2θ diffraction peaks of GaN films grown on MOCVD GaN/Si template at 690 °C (sample C), (**b**) variations of "a" (blue balls) and "c" (red balls) lattice parameters, and (**c**) edge (black triangle) and screw (blue star) threading dislocation densities with varied substrate growth temperatures in comparison with MOCVD GaN template.

Table 1. Values of estimated lattice parameters, FWHM extracted from XRD reflections, TDD, PD, RMS, and tensile stress values of MOCVD GaN and MBE GaN films grown at different growth temperatures.

Sample	Growth Temperature (°C)	Lattice Parameters		FWHM (Arcsec)		TDD (cm^{-2})		PD (cm^{-2})	RMS (nm)	Tensile Stress, σ (GPa)
		a (nm)	c (nm)	(002)	(102)	STD	ETD			
X	MOCVD GaN	0.3283	0.5171	596	966	7.14×10^8	3.03×10^9	-	0.51	0.36
A	670	0.3278	0.5174	590	939	7.00×10^8	2.84×10^9	2.12×10^9	0.18	0.31
B	680	0.3277	0.5173	572	889	6.58×10^8	2.74×10^9	1.11×10^9	0.28	0.27
C	690	0.3276	0.5172	570	880	6.53×10^8	2.47×10^9	5.58×10^8	0.30	0.12

The structural quality and reduction in dislocations of GaN films relative to sample X were studied by the X-ray rocking curve near the (002) and (102) reflections. The rocking curves and their broadening for all samples can be found in Figure S2 and Table 1, respectively. The curve broadening is associated with dislocations in the films that can be classified as screw-type dislocation (STD) and edge-type dislocation (ETD). The densities of each type of dislocation were determined through the full-width at half maximum (FWHM) of (002) and (102) reflections from Equations (3) and (4) [10]:

$$D_{screw} = \frac{\beta^2_{(002)}}{4.35 \times b^2_{screw}} \quad (3)$$

$$D_{edge} = \frac{\beta^2_{(102)} - \beta^2_{(002)}}{4.35 \times b^2_{edge}}, \quad (4)$$

where β is the FWHM values recorded for the (002) and (102) reflections. The Burgers vector magnitudes for STD and ETD are identified as b_s = 0.5185 nm and b_e = 0.3189 nm, respectively [36]. The calculated dislocation densities for GaN films are provided in Table 1, and their variation with Tsub is plotted in Figure 4c. The results revealed a considerable reduction in both types of dislocations in MBE GaN when compared to those observed in templates. In addition, with an increasing Tsub, the dislocations decrease further. Specifically, the film grown at 690 °C exhibited the most substantial decrease in dislocation densities compared to other samples. For instance, the estimated TDD of sample C demonstrated a remarkable reduction, lowering by 8.5% for STD and an impressive 18.5% for ETD when compared to the corresponding values in the template. The observed reduction in dislocation densities is basically consistent with the decrease in surface pit densities. The decline in edge dislocations indicates the impact resulting from the growth mode transitioning from 3D to 2D, which was achieved by modulating Tsub. This finding is consistent with our previous research, where the 3D-to-2D growth mode transition was facilitated by employing SiNx interlayers [14]. Threading dislocations were observed to bend at the interface of the growth mode transition with subsequent annihilation occurring during the 2D growth of MBE-grown GaN films. However, as discussed, the use of interlayers introduced potential drawbacks to device performance. Thus, the promoted 3D-to-2D growth mode transition at high growth temperatures via MBE presents a promising approach to minimize further threading dislocations achieved in cases of thin GaN thickness (<6 μm). Furthermore, STD arises from the inclusion of unexpected impurities, which work as a vertical leakage path caused by the charge-trapping phenomenon [9,37]. Meanwhile, ETD works as a Coulomb scattering center, which lessens the carrier mobility and builds up the sheet resistance [8]. Therefore, the significantly decreasing dislocation densities in our work hold promise for meeting the current state-of-the-art power devices for high-efficiency devices [38].

3.3. Optical and Electrical Characteristics

The effectively lowered dislocations are obtained from the initial in-plane stress relaxation, which is further investigated using Raman spectroscopy. Figure 5a illustrates the Raman spectra of GaN films that reveal two identifiable E2(H) and A1(LO) active phonons [10]. The shift in the E2(H) phonon with respect to the stress-free value of 567.5 cm^{-1} (see Figure 5b) is sensitive to the stress existent in GaN films. The E2(H) peak positions of samples X, A, B, and C were 565.92, 566.18, 566.37, and 566.96 cm^{-1}, respectively. The shift in the E2(H) phonon peak to the stress-free direction reflects the relaxation of the endured stress in GaN films, which agrees with the XRD result. The in-plane stress was computed from Equation (5) [10]:

$$\sigma = \frac{\Delta\omega}{4.3}\text{GPa}, \tag{5}$$

where $\Delta\omega$ is the difference between the stress-free peak position and the empirically observed E2(H) positions, and σ is the denotable in-plane stress expressed in GPa. The computed stress values are given in Table 1 and plotted in Figure 5c with the negative sign indicating tensile stress magnitudes. The observed stress relaxation trend (Figure 5c) revealed a gradual alleviation of residual tensile stress in GaN templates as compared to MBE-grown GaN. The effectiveness of stress release significantly improved with higher Tsub. Particularly notable is the substantial residual stress relaxation observed in GaN grown at 690 °C, resulting in a noteworthy reduction to 0.12 GPa. This suggests that the precise control of growth temperatures significantly contributes to stress relaxation during the 2D coalescence of MBE GaN films, consequently leading to decreased dislocation densities. The manifestation of this characteristic was evident in the improved surface morphologies observed through AFM along with low surface pit densities in MBE GaN. The stress relaxation also contributes to minimizing wafer cracking and bowing, especially in large-wafer-size manufacturing, ultimately improving device performance in GaN-based HEMT applications.

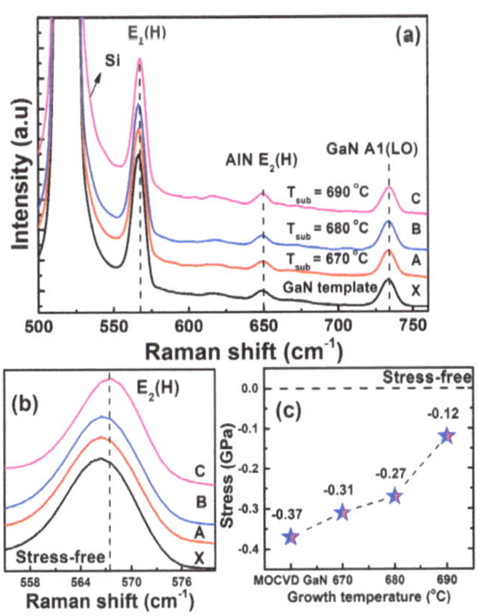

Figure 5. (**a**) Raman spectra of all GaN films; (**b**) Raman spectra in E2(H) regions; (**c**) the stress values (star-dashed line) displayed as a function of substrate growth temperature relative to the MOCVD template.

The influence of Tsub on luminescence properties was investigated through low-temperature photoluminescence spectra performed at 10 K, as shown in Figure 6. A sharp and narrow dominant near band-edge (NBE) emission associated with the exciton-bound neutral donor (D^0X) peak along with their LO-phonon replicas was observed at around 3.45 eV [39]. Samples X, A, B, and C had maximum D^0X peaks of 3.465, 3.466, 3.467, and 3.469 eV and corresponding FWHM values of 13.9, 16.1, 15.7, and 13.5 meV, respectively. The blueshift in D^0X emission peaks was toward the stress-free value of 3.471 eV for relaxed GaN film [39], which confirms the tensile stress relaxation.

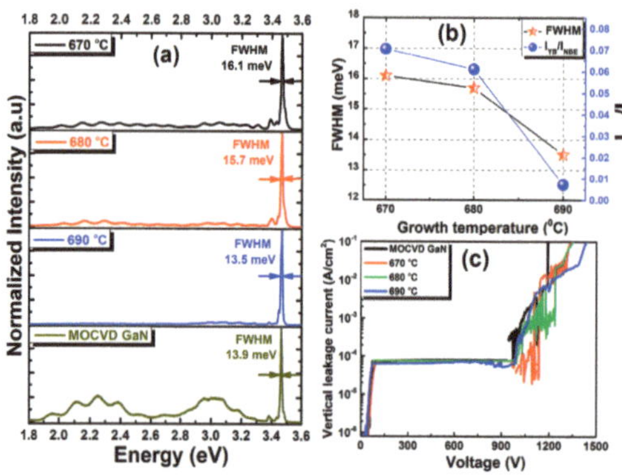

Figure 6. (**a**) The 10K PL spectra for MBE GaN grown at varying substrate temperatures in comparison with MOCVD GaN, (**b**) the full width at half maximum (FWHM) and the PL intensity ratio of I_{YB}/I_{NBE} of MBE GaN plotted as a function of growth temperatures. (**c**) Vertical leakage current versus voltage curves of GaN films.

Figure 6a shows that the MOCVD GaN exhibited the pronounced emissions of defect-related broad yellow luminescence (YL) and blue luminescence (BL) bands that were observed at around 2.2 eV and 2.8–3.0 eV, respectively [40,41]. The origin of YL and BL emissions is widely established as the attributions of edge-type dislocation [42], gallium vacancy (V_{Ga}) [43], and impurity-related complex defects [44]. It is noteworthy that the MBE GaN exhibited a remarkable decrease in both YL and BL emissions. In particular, the film grown at 690 °C completely eliminated the YL intensity and exhibited a sharper NBE emission (see Figure 6b). The findings demonstrated that the 3D to 2D growth mode transition established by modulating Tsub is advantageous in improving both their structure quality and optical properties. These characteristics were further investigated by evaluating the relative intensity ratio of the YL band to NBE emission (I_{YB}/I_{NBE}). Figure 6b plots the FWHM values of NBE emissions and I_{YB}/I_{NBE} as a function of Tsub. The results revealed that the I_{YB}/I_{NBE} ratio went down and NBE emissions became narrower as Tsub increased, affirming an improvement in film quality. Clearly, the XRD results indicated that MBE GaN films have exhibited a notable decrease in both types of dislocations with a more significant drop observed in ETD. However, this study illustrates that GaN films with the best surface morphology do not always result in the best luminescence properties. It proves the importance of controlling the growth temperature even when achieving the best-attained surface smoothness. In particular, the smoothest surface morphology found in GaN film grown at 670 °C, due to the residual Ga atoms, did not demonstrate the promising luminescence properties found in GaN grown at 690 °C. The slight remaining Ga atoms on the surface might potentially be ascribed to the YL emission. Therefore, the reduced YL intensity with the increase in Tsub can be a result of efficiently desorbing the remaining Ga atoms on the film surface by the inherent substrate thermal energy. Previously, Turkulets et al. [45] also identified the contribution of a certain molecule surface to the YL emission. The collective findings suggest that the Tsub of 690 °C can be considered the optimal growth temperature for high-quality MBE films with anticipated improvements in electrical properties.

It should be mentioned that MOCVD-grown GaN films frequently come with a considerable background of undesirable impurities [7]. The incorporation of impurities into the grown film and their buildup to the 2DEG region were proven to be accountable for the leakage current paths and reduced BV [46]. Hence, the impressively diminished defect-related YL emissions observed in MBE films denote an improvement in electrical properties that was examined using I-V measurement.

Figure 6c illustrates the vertical leakage current against the breakdown voltage curves. It is seen that the I–V characteristics observed in the 0–950 V range are attributed to the responses of C-doped GaN/AlGaN/AlN/Si [47]. Within a low applied voltage range of 0–75 V, the vertical leakage current climbed from 10^{-6} to 10^{-4} A/cm^2. Subsequently, once the applied voltage was raised to ~950 V, the leakage current value remained constant at 10^{-4} A/cm^2. Then, a rise in the vertical leakage current was observed when the applied voltage progressed from ~950 V to the highest BV value. The I–V characteristic in this region predominantly reflects the response of the top UID GaN layers. Hence, we primarily discuss the I–V behaviors in this region to assess the improved electrical properties of MBE-grown GaN layers. The applied voltage at which the vertical leakage current was 0.1 A/cm^2 is referred to as the BV value for all samples. The values of BV were found to be 1180, 1320, 1350, and 1450 V for samples X, A, B, and C, respectively. Clearly, all MBE-grown GaN samples exhibited a significant BV enhancement compared to the template. The greater breakdown voltage in MBE GaN is a result of both increased growth thickness and improved crystal quality [48]. In addition, the distinctly noisy vertical leakage characteristics observed in MBE GaN films grown at 670 °C and 680 °C, as well as in the template, could be attributed to the high densities of dislocations and surface pits [49,50]. Moreover, the origin of the failing electrical properties in samples A and B can be traced to the remaining surface Ga atoms and the nitrogen vacancy that act as interfacial defects and buffer traps. Notably, the noise-free vertical leakage current observed in the GaN grown at 690 °C resulted from a better surface morphology contributed by a tailor growth temperature.

It has been found that MBE-grown GaN films unveil a remarkable improvement in both vertical leakage current and BV compared to our prior study on MOCVD-grown GaN-based HEMTs [19]. The MBE GaN films exhibited excellent electrical properties, illustrating a current leakage of 0.1 A/cm^2 at an impressive BV of 1450 V. It showed a substantial enhancement over the previous MOCVD GaN structure, which exhibited a current leakage of 10 A/cm^2 at a BV of 1000 V. Remarkably, the growth mode transition technique significantly enhances the buffer breakdown voltage in thin MBE GaN (1.9 µm) compared to that of standard-sized GaN HEMT structures (5 µm). For instance, the breakdown voltages reported in different GaN structural designs were approximately 800 V [51], 900 V [52], and 1000 V [53] measured at a current density of 0.1 A/cm^2. The marked improvement is attributed to reduced edge threading dislocations contributed by 3D-to-2D growth mode transition in MBE GaN and a lower incorporation of donor-like impurities into film during coalescence. It not only highlights the advantages of using the MBE approach to perform 3D-to-2D growth mode transition and overcome impurity issues but also establishes it as the superior choice for enhanced electrical performance in GaN-based devices.

4. Conclusions

Our investigation into the MBE-grown UID GaN channel layer on MOCVD GaN templates with precise control over substrate growth temperatures has revealed significant advancements in crystal quality as well as morphological, optical, and electrical properties during the transition from 3D island to 2D growth mode. The optimal film quality was achieved at a growth temperature of 690 °C. The resulting relaxed strained MBE GaN film quality exhibited notably reduced edge dislocations and surface pit densities, measuring as low as 2.47 × 10^9 cm^{-2} and 5.58 × 10^8 cm^{-2}, respectively. Moreover, the tailored growth temperature conditions delivered an improved surface roughness (Rq = 0.3 nm), indicating an efficient prevention of surface desorption at elevated Tsub. The enhanced crystalline quality resulted in improved luminescence properties characterized by a narrower NBE emission band and a pronounced reduction in trap-related YL emission intensity. Furthermore, the optimized MBE-grown sample exhibited highly attractive electrical properties, featuring noise-free vertical leakage current characteristics under high applied voltage operations (BV~1450 V). The findings emphasize the efficacy of controlling MBE growth temperature to promote growth mode transition for optimizing GaN channel layer properties. It holds immense promise for achieving high efficiency and power for GaN-on-Si transistor applications. Our work collectively contributes to advancing the understanding of growth parameters in pursuing optimal GaN-based HEMT for next-generation semiconductor devices.

Supplementary Materials: The following supporting information can be downloaded at https://www.mdpi.com/article/10.3390/coatings14020191/s1, Figure S1: The 2-theta diffraction patterns of MBE-grown GaN films and the MOCVD GaN/Si template; Figure S2: X-ray rocking curves of all GaN films scanned on (a) (002) and (b) (102) reflections.

Author Contributions: T.T.M. conceived and designed the experiments, performed the epitaxial growth, analyzed the results, and wrote the manuscript. J.-J.D. supported the MOCVD GaN templates and analyzed the results. L.T.H. supported the I–V characterization. H.-C.W. supported the morphology observation. W.-C.C. is the advisor who supervised the experiments and revised the manuscript. H.H.L. discussed the results and revised the manuscript. All authors have analyzed the data, discussed the data, and revised the manuscript. All authors have read and agreed to the published version of the manuscript.

Funding: This research was funded by the National Science and Technology Council, Taiwan, under grant No. NSTC 112-2112-M-A49-399-043 and NSTC 112-2218-E-A49-018.

Institutional Review Board Statement: Not applicable.

Informed Consent Statement: Not applicable.

Data Availability Statement: Data are contained within the article or Supplementary Materials.

Conflicts of Interest: The authors declare no conflicts of interest.

References

1. Kim, J.-G.; Cho, C.; Kim, E.; Hwang, J.S.; Park, K.-H.; Lee, J.-H. High breakdown voltage and low-current dispersion in AlGaN/GaN HEMTs with high-quality AlN buffer layer. *IEEE Trans. Electron Devices* **2021**, *68*, 1513–1517. [CrossRef]
2. Ma, J.; Matioli, E. High-voltage and low-leakage AlGaN/GaN tri-anode Schottky diodes with integrated tri-gate transistors. *IEEE Electron Device Lett.* **2016**, *38*, 83–86. [CrossRef]
3. Romanitan, C.; Mihalache, I.; Tutunaru, O.; Pachiu, C. Effect of the lattice mismatch on threading dislocations in heteroepitaxial GaN layers revealed by X-ray diffraction. *J. Alloys Compd.* **2021**, *858*, 157723. [CrossRef]
4. Buffolo, M.; Caria, A.; Piva, F.; Roccato, N.; Casu, C.; De Santi, C.; Trivellin, N.; Meneghesso, G.; Zanoni, E.; Meneghini, M. Defects and reliability of GaN-based LEDs: Review and perspectives. *Phys. Status Solidi* **2022**, *219*, 2100727. [CrossRef]
5. Chatterjee, A.; Khamari, S.K.; Kumar, R.; Porwal, S.; Bose, A.; Sharma, T. Role of threading dislocations and point defects in the performance of GaN-based metal-semiconductor-metal ultraviolet photodetectors. *Superlattices Microstruct.* **2020**, *148*, 106733. [CrossRef]
6. Joh, J.; Del Alamo, J.A. A current-transient methodology for trap analysis for GaN high electron mobility transistors. *IEEE Trans. Electron. Devices* **2010**, *58*, 132–140. [CrossRef]
7. Reshchikov, M.; Vorobiov, M.; Andrieiev, O.; Ding, K.; Izyumskaya, N.; Avrutin, V.; Usikov, A.; Helava, H.; Makarov, Y. Determination of the concentration of impurities in GaN from photoluminescence and secondary-ion mass spectrometry. *Sci. Rep.* **2020**, *10*, 2223. [CrossRef] [PubMed]
8. Wong, Y.-Y.; Chang, E.Y.; Yang, T.-H.; Chang, J.-R.; Ku, J.-T.; Hudait, M.K.; Chou, W.-C.; Chen, M.; Lin, K.-L. The roles of threading dislocations on electrical properties of AlGaN/GaN heterostructure grown by MBE. *J. Electrochem. Soc.* **2010**, *157*, H746. [CrossRef]
9. Jarndal, A.; Arivazhagan, L.; Nirmal, D. On the performance of GaN-on-silicon, silicon-carbide, and diamond substrates. *Int. J. RF Microw. Comput.-Aided Eng.* **2020**, *30*, e22196. [CrossRef]

10. Lee, H.-P.; Perozek, J.; Rosario, L.; Bayram, C. Investigation of AlGaN/GaN high electron mobility transistor structures on 200-mm silicon (111) substrates employing different buffer layer configurations. *Sci. Rep.* **2016**, *6*, 37588. [CrossRef]
11. Dalapati, P.; Urata, S.; Egawa, T. Investigation of AlGaN/GaN high electron mobility transistors on Silicon (111) substrates employing multi-stacked strained layer superlattice structures. *Superlattices Microstruct.* **2020**, *147*, 106709. [CrossRef]
12. Liu, J.; Huang, Y.; Sun, X.; Zhan, X.; Sun, Q.; Gao, H.; Feng, M.; Zhou, Y.; Ikeda, M.; Yang, H. Wafer-scale crack-free 10 μm-thick GaN with a dislocation density of 5.8×10^7 cm^{-2} grown on Si. *J. Phys. D Appl. Phys.* **2019**, *52*, 425102. [CrossRef]
13. Amano, H.; Iwaya, M.; Kashima, T.; Katsuragawa, M.; Akasaki, I.; Han, J.; Hearne, S.; Floro, J.A.; Chason, E.; Figiel, J. Stress and defect control in GaN using low temperature interlayers. *Jpn. J. Appl. Phys.* **1998**, *37*, L1540. [CrossRef]
14. Dai, J.-J.; Liu, C.-W.; Wu, S.-K.; Huynh, S.-H.; Jiang, J.-G.; Yen, S.-A.; Mai, T.T.; Wen, H.-C.; Chou, W.-C.; Hu, C.-W. Improving Transport Properties of GaN-Based HEMT on Si (111) by Controlling SiH4 Flow Rate of the SiN x Nano-Mask. *Coatings* **2020**, *11*, 16. [CrossRef]
15. Chang, S.; Lung Wei, L.; Tung Luong, T.; Chang, C.; Chang, L. Threading dislocation reduction in three-dimensionally grown GaN islands on Si (111) substrate with AlN/AlGaN buffer layers. *J. Appl. Phys.* **2017**, *122*, 105306. [CrossRef]
16. Tanaka, S.; Honda, Y.; Sawaki, N.; Hibino, M. Structural characterization of GaN laterally overgrown on a (111) Si substrate. *Appl. Phys. Lett.* **2001**, *79*, 955–957. [CrossRef]
17. Strittmatter, A.; Rodt, S.; Reißmann, L.; Bimberg, D.; Schröder, H.; Obermeier, E.; Riemann, T.; Christen, J.; Krost, A. Maskless epitaxial lateral overgrowth of GaN layers on structured Si (111) substrates. *Appl. Phys. Lett.* **2001**, *78*, 727–729. [CrossRef]
18. Zhong, Y.; Zhang, J.; Wu, S.; Jia, L.; Yang, X.; Liu, Y.; Zhang, Y.; Sun, Q. A Review on the GaN-on-Si Power Electronic Devices. *Fundam. Res.* **2021**, *2*, 462–475. [CrossRef]
19. Dai, J.-J.; Mai, T.T.; Nallasani, U.R.; Chang, S.-C.; Hsiao, H.-I.; Wu, S.-K.; Liu, C.-W.; Wen, H.-C.; Chou, W.-C.; Wang, C.-P. The effect of heavy Fe-doping on 3D growth mode and Fe diffusion in GaN for high power HEMT application. *Materials* **2022**, *15*, 2058. [CrossRef] [PubMed]
20. Zhang, H.; Xue, J.; Fu, Y.; Yang, M.; Zhang, Y.; Duan, X.; Qiang, W.; Li, L.; Sun, Z.; Ma, X. Study on the effects of growth rate on GaN films properties grown by plasma-assisted molecular beam epitaxy. *J. Cryst. Growth* **2020**, *535*, 125539. [CrossRef]
21. Wu, Y.-Z.; Liu, B.; Li, Z.-H.; Tao, T.; Xie, Z.-L.; Xiu, X.-Q.; Chen, P.; Chen, D.-J.; Lu, H.; Shi, Y. Homo-epitaxial growth of high crystal quality GaN thin films by plasma assisted–molecular beam epitaxy. *J. Cryst. Growth* **2019**, *506*, 30–35. [CrossRef]
22. Heying, B.; Averbeck, R.; Chen, L.; Haus, E.; Riechert, H.; Speck, J. Control of GaN surface morphologies using plasma-assisted molecular beam epitaxy. *J. Appl. Phys.* **2000**, *88*, 1855–1860. [CrossRef]
23. Liang, K.; Sun, X.; Wu, G.; Zhang, L.; Liu, S.; Gan, Z. The investigation of molecular beam epitaxy growth of GaN by molecular dynamics simulation. *Comput. Mater. Sci.* **2020**, *173*, 109426. [CrossRef]
24. Shrestha, N.M.; Li, Y.; Suemitsu, T.; Samukawa, S. Electrical characteristic of AlGaN/GaN high-electron-mobility transistors with recess gate structure. *IEEE Trans. Electron Devices* **2019**, *66*, 1694–1698. [CrossRef]
25. Cao, Y.; Xing, H.; Jena, D. Polarization-mediated remote surface roughness scattering in ultrathin barrier GaN high-electron mobility transistors. *Appl. Phys. Lett.* **2010**, *97*, 222116. [CrossRef]
26. Kruse, C.; Einfeldt, S.; Böttcher, T.; Hommel, D.; Rudloff, D.; Christen, J. Spatially modified layer properties related to the formation of gallium droplets on GaN (0001) surfaces during plasma-assisted molecular-beam epitaxy. *Appl. Phys. Lett.* **2001**, *78*, 3827–3829. [CrossRef]
27. Sampath, A.; Garrett, G.; Collins, C.; Boyd, P.; Choe, J.; Newman, P.; Shen, H.; Wraback, M.; Molnar, R.; Caissie, J. Effect of Ga-rich growth conditions on the optical properties of GaN films grown by plasma-assisted molecular beam epitaxy. *J. Vac. Sci. Technol. B Microelectron. Nanometer Struct. Process. Meas. Phenom.* **2004**, *22*, 1487–1490. [CrossRef]
28. Krishna, S.; Aggarwal, N.; Mishra, M.; Maurya, K.; Singh, S.; Dilawar, N.; Nagarajan, S.; Gupta, G. Correlation of growth temperature with stress, defect states and electronic structure in an epitaxial GaN film grown on c-sapphire via plasma MBE. *Phys. Chem. Chem. Phys.* **2016**, *18*, 8005–8014. [CrossRef] [PubMed]
29. Adelmann, C.; Brault, J.; Jalabert, D.; Gentile, P.; Mariette, H.; Mula, G.; Daudin, B. Dynamically stable gallium surface coverages during plasma-assisted molecular-beam epitaxy of (0001) GaN. *J. Appl. Phys.* **2002**, *91*, 9638–9645. [CrossRef]
30. Able, A.; Wegscheider, W.; Engl, K.; Zweck, J. Growth of crack-free GaN on Si (1 1 1) with graded AlGaN buffer layers. *J. Cryst. Growth* **2005**, *276*, 415–418. [CrossRef]
31. Smith, A.; Ramachandran, V.; Feenstra, R.; Greve, D.; Ptak, A.; Myers, T.; Sarney, W.; Salamanca-Riba, L.; Shin, M.; Skowronski, M. Surface reconstruction during molecular beam epitaxial growth of GaN (0001). *Mater. Res. Soc. Internet J. Nitride Semicond. Res.* **1998**, *3*, e12. [CrossRef]
32. Krishna, S.; Aggarwal, N.; Gundimeda, A.; Sharma, A.; Husale, S.; Maurya, K.; Gupta, G. Correlation of donor-acceptor pair emission on the performance of GaN-based UV photodetector. *Mater. Sci. Semicond. Process.* **2019**, *98*, 59–64. [CrossRef]
33. Darakchieva, V.; Monemar, B.; Usui, A. On the lattice parameters of GaN. *Appl. Phys. Lett.* **2007**, *91*, 031911. [CrossRef]
34. Van de Walle, C.G. Effects of impurities on the lattice parameters of GaN. *Phys. Rev. B* **2003**, *68*, 165209. [CrossRef]
35. Moram, M.; Vickers, M. X-ray diffraction of III-nitrides. *Rep. Prog. Phys.* **2009**, *72*, 036502. [CrossRef]
36. Weyher, J.v.L.; Lazar, S.; Macht, L.; Liliental-Weber, Z.; Molnar, R.; Müller, S.; Sivel, V.; Nowak, G.; Grzegory, I. Orthodox etching of HVPE-grown GaN. *J. Cryst. Growth* **2007**, *305*, 384–392. [CrossRef]

37. Besendörfer, S.; Meissner, E.; Lesnik, A.; Friedrich, J.; Dadgar, A.; Erlbacher, T. Methodology for the investigation of threading dislocations as a source of vertical leakage in AlGaN/GaN-HEMT heterostructures for power devices. *J. Appl. Phys.* **2019**, *125*, 095704. [CrossRef]
38. Heuken, L.; Kortemeyer, M.; Ottaviani, A.; Schröder, M.; Alomari, M.; Fahle, D.; Marx, M.; Heuken, M.; Kalisch, H.; Vescan, A. Analysis of an AlGaN/AlN super-lattice buffer concept for 650-V low-dispersion and high-reliability GaN HEMTs. *IEEE Trans. Electron Devices* **2020**, *67*, 1113–1119. [CrossRef]
39. Kornitzer, K.; Ebner, T.; Thonke, K.; Sauer, R.; Kirchner, C.; Schwegler, V.; Kamp, M.; Leszczynski, M.; Grzegory, I.; Porowski, S. Photoluminescence and reflectance spectroscopy of excitonic transitions in high-quality homoepitaxial GaN films. *Phys. Rev. B* **1999**, *60*, 1471. [CrossRef]
40. Reshchikov, M.; Vorobiov, M.; Demchenko, D.; Özgür, Ü.; Morkoç, H.; Lesnik, A.; Hoffmann, M.; Hörich, F.; Dadgar, A.; Strittmatter, A. Two charge states of the C N acceptor in GaN: Evidence from photoluminescence. *Phys. Rev. B* **2018**, *98*, 125207. [CrossRef]
41. Calleja, E.; Sánchez, F.; Basak, D.; Sanchez-Garcia, M.; Munoz, E.; Izpura, I.; Calle, F.; Tijero, J.; Sánchez-Rojas, J.; Beaumont, B. Yellow luminescence and related deep states in undoped GaN. *Phys. Rev. B* **1997**, *55*, 4689. [CrossRef]
42. Zhao, D.; Jiang, D.; Yang, H.; Zhu, J.; Liu, Z.; Zhang, S.; Liang, J.; Li, X.; Li, X.; Gong, H. Role of edge dislocations in enhancing the yellow luminescence of n-type GaN. *Appl. Phys. Lett.* **2006**, *88*, 241917. [CrossRef]
43. Xie, Z.; Sui, Y.; Buckeridge, J.; Sokol, A.A.; Keal, T.W.; Walsh, A. Prediction of multiband luminescence due to the gallium vacancy–oxygen defect complex in GaN. *Appl. Phys. Lett.* **2018**, *112*, 262104. [CrossRef]
44. Lyons, J.; Janotti, A.; Van de Walle, C. Carbon impurities and the yellow luminescence in GaN. *Appl. Phys. Lett.* **2010**, *97*, 152108. [CrossRef]
45. Turkulets, Y.; Shauloff, N.; Chaulker, O.H.; Shapira, Y.; Jelinek, R.; Shalish, I. The GaN (0001) yellow-luminescence-related surface state and its interaction with air. *Surf. Interfaces* **2023**, *38*, 102834. [CrossRef]
46. Borga, M.; Meneghini, M.; Stoffels, S.; Li, X.; Posthuma, N.; Van Hove, M.; Decoutere, S.; Meneghesso, G.; Zanoni, E. Impact of substrate resistivity on the vertical leakage, breakdown, and trapping in GaN-on-Si E-mode HEMTs. *IEEE Trans. Electron. Devices* **2018**, *65*, 2765–2770. [CrossRef]
47. Song, C.; Yang, X.; Ji, P.; Tang, J.; Wu, S.; Xu, Y.; Imran, A.; Wang, M.; Yang, Z.; Xu, F. Impact of Silicon Substrate with Low Resistivity on Vertical Leakage Current in AlGaN/GaN HEMTs. *Appl. Sci.* **2019**, *9*, 2373. [CrossRef]
48. Rowena, I.B.; Selvaraj, S.L.; Egawa, T. Buffer thickness contribution to suppress vertical leakage current with high breakdown field (2.3 MV/cm) for GaN on Si. *IEEE Electron. Device Lett.* **2011**, *32*, 1534–1536. [CrossRef]
49. Tajalli, A.; Borga, M.; Meneghini, M.; De Santi, C.; Benazzi, D.; Besendörfer, S.; Püsche, R.; Derluyn, J.; Degroote, S.; Germain, M. Vertical leakage in GaN-on-Si stacks investigated by a buffer decomposition experiment. *Micromachines* **2020**, *11*, 101. [CrossRef]
50. Choi, F.; Griffiths, J.; Ren, C.; Lee, K.; Zaidi, Z.; Houston, P.; Guiney, I.; Humphreys, C.; Oliver, R.; Wallis, D. Vertical leakage mechanism in GaN on Si high electron mobility transistor buffer layers. *J. Appl. Phys.* **2018**, *124*, 055702. [CrossRef]
51. Borga, M.; Meneghini, M.; Benazzi, D.; Canato, E.; Püsche, R.; Derluyn, J.; Abid, I.; Medjdoub, F.; Meneghesso, G.; Zanoni, E. Buffer breakdown in GaN-on-Si HEMTs: A comprehensive study based on a sequential growth experiment. *Microelectron. Reliab.* **2019**, *100*, 113461. [CrossRef]
52. Chen, X.; Zhong, Y.; Yan, S.; Guo, X.; Gao, H.; Sun, X.; Wang, H.; Li, F.; Zhou, Y.; Feng, M. Enhanced breakdown voltage and dynamic performance of GaN HEMTs with AlN/GaN superlattice buffer. *J. Phys. D Appl. Phys.* **2023**, *56*, 355101. [CrossRef]
53. Tajalli, A.; Meneghini, M.; Besendörfer, S.; Kabouche, R.; Abid, I.; Püsche, R.; Derluyn, J.; Degroote, S.; Germain, M.; Meissner, E. High breakdown voltage and low buffer trapping in superlattice gan-on-silicon heterostructures for high voltage applications. *Materials* **2020**, *13*, 4271. [CrossRef]

coatings

MDPI

Article

Epitaxial Growth of Cobalt Oxide Thin Films on Sapphire Substrates Using Atmospheric Pressure Mist Chemical Vapor Deposition

Hou-Guang Chen *, Huei-Sen Wang *, Sheng-Rui Jian, Tung-Lun Yeh and Jing-Yi Feng

Department of Materials Science and Engineering, I-Shou University, Kaohsiung City 84001, Taiwan
* Correspondence: houguang@isu.edu.tw (H.-G.C.); huei@isu.edu.tw (H.-S.W.)

Abstract: This study demonstrated the epitaxial growth of single-phase (111) CoO and (111) Co_3O_4 thin films on *a*-plane sapphire substrates using an atmospheric pressure mist chemical vapor deposition (mist-CVD) process. The phase structure of the grown cobalt oxide films was manipulated by controlling the growth temperature and process ambient, confirmed through X-ray diffraction, Raman spectroscopy, and X-ray photoelectron spectroscopy. Furthermore, the electrical properties of Co_3O_4 films were significantly improved after thermal annealing in oxygen ambient, exhibiting a stable p-type conductivity with an electrical resistivity of 8.35 Ohm cm and a carrier concentration of 4.19×10^{16} cm^{-3}. While annealing CoO in oxygen atmosphere, the Co_3O_4 films were found to be most readily formed on the CoO surface due to the oxidation reaction. The orientation of the atomic arrangement of formed Co_3O_4 was epitaxially constrained by the underlying CoO epitaxial layer. The oxidation of CoO to Co_3O_4 was largely driven by outward diffusion of cobalt cations, resulting in the formation of pores in the interior of formed Co_3O_4 films.

Keywords: cobalt oxide films; epitaxial growth; p-type metal oxide semiconductor; mist chemical vapor deposition

1. Introduction

Metal oxide semiconductors are regarded as promising, next-generation, functional materials due to their various functionalities, including transparency, electrical conductivity, ferroelectricity, ferromagnetism, and superconductivity [1–4]. Thus, they are in demand for a wide range of applications, including transparent electronics, gas sensors, energy storage, thin film transistors, photodetectors, solar cells, catalysis, memory devices, etc. Most metal oxide semiconductors exhibit n-type conductivity, such as Al-doped ZnO, Sn-doped In_2O_3, and Sn-doped Ga_2O_3. However, several metal oxide semiconductors demonstrate p-type conductivity, including cobalt oxide, nickel oxide, and copper oxide [5]. Among these materials, cobalt oxide, studied in this work, has attracted intensive attention in the view of its applications in various fields, including photovoltaics devices [6,7], catalysis [8], gas sensors [9], water splitting [10], resistive switching [11], rechargeable batteries [12], supercapacitors [13], photodetectors [14], and spintronic devices [15]. There are two stable cobalt oxide phases: CoO (rock-salt structure, *a* = 0.426 nm) and Co_3O_4 (normal spinel structure, *a* = 0.808 nm). Both phases are direct band gap p-type semiconductors. CoO has a cubic rock-salt structure with high-spin Co^{2+} cations occupying the octahedral sites. Co_3O_4, on the other hand, has a normal spinel structure, with the mixed Co^{2+} and Co^{3+} cations occupying the tetrahedral and octahedral sites, respectively.

To date, cobalt oxide thin films have been successfully prepared through various methods, such as radio frequency reactive sputtering [7], spray pyrolysis [16], sol–gel [17], pulsed laser deposition (PLD) [18], molecular beam epitaxy (MBE) [19], atomic layer deposition (ALD) [20], and metalorganic chemical vapor deposition (MOCVD) [21]. The cobalt oxide thin films of both CoO and Co_3O_4 can be synthesized selectively by various thin film deposition technologies by controlling the process temperature and the composition of the reactive atmosphere during thin film deposition [18,22]. Furthermore, the increasing interest in the cobalt oxide thin films has prompted the development of epitaxial growth technology for obtaining high-quality single-crystalline films. In contrast to the polycrystalline films comprising randomly oriented grains, epitaxial film having well-ordered structures is essential to developing high-performance cobalt-oxide-based spintronic or optoelectronic devices. Although the epitaxial growth of cobalt oxide thin film can be implemented through PLD [18], MBE [19], ALD [20], and MOCVD [21], the processes mentioned above still need to adopt a high vacuum pumping system, complex gas transport system, or flammable and toxic organometallic precursors. In this study, we employed atmospheric pressure solution-processed mist chemical vapor deposition (mist-CVD) to perform the epitaxial growth of cobalt oxide films on sapphire substrates. Mist-CVD possesses the advantages of a non-vacuum system, low cost, low energy consumption, and non-toxicity. Furthermore, different works have shown that various functional metal oxide epitaxial films, including ZnO, Ga_2O_3, $ZnGa_2O_4$, In_2O_3, and SnO_2, can be successfully grown using mist-CVD [23–27]. Therefore, the mist-CVD technique has great potential for application in the epitaxial growth of high-quality metal oxide thin films. However, to the best of our knowledge, less attention has been given to the study of the mist-CVD growth of cobalt oxide epitaxial films. Moreover, the selective epitaxial growth of single-phase CoO and Co_3O_4 films using mist-CVD by controlling growth conditions, such as temperature and ambient, has not been explored yet. This work demonstrated selective epitaxial growth of single-phase CoO and Co_3O_4 films on sapphire substrates via the atmospheric pressure mist-CVD process. In addition, the influence of post-annealing treatment on the electrical and structural properties of grown cobalt oxide films was also investigated in this work.

2. Materials and Methods

The epitaxial growth of cobalt oxide thin films was performed using an atmospheric pressure mist-CVD system. The details of the mist-CVD system can be found in our previous works [28]. Prior to the growth of cobalt oxide thin films, as received, *a*-plane sapphire substrates (12 mm × 12 mm) were cleaned ultrasonically in acetone for 10 min, rinsed with deionized water, and then blown with dry nitrogen. For the mist-CVD growth of cobalt oxide thin films, a precursor solution was prepared in a solution of 0.05 M cobalt (II) acetate tetrahydrate ($Co(CH_3COO)_2\,4H_2O$, 98%, Showa) dissolved into a solution of deionized water and acetic acid mixed in a 7:3 ratio. The precursor solution was atomized to produce mist droplets using ultrasonic transducers at 2.4 MHz and transferred into a reaction furnace, with a carrier gas of either nitrogen (N_2, 99.99% purity) or oxygen (O_2, 99.99%, purity), at a flow rate of 2 L/min. The growth of cobalt oxide films on sapphire substrates was implemented at the temperature of 450–500 °C for 60 min. The post-annealing treatment was implemented using an ULVAC MILA-5000 annealing system. The cobalt oxide films were annealed at 650 °C for 30 min under oxygen ambient.

The crystal structure of cobalt oxide films was analyzed using X-ray diffraction (XRD). A X'Pert PRO-MPD (PANalytical) diffractometer, equipped with a hybrid monochromator, was employed for conducting 2θ-ω scan measurements. Additionally, we used the D8 Discover (Bruker, Billerica, MA, USA) diffractometer for X-ray rocking curve (XRC) and φ-scan measurements. The Raman spectra were recorded using a Horiba Jobin Yvon HR800 system, using a 633 nm He-Ne laser as an excitation source. The surface morphologies of the grown films were examined through scanning electron microscopy (SEM, Hitachi, Tokyo, Japan, S3400). The chemical state of the cobalt oxide films was analyzed through X-ray photoelectron spectroscopy (XPS) using a JEOL JAMP 9500 F system. The electrical properties of the cobalt oxide films were examined through Hall-effect measurement (Ecopia, Anyang City, Republic of Korea, HMS-3000). The optical transmission spectrum was obtained using a UV-visible spectrophotometer (UV-Vis, Thermo Scientific, Waltham, MA, USA, Evolution 201). The microstructure of the cobalt oxide–sapphire interface was characterized using cross-sectional transmission electron microscopy (XTEM, JEOL

JEM-2100F). The XTEM specimen was prepared using a focused ion beam (FIB, FEI, Nova-200 NanoLab). The high-resolution TEM (HRTEM) images were analyzed using Gatan Digital Micrograph^{TM} version 3.10.1 (Pleasanton, CA, USA) software.

3. Results

3.1. Epitaxial Growth of CoO and Co_3O_4 Thin Films on Sapphire Substrates

Based on previous successful work on the growth of ZnO epitaxial films using mist-CVD using the solution of zinc acetate dissolved into a solution of deionized water and acetate acid [28,29], we replaced zinc acetate with cobalt acetate in this work. The cobalt oxide thin films were grown at 450 and 500 °C using mist-CVD under nitrogen and oxygen ambient, respectively. We performed XRD to identify the phase of the grown films. Figure 1 depicts the 2θ-ω scan profiles of cobalt oxide thin films grown under various conditions. Figure 1b,c show the XRD profiles of the films grown under nitrogen ambient at 500 °C (Figure 1b) and oxygen ambient at the temperature of 450 °C (Figure 1c), respectively, revealing that the diffracted peaks at 2θ positions of 36.62° and 77.83° correspond to the (111) and (222) planes of the CoO phase (JCPDS No. 48-1719) and the diffracted peak at 38.59° and 59.42° corresponds to the (222) and (333) plane of the Co_3O_4 phase (JCPDS No. 42-1467), together with intense peaks of sapphire substrate (2θ = 37.7°). It is shown that CoO and Co_3O_4 phases coexisted in both grown cobalt oxide films. Notably, the cobalt oxide film grown under nitrogen ambient at the temperature of 450 °C solely exhibited a single CoO phase, as shown in Figure 1a. On the other hand, only the Co_3O_4 phase was detected in the cobalt oxide film grown at 500 °C under oxygen ambient, as shown in Figure 1d. Nevertheless, all of the cobalt oxide thin films grown on sapphire substrates under various conditions exhibited the <111> texture. The formation of CoO and Co_3O_4 significantly depends on the growth temperature and ambient. In general, the usage of metalorganic precursor containing only divalent Co^{2+} tends to favor the formation of a less stable CoO phase containing cobalt cations in a lower oxidation state (Co^{2+}) [21,30]. However, they can still be used to grow the single-phase Co_3O_4 films comprising cobalt cations in two different oxidation states, Co^{2+} and Co^{3+}, by adjusting CVD growth conditions (temperature and ambience). According to the abovementioned results, the use of divalent cobalt (cobalt (II) acetate) precursors for mist-CVD growth facilitated the growth of CoO films at relatively low temperatures in nitrogen (oxygen-deficient) atmosphere. On the other hand, the relatively high temperature and oxygen-rich atmosphere were beneficial for maintaining cobalt in its higher oxidation state, leading to the growth of single-phase Co_3O_4. The crystallite sizes of grown CoO and Co_3O_4 films can be estimated using the Scherrer method based on the results of the 2θ-ω scan [18,31]. Utilizing the Scherrer equation and the full-width at half-maximum (FWHM) of diffracted peaks corresponding to CoO (222) and Co_3O_4 (333), the average crystallite sizes of CoO and Co_3O_4 were determined to be 53 nm and 30 nm, respectively. In addition, the out-of-plane strains in both films were calculated from the 2θ positions of the diffracted peaks corresponding to CoO (222) and Co_3O_4 (333), and they were found to be −0.32% and −0.1%, respectively. These negative strains along the out-of-plane direction resulted from the presence of tensile in-plane strain in both of the grown epitaxial films. Furthermore, we carried out Raman analysis to explore the structural details in both of the cobalt oxide films grown at 450 °C in nitrogen ambient for CoO and 500 °C in oxygen ambient for Co_3O_4, respectively. Figure 2a,b display the corresponding Raman spectra of CoO and Co_3O_4 films. In Figure 2a, due to the NaCl-type centrosymmetric lattice structure, CoO exhibited a weak Raman scattering [32,33]. Except for the sharp peaks of sapphire substrates, as indicated by an asterisk, a broad asymmetric band with weak intensity appeared at 530 cm^{-1}, indicated by a square mark. Although the first-order phonon scattering is forbidden based on the selection rules for the NaCl-type lattice structure, the presence of cobalt vacancy or other structural defects in the CoO lattice can lead to the emergence of the first-order phonon scattering. The broad band observed at 530 cm^{-1} can be attributed to this defect-induced one-phonon longitudinal optical (LO) Raman scatter [32]. The Raman spectrum of the Co_3O_4 film (Figure 2b) reveals the presence of five sharp peaks, which are located at 194, 482, 521, 619, and 691 cm^{-1}. These peaks are in good agreement with the five Raman active phonon modes (A_{1g} + $3F_{2g}$ + E_g) associated with the spinel Co_3O_4 reported by Hadjiev et al. [34]. In comparison to CoO, the Co_3O_4 film exhibits lower optical transmittance at 633 nm, leading to the absorption of most incident lasers. However, CoO has higher optical transmittance and weaker Raman scattering due to its NaCl-type centrosymmetric lattice structure. This behavior is likely responsible for the appearance of Raman scattering corresponding to the underlying sapphire substrate. Based on the results of the XRD 2θ-ω scan and Raman scatter analysis, single-phase CoO and Co_3O_4 films can be obtained by controlling growth temperature and ambience during the mist-CVD process.

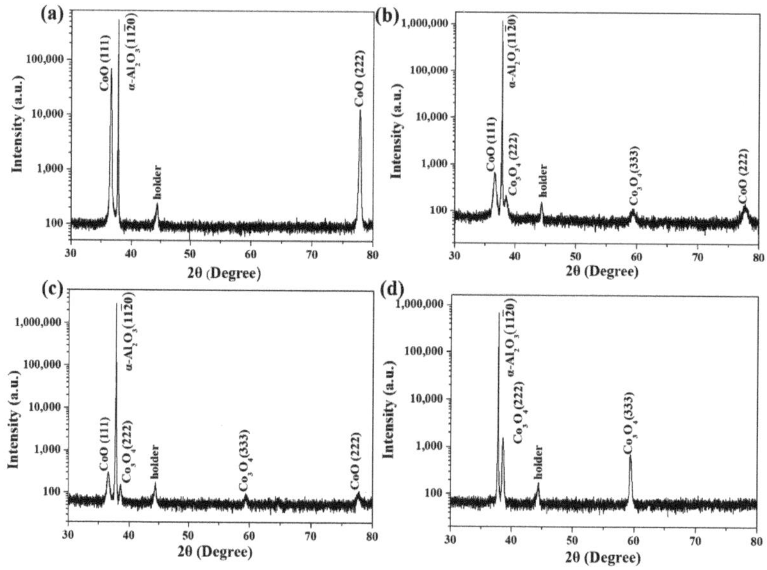

Figure 1. X-ray diffraction 2θ-ω scan profiles of the films grown under nitrogen ambient at (**a**) 450 °C and (**b**) 500 °C and under oxygen ambient at temperatures of (**c**) 450 °C and (**d**) 500 °C, respectively.

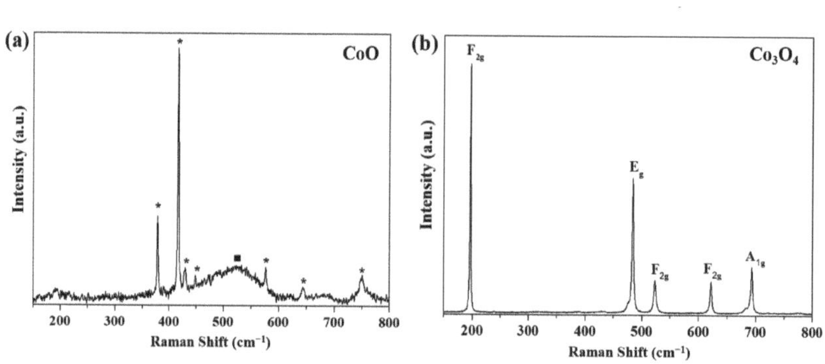

Figure 2. The Raman spectra of (**a**) CoO and (**b**) Co_3O_4 films. The asterisk and square in (**a**) indicate Raman scattering peaks of sapphire substrate and CoO film, respectively.

The corresponding surface morphologies of the CoO and Co_3O_4 films were observed through SEM, as shown in Figure 3a,b; both as-grown cobalt oxide films possessed uniform surface morphologies. The CoO film exhibited a flat and smooth surface morphology;

however, the triangular-shaped grains could be observed on the surfaces of the Co_3O_4 films. The thicknesses of the CoO and Co_3O_4 films, which were grown for one hour, were approximately 294 nm and 180 nm, respectively, as shown in Figure 3c,d. In this work, the growth rates of single-phase CoO and Co_3O_4 epitaxial films were 294 nm and 180 nm/h, respectively, which were slower than those grown using other CVD processes (230–920 nm/h) [21,35]. The cross-sectional SEM image shown in Figure 3d reveals the presence of the columnar grain growth of Co_3O_4 on the sapphire substrate. The grain sizes measured from the cross-sectional SEM image ranged from 57 to 150 nm. Remarkably, the grain sizes observed through SEM are not equivalent to crystallite sizes estimated from XRD peak widths using the Scherrer equation because a grain may be composed of two or more crystallites.

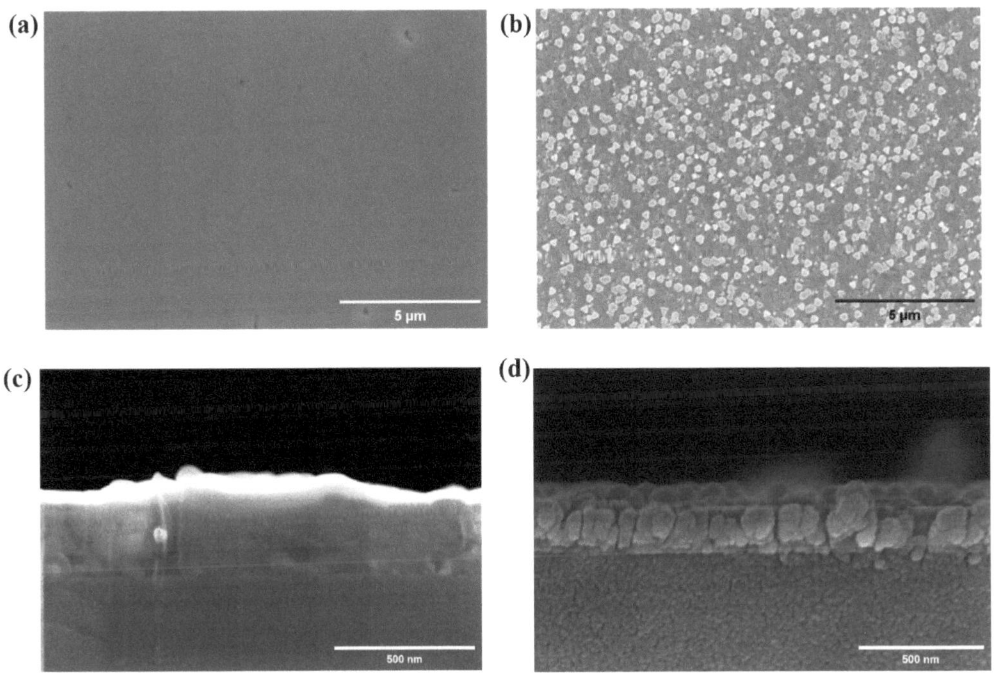

Figure 3. Top-view and cross-sectional SEM micrographs of the films: (**a**,**c**) CoO; (**b**,**d**) Co_3O_4.

Based on the results of XRD 2θ-ω scans, regardless of the growth conditions (growth temperature or ambient), the cobalt oxide films grown on *a*-plane sapphire substrates using mist-CVD exhibited <111> preferred orientation. To characterize the orientation relationship between cobalt oxide films and *a*-plane sapphire substrates, the XRD φ-scan measurement of CoO and Co_3O_4 films was performed, respectively. Figure 4a,b show the XRD φ-scan profiles of CoO and Co_3O_4 films grown on *a*-plane sapphire substrates. Remarkably, the XRD φ-scan profiles of {200} reflection of CoO film as well as {400} reflection of Co_3O_4 film exhibited six peaks separated by 60°, consisting of two sets of three-fold symmetric peaks separated by 60°, due to the crystal structures of rock salt CoO and spinel Co_3O_4 having three-fold symmetry along the <111> direction. This confirms the presence of 60°-rotated twin domains in either CoO or Co_3O_4 films grown on *a*-plane sapphire substrates. A single grain of CoO or Co_3O_4 can contain multiple twin domains, resulting in the small crystallite sizes determined using the Scherrer equation. Based on the results of XRD φ-scans, the obtained CoO films were epitaxially grown on sapphire substrate with the following epitaxial orientation relationship of $(111)[1\bar{1}0]_{CoO} || (11\bar{2}0)[0001]_{Al2O3}$;

however, the epitaxial relationship between Co_3O_4 film and sapphire substrate was determined as $(111)[11\bar{2}]_{Co3O4} || (11\bar{2}0)[0001]_{Al2O3}$, indicating the rotation of the in-plane axis of Co_3O_4 film on sapphire substrate through an angle of 30° with respect to that of the CoO film. Moreover, the epitaxial qualities of cobalt oxide films were evaluated through the XRD rocking curve (XRC). Figure 4c,d show the XRC profiles of the CoO (111) plane and Co_3O_4 (222) plane, respectively. It was found that the FWHM values of both XRCs were 0.25°, determined by fitting a single pseudo-Voigt function, indicating that the epitaxial quality of the cobalt oxide films grown using atmospheric pressure mist-CVD was comparable to those grown using the conventional vacuum-based processes [18,20,35–38]. The threading dislocation densities of epitaxial films can be determined from the XRD rocking curves. The dislocation density (*D*) can be calculated using the measured FWHM of the XRC with the following equation [39]:

$$D = FWHM^2 / \left(2\pi ln(2) \cdot b^2\right) \quad (1)$$

where *b* is the Burgers vector of dislocation. The Burgers vectors in both rock-salt and spinel structures are $a_0/2\langle 110 \rangle$ [40,41]; hence, the Burgers vectors of CoO and Co_3O_4 are 0.304 nm and 0.5716 nm, respectively. Thus, the dislocation densities in the CoO and Co_3O_4 epitaxial films were estimated to be 4.7×10^9 cm^{-2} and 1.3×10^9 cm^{-2}, respectively.

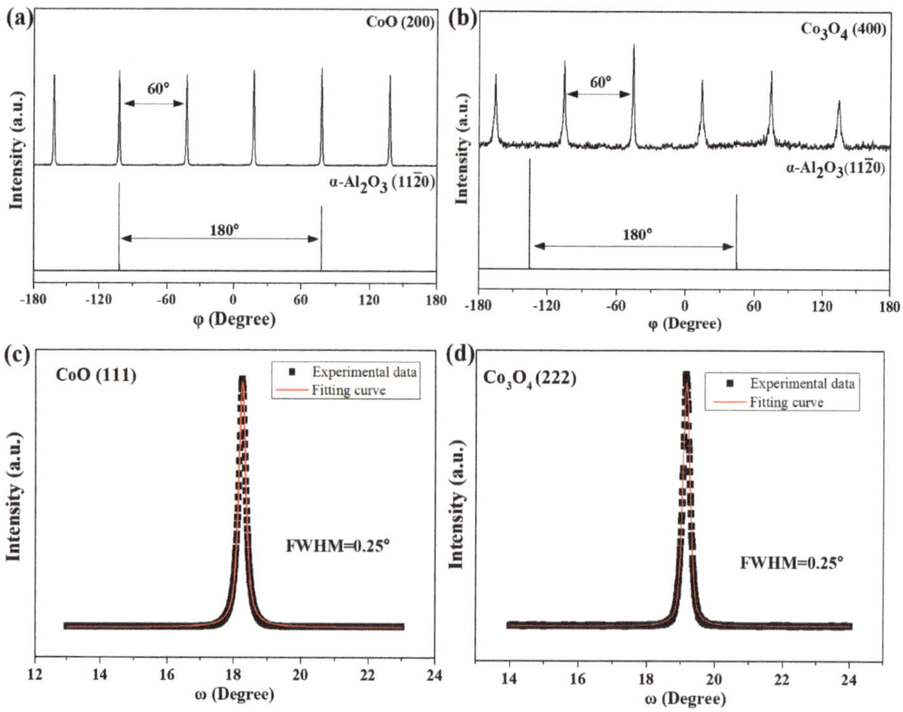

Figure 4. X-ray diffraction φ-scan profiles of (**a**) CoO {200} and sapphire $\{11\bar{2}0\}$ reflections for the CoO sample, and (**b**) Co_3O_4 {400} and sapphire $\{11\bar{2}0\}$ reflections for the Co_3O_4 sample. Rocking curve profiles of (**c**) CoO (111) and (**d**) Co_3O_4 (222) reflections.

The optical properties of CoO and Co_3O_4 epitaxial films were examined through UV-visible (UV-Vis) spectroscopy. Figure 5a,c are the UV-Vis transmission spectra of CoO and Co_3O_4 epitaxial films. The transmittance spectrum of the CoO film shows an absorption edge between approximately 400 and 500 nm. The average transmittance of

CoO film in the visible region was in the range of 30–40%. For the Co_3O_4 epitaxial film, the corresponding UV-Vis spectrum (Figure 5c) shows an increase in the transmittance as the wavelength increases, with two absorptions at approximately 600 nm and 800 nm, respectively. The optical absorption coefficient (α) can be estimated from the transmittance (T) by the following equation:

$$\alpha = -\ln(T)/d \quad (2)$$

where d is the film thickness. The optical bandgap (E_g) can be determined from $(\alpha h\nu)^2$ versus the $h\nu$ plot (Tauc Plot), as shown in Figure 5b,d, where $h\nu$ is the photon energy. The extrapolation of the linear region of the Tauc plot led to the optical bandgaps of the thin films. Based on Figure 5b, the optical bandgap of the CoO film was estimated to be 2.81 eV, consistent with that previously reported [18,22]. In contrast, the linearly fitted segments in the Tauc plot of the Co_3O_4 film (Figure 5d) prove the presence of two energy levels of the direct allowed transitions at 1.41 eV and 2.03 eV, respectively. As mixed Co^{2+} and Co^{3+} cations occupy the tetrahedral and octahedral sites in the spinel Co_3O_4 lattice, the charge transfer process associated with the Co^{3+}–Co^{2+} and an oxide ligand (O^{2-}) to metal (Co^{2+}) transition, respectively, were identified. The bandgap of 1.41 eV corresponds to the Co^{3+}–Co^{2+} charge transfer transition, reported within the range of 1.4–1.5 eV [16,21,22,42]. The bandgap of 2.03 eV is interpreted as an oxide ligand (O^{2-}) to metal (Co^{2+}) charge transfer transition, in agreement with the values (1.9–2.4 eV) reported in earlier studies [16,21,22,42].

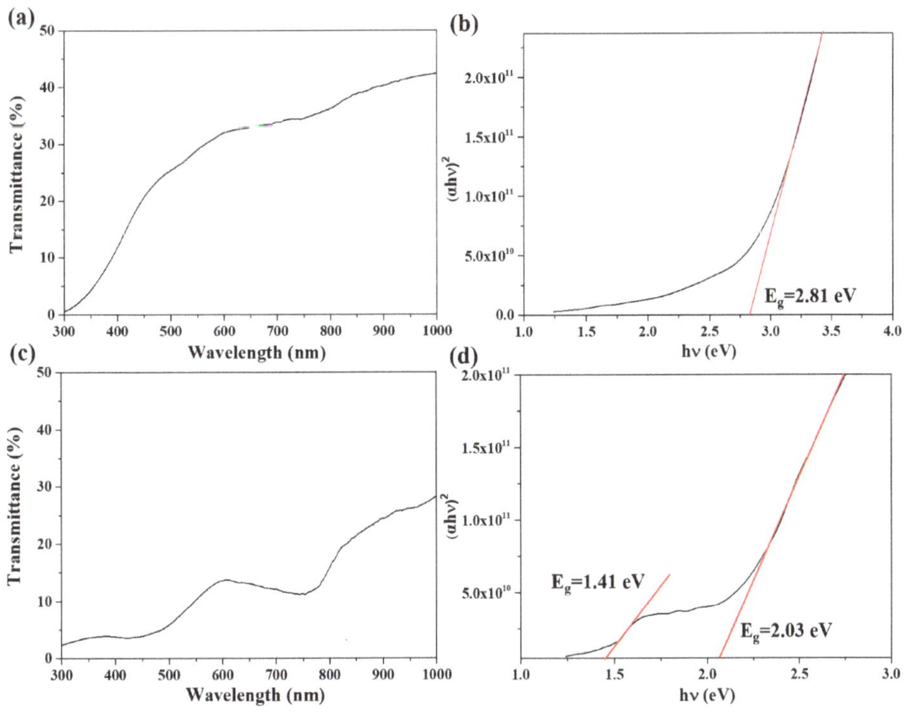

Figure 5. UV-visible transmission spectra of (**a**) CoO and (**c**) Co_3O_4 films; corresponding Tauc plots of $(\alpha h\nu)^2$ versus photo energy for (**b**) CoO and (**d**) Co_3O_4 films.

The electrical properties of CoO films were evaluated using the Hall-effect measurement. The electrical resistivities of the as-grown CoO and Co_3O_4 films were 1.8×10^5 Ω-cm and 2.3×10^2 Ω-cm, respectively, which are consistent with the values reported by other studies (>10^4 Ohm cm for CoO; 1~10^3 Ohm cm for Co_3O_4) [22]. Based on the results reported by previous studies, the electrical performance of various metal-oxide semiconductor films can be further improved via annealing at the temperature higher than that used in the film growth process [43,44]. Therefore, the annealing temperature of 650 °C, being higher than that used in the mist-CVD growth process, was chosen to further modify the electrical properties of the grown films. Notably, the electrical properties of CoO and Co_3O_4 films can be further improved using a post-thermal annealing treatment at 650 °C in oxygen ambient for 30 min. After thermal annealing, the electrical properties of annealed Co_3O_4 film were significantly enhanced, showing that the electrical resistivity was 8.35 Ohm cm, and the carrier concentration, with a positive Hall coefficient sign, increased from 1.5×10^{15} cm^{-3} to 4.19×10^{16} cm^{-3}. In contrast, the as-grown CoO film behaved almost as an electrical insulator due to its extremely high electrical resistivity and low carrier concentration (2.3×10^{12} cm^{-3}). When annealing the sample of CoO film in oxygen ambient, the sample showed p-type conductivity with the electrical resistivity and carrier concentration of 14.2 Ohm cm and 4.37×10^{16} cm^{-3}. Annealing in an oxygen atmosphere significantly improves the electrical properties of both cobalt oxide thin films, which can increase carrier concentration to the order of magnitude of 10^{16} cm^{-3}. When annealing in an oxygen-rich environment, the cobalt vacancies are the dominant point defects in cobalt oxide lattices and act as electron acceptors, which are the origin of hole carriers and p-type conductivity [45]. For the annealing of CoO film, the oxidation of CoO to Co_3O_4 readily occurred while heating at 300–650 °C in the presence of oxygen [46]. Therefore, the possible reason for improved electrical conductivity in the samples of annealed CoO film is attributed to the transformation into the Co_3O_4 phase through an oxidation process.

3.2. Thermal Annealing of CoO Epitaxial Films

Although improved electrical conductivity in the samples of annealed CoO film can possibly be attributed to the formation of the Co_3O_4 phase, the details of microstructural and chemical information of Co_3O_4 converted from CoO was unknown. Thus, the main aim of this study in this section is to explore the microstructural evolution during phase transition from CoO to Co_3O_4. To further prove the phase transition of CoO to Co_3O_4 during thermal annealing, XRD and Raman measurements were implemented, respectively, as shown in Figure 6. The result of the 2θ-ω scan profile revealed that the phase corresponding to the annealed sample of CoO film was Co_3O_4 (Figure 6a). The average crystallite size of the corresponding Co_3O_4 film was 35 nm, which was estimated from the diffraction peak width using the Scherrer equation. In addition, the Raman spectrum recorded from the annealed sample was consistent with the typical characteristic Raman peaks of spinel Co_3O_4 (Figure 6b). From the above results, it is shown that cobalt monoxide (CoO) was fully converted to tricobalt tetroxide (Co_3O_4) through an oxidation reaction, labeled as $Co_3O_{4(Oxid)}$, during the annealing process in oxygen ambient. Figure 7 shows the top-view and cross-sectional SEM images of the annealed CoO film ($Co_3O_{4(Oxid)}$). In contrast to the flat surface of the CoO film before the annealing process (Figure 3a), the surface of the annealed CoO film exhibited an undulating and uneven morphology (Figure 7a). The average thickness of the annealed CoO film was approximately 300 nm. The density of CoO (6.45 g/cm^3) is higher than that of Co_3O_4 (6.11 g/cm^3). Thus, while CoO films were transformed into Co_3O_4 during thermal annealing, the volume (thickness) of the formed Co_3O_4 films was slightly increased. Remarkably, the pore-like structures present in the interiors of the films can be observed from the cross-sectional SEM image (Figure 7b). The details regarding the microstructure need to be further characterized through TEM and will be discussed in the following paragraphs.

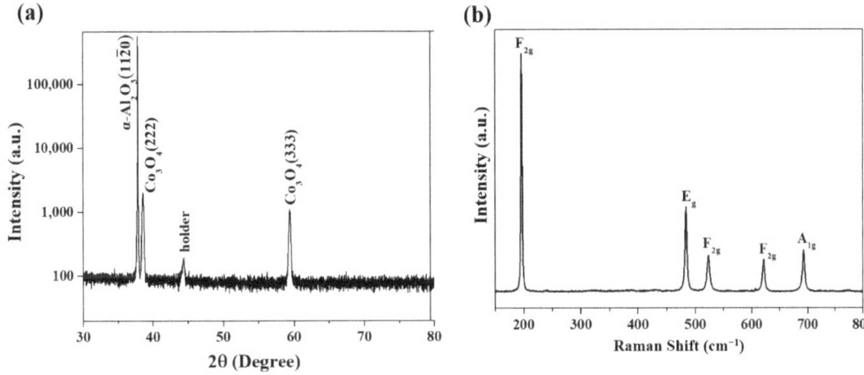

Figure 6. (**a**) X-ray diffraction 2θ-ω scan profile and (**b**) Raman spectrum of the CoO film sample annealed at 650 °C in oxygen ambience.

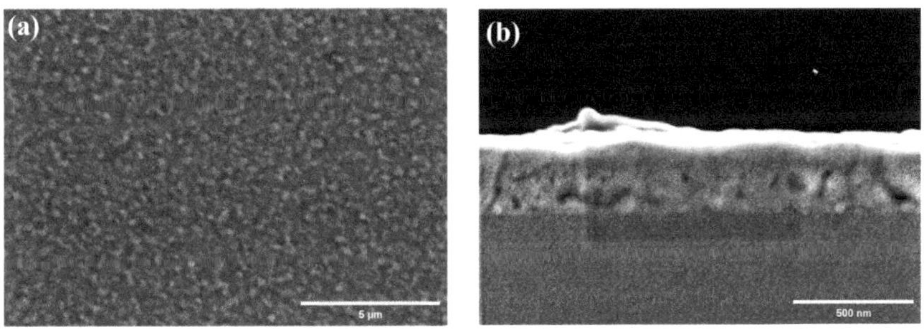

Figure 7. (**a**) Top-view and (**b**) cross-sectional SEM micrographs of the CoO film annealed at 650 °C in oxygen ambience.

To further investigate the valence and chemical state, X-ray photoelectron spectroscopy was used to characterize as-grown CoO, Co_3O_4 converted from CoO ($Co_3O_{4(Oxid)}$), and CVD-grown Co_3O_4 films. Figure 8 shows survey spectra of as-grown CoO, Co_3O_4 converted from CoO ($Co_3O_{4(Oxid)}$), and CVD-grown Co_3O_4 films. These XPS spectra dominantly show the element of Co and O along with the signal of carbon impurity (C 1s). The corresponding atomic concentrations (at.%) of cobalt were 48.3, 35.7, and 35.1 at.%, respectively, which were lower than those of cobalt oxides with stoichiometry (CoO: 50 at.%; Co_3O_4: 42.9 at.%). This confirms that all of the cobalt oxide films were metal-deficient-type oxides, indicating that cobalt vacancies were the dominant point defects in cobalt oxide lattices. The Co 2p spectra of CoO, $Co_3O_{4(Oxid)}$, and CVD-grown Co_3O_4, are shown in Figure 8a–c. The Co 2p core level splits into $2p^{3/2}$ and $2p^{1/2}$ due to spin–orbit coupling. For as-grown CoO film (Figure 9a), the spin–orbit doublet was observed at a binding energy (B.E.) of 780 and 795.8 eV with a spin–orbit splitting of 15.8 eV between Co $2p^{3/2}$ and $2p^{1/2}$ core levels, along with an intense shake-up satellite peak at 786 eV, consistent with the characteristic XPS spectrum of CoO reported by other literatures [22,47]. The Co 2p spectrum of the $Co_3O_{4(Oxid)}$ film (Figure 9b) was similar to that of the CVD-grown Co_3O_4 film (Figure 9c). The two main peaks corresponding to Co $2p^{3/2}$ and Co $2p^{1/2}$ core levels were located at 780 and 795.1 eV with spin–orbit splitting of 15.1 eV, together with a weak and broadened shake-up satellite peak located at 789 eV, consistent with the characteristic XPS spectrum of Co_3O_4 reported by other studies [10,22]. The Co^{2+} and Co^{3+} cations coexist in Co_3O_4; therefore, the Co 2p doublet peaks can be deconvoluted into four subpeaks, where the fitting peaks at 779.6 and 794.8 eV are assigned as the Co^{3+}

state and the fitting peaks at 781 and 796 eV are ascribed to the Co^{2+} state. The signals of Co^{3+} and Co^{2+} in the Co 2p spectra were numerically integrated to obtain the ratios of Co^{3+}/Co^{2+} cations of $Co_3O_{4(Oxid)}$ and CVD-grown Co_3O_4 films, respectively, estimated to be 0.50 and 0.89, respectively, which are both lower than the theoretical ratio of Co^{3+}/Co^{2+} in Co_3O_4 (2:1). The Co^{3+} vacancies (V_{Co3+}) at the octahedral sites in spinel Co_3O_4 were indicated to dominate in both samples [45]. Remarkably, compared to CVD-grown Co_3O_4, $Co_3O_{4(Oxid)}$ thin films, converted from CoO to Co_3O_4, revealed a lower Co^{3+}/Co^{2+}ratio, possibly due to the initial CoO phase solely comprising divalent cobalt. Theoretical and experimental studies have proven that the formation of the cation vacancies in metal oxides is favorable in an oxygen-rich environment [7,45,48]. Thus, the Co vacancies in Co_3O_4 were regarded as the major origin of their p-type semiconducting properties. The O 1s spectra of CoO, $Co_3O_{4(Oxid)}$, and CVD-grown Co_3O_4, as shown in Figure 9d, are deconvoluted into three components at approximately 530, 531.2, and 532.2 eV, which is attributed to oxygen bonded with cobalt in the lattice (O_{lat}), hydroxyl groups bonded to surface defects, such as oxygen vacancies (O_V), and weakly absorbed molecular water (O_{abs}), respectively. The difference between the three O 1s core level XPS spectra recorded from CoO, $Co_3O_{4(Oxid)}$, and CVD-grown Co_3O_4 films was not obvious. The contribution of oxygen bonded to cobalt in the lattice was estimated to be 72.8%, 72.0%, and 75.6%, respectively, indicating that all of the cobalt oxide films retain more structural oxygen in the lattice. Considering a water/acetate solution of cobalt acetate used as the cobalt source, the incorporation of carbon from the source is still possible [49]. Therefore, the residual impurities, such as carbon, in grown films were also detected through XPS. Although the signal of carbon 1s core level emission was not detected from the sample after argon ion sputtering, as shown in Figure S1 in Supplementary information, it is possible that the carbon content was below the instrument detection limit. In future work, secondary ion mass spectroscopy (SIMS) is needed for the detection of the residual impurity with low concentration.

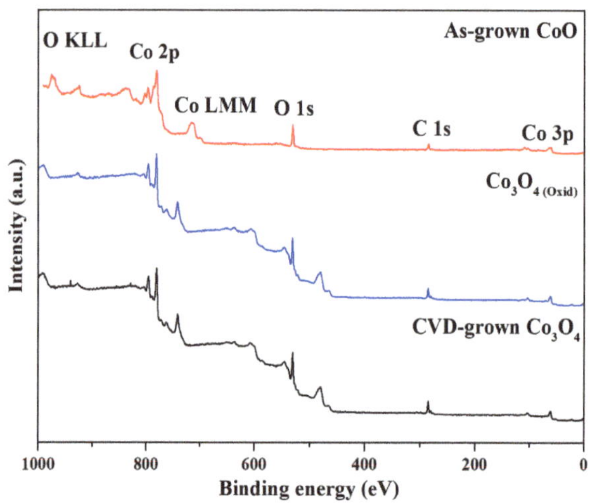

Figure 8. X-ray photoelectron spectroscopy (XPS) survey spectra of as-grown CoO, Co_3O_4 converted from CoO ($Co_3O_{4(Oxid)}$), and CVD-grown Co_3O_4 films.

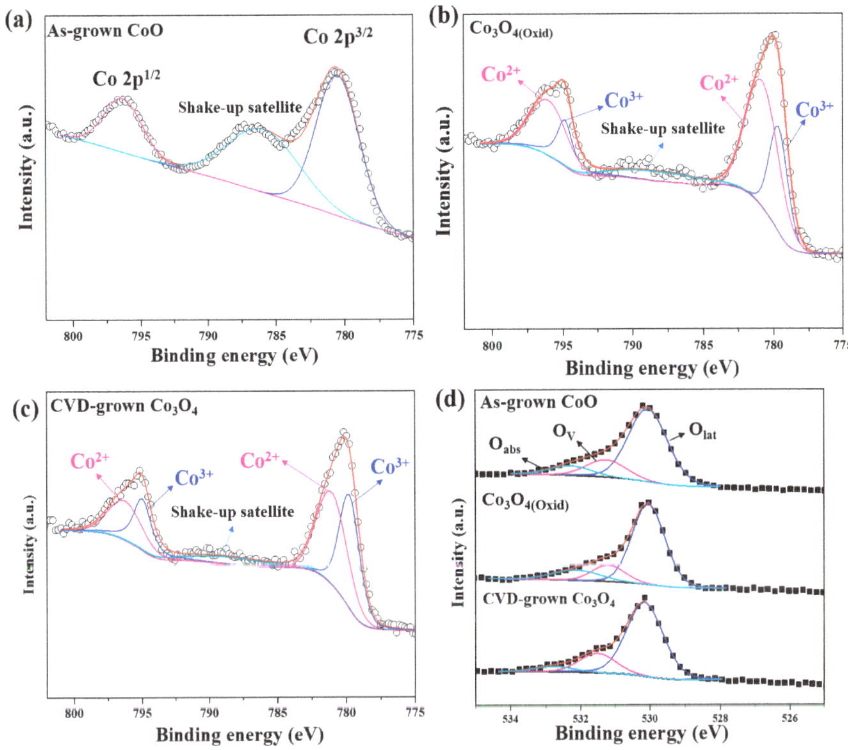

Figure 9. Core-level photoemission spectra of the Co 2p recorded from (**a**) as-grown CoO, (**b**) Co_3O_4 converted from CoO ($Co_3O_{4(Oxid)}$), (**c**) CVD-grown Co_3O_4 films, and (**d**) corresponding core-level photoemission spectra of O 1s.

According to the studies reported on the oxidation of CoO to Co_3O_4, the formation of the $Co_3O_{4(Oxid)}$ layer initially occurs on the surface of CoO [50,51]. Therefore, the crystallographic alignment of the $Co_3O_{4(Oxid)}$ film formed through the oxidation reaction is closely related to the orientation of the underlying CoO epitaxial film [50,51]. The X-ray diffraction φ-scan and XRC analysis were employed to assess the relationship and the crystalline quality of the formed $Co_3O_{4(Oxid)}$ film. Figure 10a shows the XRD φ-scan profiles for the $Co_3O_{4(Oxid)}$ film detecting Co_3O_4 (400) and Al_2O_3 (11$\bar{2}$0) reflections. The profile of the $Co_3O_{4(Oxid)}$ film revealed six-fold symmetry with a spacing of 60°, indicating that the $Co_3O_{4(Oxid)}$ film was in epitaxy with the sapphire substrate. Similarly, the presence of 60°-rotated twin domains in the $Co_3O_{4(Oxid)}$ film was detected from the X-ray diffraction φ-scan profile. Interestingly, the crystallographic relationship between the $Co_3O_{4(Oxid)}$ film and sapphire is (111)[1$\bar{1}$0]$Co_3O_{4(Oxid)}$ | | (11$\bar{2}$0)[0001]$_{Al2O3}$, rotated through an angle of 30° with respect to that of CVD-grown Co_3O_4 film on sapphire substrate ((111)[11$\bar{2}$]$_{Co3O4}$ | | (11$\bar{2}$0)[0001]$_{Al2O3}$), as shown in Figure 4b. In Figure 10b, the $Co_3O_{4(Oxid)}$ presents a broader FWHM (0.62°) of XRC for Co_3O_4 (222) diffraction. According to the cross-sectional SEM observation (Figure 7b), there were several pore-like structures in the interiors of the $Co_3O_{4(Oxid)}$ films. These pore-like structures affected the growth of crystallites during the annealing process and were possibly responsible for the broader width of the XRC. The formation of the Co_3O_4 layer initially occurred on the surface of CoO. Sequentially, the entire CoO layer was gradually converted into Co_3O_4 during the oxidation process, as illustrated in Figure 11. The oxygen sublattices of the CoO and Co_3O_4, both being fcc sublattices, are nearly identical, with a lattice mismatch of about

5% [50]. Therefore, the oxygen sublattice of newly formed $Co_3O_{4(Oxid)}$ inherited the original framework of CoO and was epitaxially constrained by the lattice planes of the underlying CoO, as illustrated in Figure 10.

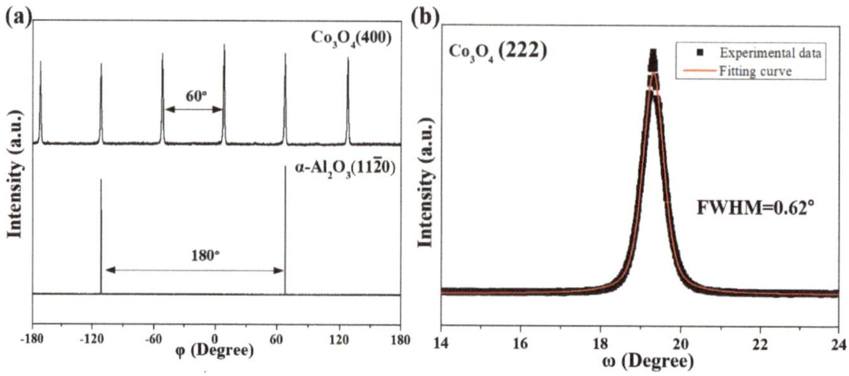

Figure 10. (**a**) X-ray diffraction φ-scan profiles of Co_3O_4 {400} and sapphire {11$\bar{2}$0} reflections, and (**b**) rocking curve profiles of Co_3O_4 (222) reflections recorded from samples of the $Co_3O_{4(Oxid)}$ film.

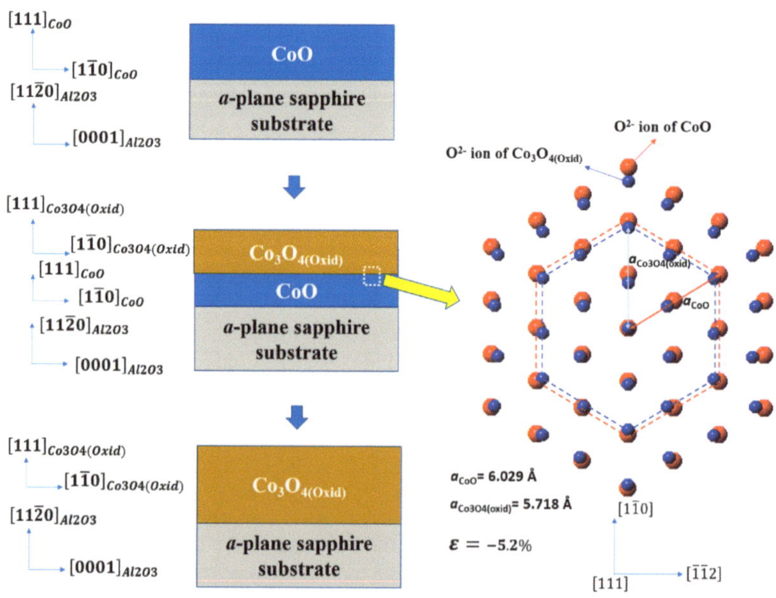

Figure 11. Schematic representation $Co_3O_{4(Oxid)}$ layer formation through the oxidation of the CoO epitaxial layer and lattice match between (111) planes of the CoO and Co_3O_4 oxygen sublattices.

We performed cross-sectional TEM to gain insight into the microstructure of the $Co_3O_{4(Oxid)}$ film converted from CoO. Importantly, several pores present in the interior of $Co_3O_{4(Oxid)}$ films were observed from the low-magnification bright-field TEM image (Figure 12a). During the oxidation of CoO to Co_3O_4, the growth of Co_3O_4 is initiated on the surface of CoO. Then, the Co_3O_4 layer grows inward and outward simultaneously [52–54].

In the case of outward growth, the formation of Co_3O_4 occurs due to the outward diffusion of cobalt cations and electrons as a result of the following reaction [52–54]:

$$3Co_i^{\cdot\cdot} + 6e' + 2O_2 \rightarrow Co_3O_4 \tag{3}$$

where $Co_i^{\cdot\cdot}$ and e' denote the double ionized interstitial cobalt cation and a free electron, respectively. The migrating interstitial cobalt cations reacted with oxygen on the surface to form Co_3O_4. In the case of inward growth, the reaction occurred at the interface between CoO and Co_3O_4, according to the displacement reaction [52–54]:

$$4CoO \rightarrow Co_3O_4 + Co_i^{\cdot\cdot} + 2e' \tag{4}$$

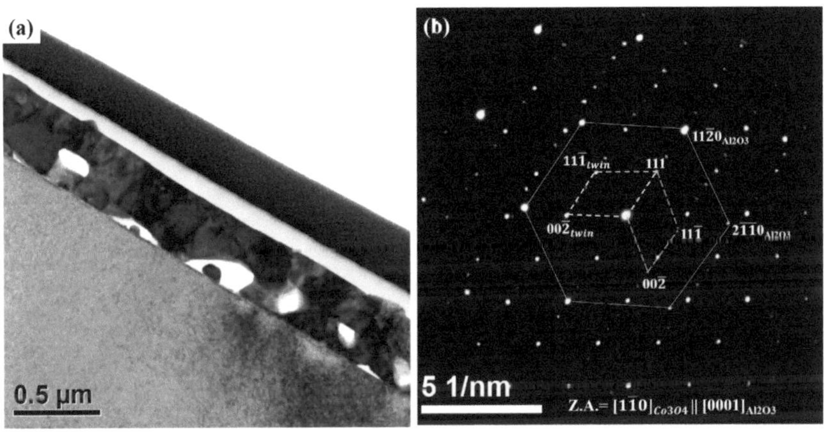

Figure 12. (**a**) Low-magnification bright-field transmission electron micrograph of the $Co_3O_{4(Oxid.)}$ film converted from the CoO epitaxial layer via oxidation reaction; (**b**) Corresponding selected area electron diffraction (SAED) patterns recorded at the interfacial regions and viewed along the sapphire [0001] zone axis. The normal and 60°-rotated twin variants are indicated by green and yellow dashed lines, respectively.

Based on the marker experiment for studying the oxidation of CoO, it was observed that cobalt cations continued to diffuse outward through the Co_3O_4 layer during the oxidation reaction [53,54]. Consequently, this resulted in the excessive concentration of cobalt vacancies inside the film layer due to the directional diffusion cobalt cations (the so-called Kirkendall effect). Thus, the thickness of formed Co_3O_4 films became thicker after the annealing process. The formation of pores was the consequence of a supersaturation of vacancies in the lattice brought about by the coalescence of cobalt vacancies [52,54]. In Figure 12b, the pattern of the selected area electron diffraction (SAED) recorded from the interface region shows the presence of two twinned variants in the formed $Co_3O_{4(Oxid)}$ film, designated by green and yellow dashed lines, both indexed as the normal and 60°-rotated twin variants, respectively. They were oriented epitaxially with respect to the sapphire substrate. The epitaxial relationship between $Co_3O_{4(Oxid)}$ and the sapphire substrate was identified as $(111)[1\bar{1}0]Co_3O_{4(Oxid)}$ | | $(11\bar{2}0)[0001]_{Al2O3}$, which is consistent with the result of the XRD φ-scan, as shown in Figure 10a. Despite the presence of numerous pore structures in the formed $Co_3O_{4(Oxid)}$ layer, the diffraction pattern of the $Co_3O_{4(Oxid)}$ film remained consistent throughout the entire film layer.

The microstructure near the film/substrate heterointerface was visualized using HRTEM, as shown in Figure 13a. This view is along the sapphire [0001] zone axis. Remarkably, the interface between the formed $Co_3O_{4(Oxid)}$ layer and the sapphire substrate is not atomically abrupt; instead, it appears intermixed. It is possible that during the annealing process at 650 °C for 30 min, the occurrence of slight interdiffusion between the formed $Co_3O_{4(Oxid)}$ film and the sapphire substrate is not ignored. Furthermore, the corresponding fast Fourier transformation pattern (Figure 13b) of the HRTEM image was consistent with the SAED pattern (Figure 12b). In Figure 13c, the Fourier-filtered image, reconstructed using the Co_3O_4 $22\overline{2}$ and sapphire $2\overline{1}\overline{1}0$ reflections, reveals a lattice-matched interface, accompanied by several of the extra-half planes of misfit dislocations located at the interface, indicated by black triangles, to accommodate the lattice mismatch. The misfit dislocations at the interface region were not arranged in a periodic or evenly spaced manner. It is evident that the paradigm of the domain match epitaxy does not seem to be applicable to this intermixed interface. Limited by the spatial resolution of conventional HRTEM, the atomic arrangement of different elements located near the interface cannot be clearly identified. To explore the explicit mechanism governing this interface, achieving sub-angstrom resolution high angle annular dark field (HAADF) imaging using scanning transmission electron microscopy (STEM) with a spherical aberration corrector is highly necessary in future research. Such imaging will help identify the position of atomic columns at the interface region. Despite the presence of pores in the films, the lattice-matched interface between the formed $Co_3O_{4(Oxid)}$ film and sapphire substrate was confirmed by the HRTEM image.

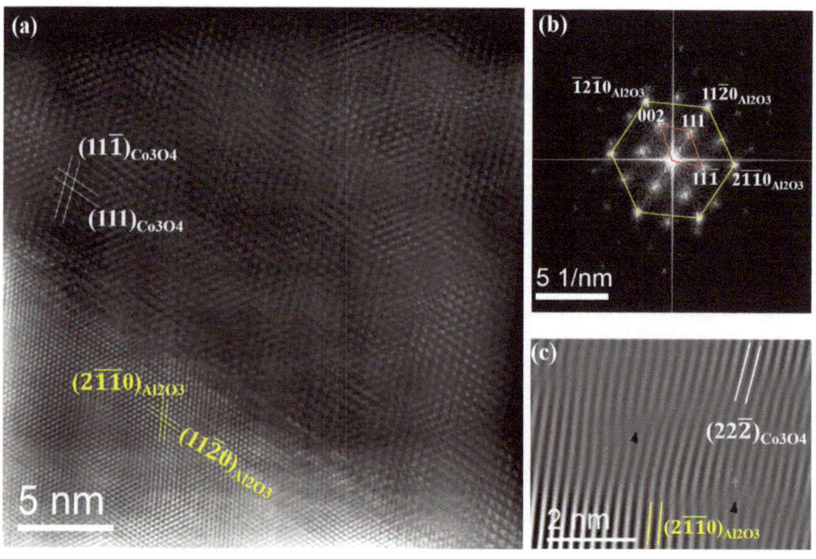

Figure 13. (**a**) High-resolution transmission electron microscopic image of the $Co_3O_{4(Oxid)}$ film–sapphire interface visualized along the sapphire [0001] zone axis. (**b**) Corresponding fast Fourier transformation pattern of (**a**). (**c**) Fourier-filtered image reconstructed using Co_3O_4 $22\overline{2}$ and sapphire $2\overline{1}\overline{1}0$ reflections. The black triangles in the Fourier-filtered image indicate the position of misfit dislocations.

4. Conclusions

In this study, we demonstrated selective epitaxial growth of the single-phase rock-salt CoO and spinel Co_3O_4 thin films on sapphire substrates using mist-CVD by controlling growth temperature and ambient, as confirmed by XRD, Raman, and XPS spectroscopy measurements. The surface morphologies of the obtained CoO films exhibited a flat and smooth surface morphology; in contrast, the triangular-shaped grains were observed on the surfaces of the Co_3O_4 films. The epitaxial relationship of CoO and Co_3O_4 on sapphire substrates was determined to be $(111)[1\overline{1}0]_{CoO} \| (11\overline{2}0)[0001]_{Al2O3}$ for CoO and $(111)[11\overline{2}]_{Co3O4} \| (11\overline{2}0)[0001]_{Al2O3}$ for Co_3O_4. The crystalline quality of both CoO and

Co_3O_4 epitaxial films grown using atmospheric pressure mist-CVD was comparable to those grown by the other vapor phase epitaxy in a vacuum environment. The CoO films had a direct bandgap of 2.81 eV, and the two direct bandgap values at 1.41 eV and 2.03 eV were observed for Co_3O_4. The electrical properties of Co_3O_4 epitaxial films could be enhanced through thermal annealing in oxygen ambient, exhibiting a stable p-type conductivity with an electrical resistivity of 8.35 Ohm cm and a carrier concentration of 4.19×10^{16} cm^{-3}. According to XPS analysis, cobalt vacancies, which functioned as acceptors and were predominantly present in Co_3O_4 films, were found to facilitate the generation of hole carriers. In addition, the phase transition of CoO to Co_3O_4 during the annealing process was evidenced by XRD, Raman, and XPS spectroscopy measurements. Therefore, the oxidation of CoO to Co_3O_4 primarily accounts for the improved electrical properties observed in the annealed samples of CoO films. The crystallographic orientation of the Co_3O_4 film formed via the oxidation reaction was epitaxially constrained by the lattice planes of the underlying CoO epitaxial film, primarily because CoO and Co_3O_4 share a nearly identical oxygen sublattice. The formation of Co_3O_4 films during the oxidation of CoO was predominantly governed by the outward diffusion of cobalt cations. Therefore, the formation of the pores within the interiors of Co_3O_4 films was attributed to the accumulation of excessive cobalt vacancies. Despite the presence of several pores within the interiors of Co_3O_4 films, the lattice-matched interface between the Co_3O_4 film and the sapphire substrate was confirmed by the HRTEM image.

Supplementary Materials: The following supporting information can be downloaded at: https://www.mdpi.com/article/10.3390/coatings13111878/s1. Figure S1: Core-level photoemission spectra of carbon 1s of CoO film before and after argon-ion clearing.

Author Contributions: Conceptualization, H.-G.C.; methodology, T.-L.Y.; validation, H.-G.C., T.-L.Y. and J.-Y.F.; formal analysis, H.-G.C., H.-S.W., S.-R.J., T.-L.Y. and J.-Y.F.; investigation, T.-L.Y. and J.-Y.F.; resources, H.-S.W. and S.-R.J.; data curation, T.-L.Y.; writing—original draft preparation, H.-G.C.; writing—review and editing, H.-G.C., H.-S.W. and S.-R.J.; visualization, H.-G.C.; supervision, H.-G.C.; project administration, H.-G.C.; funding acquisition, H.-G.C. and H.-S.W. All authors have read and agreed to the published version of the manuscript.

Funding: This research was funded by the National Science and Technology Council, Taiwan, R.O.C., grant numbers MOST 109-2221-E-214-018, MOST 110-2221-E-214-004 and NSTC 112-2221-E-214-010, and I-Shou University, grant number ISU-112-01-03A.

Institutional Review Board Statement: Not applicable.

Informed Consent Statement: Not applicable.

Data Availability Statement: Not applicable.

Acknowledgments: The authors gratefully acknowledge the Core Facility Center, National Cheng Kung University, Taiwan, for the XRD, FIB, HRTEM experiment, and thank the MANALAB at I-Shou University for the SEM.

Conflicts of Interest: The authors declare no conflict of interest.

References

1. Schlom, D.G.; Chen, L.-Q.; Pan, X.; Schmehl, A.; Zurbuchen, M.A. A Thin Film Approach to Engineering Functionality into Oxides. *J. Am. Ceramic Soc.* **2008**, *91*, 2429–2454. [CrossRef]
2. Oka, D.; Fukumura, T. Crystal engineering for novel functionalities with oxide thin film epitaxy. *CrystEngComm* **2017**, *19*, 2144–2162. [CrossRef]
3. Zhu, Y.; Sun, K.; Wu, S.; Zhou, P.; Fu, Y.; Xia, J.; Li, H.-F. A comprehensive review on the ferroelectric orthochromates: Synthesis, property, and application. *Coord. Chem. Rev.* **2023**, *475*, 214873. [CrossRef]
4. Kumah, D.P.; Ngai, J.H.; Kornblum, L. Epitaxial Oxides on Semiconductors: From Fundamentals to New Devices. *Adv. Funct. Mater.* **2020**, *30*, 1901597. [CrossRef]
5. Zhang, K.H.L.; Xi, K.; Blamire, M.G.; Egdell, R.G. P-type transparent conducting oxides. *J. Phys. Condens. Matter.* **2016**, *28*, 383002. [CrossRef]

6. Kupfer, B.; Majhi, K.; Keller, D.A.; Bouhadana, Y.; Rühle, S.; Barad, H.N.; Anderson, A.Y.; Zaban, A. Thin Film Co_3O_4/TiO_2 Heterojunction Solar Cells. *Adv. Energy Mater.* **2015**, *5*, 1401007. [CrossRef]
7. Patel, M.; Park, S.-H.; Kim, J. Rapid Thermal Treatment of Reactive Sputtering Grown Nanocrystalline Co_3O_4 for Enhanced All-Oxide Photovoltaics. *Phys. Status Solidi A* **2018**, *215*, 1800216. [CrossRef]
8. Hu, Z.; Hao, L.; Quan, F.; Guo, R. Recent developments of Co_3O_4-based materials as catalysts for the oxygen evolution reaction. *Catal. Sci. Technol.* **2022**, *12*, 436–461. [CrossRef]
9. Zhou, T.; Zhang, T.; Deng, J.; Zhang, R.; Lou, Z.; Wang, L. P-type Co_3O_4 nanomaterials-based gas sensor: Preparation and acetone sensing performance. *Sens. Actuators B Chem.* **2017**, *242*, 369–377. [CrossRef]
10. Liu, L.; Jiang, Z.; Fang, L.; Xu, H.; Zhang, H.; Gu, X.; Wang, Y. Probing the Crystal Plane Effect of Co_3O_4 for Enhanced Electrocatalytic Performance toward Efficient Overall Water Splitting. *ACS Appl. Mater. Interfaces* **2017**, *9*, 27736–27744. [CrossRef]
11. Yao, C.; Ismail, M.; Hao, A.; Thatikonda, S.K.; Huang, W.; Qin, N.; Bao, D. Annealing atmosphere effect on the resistive switching and magnetic properties of spinel Co_3O_4 thin films prepared by a sol–gel technique. *RSC Adv.* **2019**, *9*, 12615–12625. [CrossRef] [PubMed]
12. Kim, H.; Seo, D.-H.; Kim, S.-W.; Kim, J.; Kang, K. Highly reversible Co_3O_4/graphene hybrid anode for lithium rechargeable batteries. *Carbon* **2011**, *49*, 326–332. [CrossRef]
13. Zhao, X.; Mao, L.; Cheng, Q.; Li, J.; Liao, F.; Yang, G.; Xie, L.; Zhao, C.; Chen, L. Two-dimensional Spinel Structured Co-based Materials for High Performance Supercapacitors: A Critical Review. *Chem. Eng. J.* **2020**, *387*, 124081. [CrossRef]
14. Patel, M.; Kumar, M.; Kim, H.-S.; Park, W.-H.; Choi, E.H.; Kim, J. Reactive sputtering growth of Co_3O_4 thin films for all metal oxide device: A semitransparent and self-powered ultraviolet photodetector. *Mater. Sci. Semicond. Process.* **2018**, *74*, 74–79. [CrossRef]
15. Negusse, E.; Idzerda, Y.U.; Suci, P.A. Exchange biased magnetoresistance based spin-transport sensor. *J. Appl. Phys.* **2004**, *95*, 4506–4508. [CrossRef]
16. Kadam, L.D.; Patil, P.S. Thickness-dependent properties of sprayed cobalt oxide thin films. *Mater. Chem. Phys.* **2001**, *68*, 225–232. [CrossRef]
17. Tototzintle-Huitle, H.; Prokhorov, E.; Mendoza-Galván, A.; Urbina, J.E.; González-Hernández, J. Study of the formation of Co_3O_4 thin films using sol–gel method. *J. Phys. Chem. Solids* **2003**, *64*, 975–980. [CrossRef]
18. Matsuda, A.; Yamauchi, R.; Shiojiri, D.; Tan, G.; Kaneko, S.; Yoshimoto, M. Room-temperature selective epitaxial growth of CoO (111) and Co_3O_4 (111) thin films with atomic steps by pulsed laser deposition. *Appl. Surf. Sci.* **2015**, *349*, 78–82. [CrossRef]
19. Vaz, C.; Wang, H.-Q.; Ahn, C.; Henrich, V.; Baykara, M.; Schwendemann, T.; Pilet, N.; Albers, B.; Schwarz, U.; Zhang, L. Interface and electronic characterization of thin epitaxial Co_3O_4 films. *Surf. Sci.* **2009**, *603*, 291–297. [CrossRef]
20. Klepper, K.B.; Nilsen, O.; Fjellvåg, H. Epitaxial growth of cobalt oxide by atomic layer deposition. *J. Cryst. Growth* **2007**, *307*, 457–465. [CrossRef]
21. Alema, F.; Osinsky, A.; Mukhopadhyay, P.; Schoenfeld, W.V. Epitaxial growth of Co_3O_4 thin films using $Co(dpm)_3$ by MOCVD. *J. Cryst. Growth* **2019**, *525*, 125207. [CrossRef]
22. Prieto, P.; Marco, J.F.; Serrano, A.; Manso, M.; de la Figuera, J. Highly oriented (111) CoO and Co_3O_4 thin films grown by ion beam sputtering. *J. Alloys Compd.* **2019**, *810*, 151912. [CrossRef]
23. Tanoue, H.; Takenouchi, M.; Yamashita, T.; Wada, S.; Yatabe, Z.; Nagaoka, S.; Naka, Y.; Nakamura, Y. Improvement of m-plane ZnO films formed on buffer layers on sapphire substrates by mist chemical vapor deposition. *Phys. Status Solidi A* **2017**, *214*, 1600603. [CrossRef]
24. Park, S.-Y.; Ha, M.-T.; Kim, K.-H.; Van Lich, L.; Shin, Y.-J.; Jeong, S.-M.; Kwon, S.-H.; Bae, S.-Y. Enhanced thickness uniformity of large-scale α-Ga_2O_3 epilayers grown by vertical hot-wall mist chemical vapor deposition. *Ceram. Int.* **2021**, *48*, 5075–5082. [CrossRef]
25. Oshima, T.; Niwa, M.; Mukai, A.; Nagami, T.; Suyama, T.; Ohtomo, A. Epitaxial growth of wide-band-gap $ZnGa_2O_4$ films by mist chemical vapor deposition. *J. Cryst. Growth* **2014**, *386*, 190–193. [CrossRef]
26. Nishinaka, H.; Yoshimoto, M. Mist Chemical Vapor Deposition of Single-Phase Metastable Rhombohedral Indium Tin Oxide Epitaxial Thin Films with High Electrical Conductivity and Transparency on Various alpha-Al_2O_3 Substrates. *Cryst. Growth Des.* **2018**, *18*, 4022–4028. [CrossRef]
27. Yatabe, Z.; Tsuda, T.; Matsushita, J.; Sato, T.; Otabe, T.; Sue, K.; Nagaoka, S.; Nakamura, Y. Single crystalline SnO_2 thin films grown on m -plane sapphire substrate by mist chemical vapor deposition. *Phys. Status Solidi C* **2017**, *14*, 1600148. [CrossRef]
28. Chen, H.-G.; Shih, Y.-H.; Wang, H.-S.; Jian, S.-R.; Yang, T.-Y.; Chuang, S.-C. Van der Waals Epitaxial Growth of ZnO Films on Mica Substrates in Low-Temperature Aqueous Solution. *Coatings* **2022**, *12*, 706. [CrossRef]
29. Nishinaka, H.; Kamada, Y.; Kameyama, N.; Fujita, S. Epitaxial ZnO Thin Films on a-Plane Sapphire Substrates Grown by Ultrasonic Spray-Assisted Mist Chemical Vapor Deposition. *Jpn. J. Appl. Phys.* **2009**, *48*, 121103. [CrossRef]
30. Burriel, M.; Garcia, G.; Santiso, J.; Hansson, A.N.; Linderoth, S.; Figueras, A. Co_3O_4 protective coatings prepared by Pulsed Injection Metal Organic Chemical Vapour Deposition. *Thin Solid Film.* **2005**, *473*, 98–103. [CrossRef]
31. Muniz, F.T.L.; Miranda, M.A.R.; Morilla dos Santos, C.; Sasaki, J.M. The Scherrer equation and the dynamical theory of X-ray diffraction. *Acta Crystallogr. A* **2016**, *72*, 385–390. [CrossRef]
32. Rivas-Murias, B.; Salgueiriño, V. Thermodynamic CoO–Co_3O_4 crossover using Raman spectroscopy in magnetic octahedron-shaped nanocrystals. *J. Raman Spectrosc.* **2017**, *48*, 837–841. [CrossRef]

33. Mironova-Ulmane, N.; Kuzmin, A.; Grube, M. Raman and infrared spectromicroscopy of manganese oxides. *J. Alloys Compd.* **2009**, *480*, 97–99. [CrossRef]
34. Hadjiev, V.G.; Iliev, M.N.; Vergilov, I.V. The Raman spectra of Co_3O_4. *J. Phys. C Solid State Phys.* **1988**, *21*, L199. [CrossRef]
35. Shalini, K.; Mane, A.U.; Shivashankar, S.A.; Rajeswari, M.; Choopun, S. Epitaxial growth of Co_3O_4 films by low temperature, low pressure chemical vapour deposition. *J. Cryst. Growth* **2001**, *231*, 242–247. [CrossRef]
36. Forsythe, J.; Leiko, O.; Uvarov, V.; Škvára, J.; Mysliveček, J.; Mašek, K.; Johánek, V. CoO epitaxial growth on Cu(111) by reactive PVD and plasma oxidation. *Surf. Interfaces* **2022**, *34*, 102364. [CrossRef]
37. Entani, S.; Kiguchi, M.; Saiki, K.; Koma, A. Epitaxial growth of CoO films on semiconductor and metal substrates by constructing a complex heterostructure. *J. Cryst. Growth* **2003**, *247*, 110–118. [CrossRef]
38. Feeler, D.; Williams, R.S. Epitaxial Growth of CoO (001) and Co_3O_4 (001) Films on MgO (001) by Pulsed Laser Deposition. *MRS Online Proc. Libr.* **1994**, *341*, 29–34. [CrossRef]
39. Ayers, J.E. The measurement of threading dislocation densities in semiconductor crystals by X-ray diffraction. *J. Cryst. Growth* **1994**, *135*, 71–77. [CrossRef]
40. Amodeo, J.; Merkel, S.; Tromas, C.; Carrez, P.; Korte-Kerzel, S.; Cordier, P.; Chevalier, J. Dislocations and Plastic Deformation in MgO Crystals: A Review. *Crystals* **2018**, *8*, 240. [CrossRef]
41. Mitchell, T.E.; Hwang, L.; Heuer, A.H. Deformation in spinel. *J. Mater. Sci.* **1976**, *11*, 264–272. [CrossRef]
42. Patil, P.S.; Kadam, L.D.; Lokhande, C.D. Preparation and characterization of spray pyrolysed cobalt oxide thin films. *Thin Solid Film.* **1996**, *272*, 29–32. [CrossRef]
43. Tsai, S.-H.; Basu, S.; Huang, C.-Y.; Hsu, L.-C.; Lin, Y.-G.; Horng, R.-H. Deep-Ultraviolet Photodetectors Based on Epitaxial $ZnGa_2O_4$ Thin Films. *Sci. Rep.* **2018**, *8*, 14056. [CrossRef] [PubMed]
44. Chen, H.-G.; Hung, S.-P. Epitaxial growth of Sb-doped nonpolar a-plane ZnO thin films on r-plane sapphire substrates by RF magnetron sputtering. *J. Alloys Compd.* **2014**, *586*, S339–S342. [CrossRef]
45. Wang, P.; Jin, C.; Zheng, D.; Yang, T.; Wang, Y.; Zheng, R.; Bai, H. Engineering Co Vacancies for Tuning Electrical Properties of p-Type Semiconducting Co_3O_4 Films. *ACS Appl. Mater. Interfaces* **2021**, *13*, 26621–26629. [CrossRef] [PubMed]
46. Wang, S.; Zhang, B.; Zhao, C.; Li, S.; Zhang, M.; Yan, L. Valence control of cobalt oxide thin films by annealing atmosphere. *Appl. Surf. Sci.* **2011**, *257*, 3358–3362. [CrossRef]
47. Jena, A.; Penki, T.R.; Munichandraiah, N.; Shivashankar, S.A. Flower-like porous cobalt(II) monoxide nanostructures as anode material for Li-ion batteries. *J. Electroanal. Chem.* **2016**, *761*, 21–27. [CrossRef]
48. Perekrestov, R.; Spesyvyi, A.; Maixner, J.; Mašek, K.; Leiko, O.; Khalakhan, I.; Maňák, J.; Kšírová, P.; Hubička, Z.; Čada, M. The comparative study of electrical, optical and catalytic properties of Co_3O_4 thin nanocrystalline films prepared by reactive high-power impulse and radio frequency magnetron sputtering. *Thin Solid Film.* **2019**, *686*, 137427. [CrossRef]
49. Masaya, O.; Rie, T.; Hitoshi, K.; Tomochika, T.; Takahiro, S.; Toshimi, H. Schottky barrier diodes of corundum-structured gallium oxide showing on-resistance of 0.1 $m\Omega \cdot cm^2$ grown by MIST EPITAXY®. *Appl. Phys. Express* **2016**, *9*, 021101.
50. Carson, G.A.; Nassir, M.H.; Langell, M.A. Epitaxial growth of Co_3O_4 on CoO (100). *J. Vac. Sci. Technol. A* **1996**, *14*, 1637–1642. [CrossRef]
51. Carson, G.A.; Nassir, M.H.; Langell, M.A. Co_3O_4 epitaxial formation on CoO (100). *Mater. Res. Soc.* **1995**, *355*, 163–168. [CrossRef]
52. Żyła, M.; Smoła, G.; Knapik, A.; Rysz, J.; Sitarz, M.; Grzesik, Z. The formation of the Co_3O_4 cobalt oxide within CoO substrate. *Corros. Sci.* **2016**, *112*, 536–541. [CrossRef]
53. Kaczmarska, A.; Grzesik, Z.; Mrowec, S. On the Defect Structure and Transport Properties of Co_3O_4 Spinel Oxide. *High Temp. Mater. Process.* **2012**, *31*, 371–379. [CrossRef]
54. Przybylski, K.; Smeltzer, W.W. High Temperature Oxidation Mechanism of CoO to Co_3O_4. *J. Electrochem. Soc.* **1981**, *128*, 897–902. [CrossRef]

coatings

Article

Collision Cascade in a Silicon-Based Device under Energetic Ar Ions Irradiation

Guoying Liang [1,*], Baoming Xu [1] and Xiaoyun Wei [2]

1 Shandong Provincial Key Laboratory of Biophysics, Institute of Biophysics, Dezhou University, Dezhou 253023, China; xubm2018@163.com
2 Laboratory of Research & Design Center, Dezhou Shihua Chemical Company Limited, Dezhou 253023, China; mysky1980@163.com
* Correspondence: gyliang@buaa.edu.cn; Tel.: +86-0534-8985884

Abstract: Silicon, as the basic material of biochips and electronic devices, is often exposed to irradiation environments, and its radiation resistance has attracted much attention in recent decades. We calculated collision cascade in a silicon-based device under energetic Ar ions irradiation by using Monte Carlo and molecular dynamics simulations. The difference in vacancy probability density under different energetic incident ion irradiation is caused by the penetrating power and the straggling power of incident ions. The kinetic energy of an incident ion determines the size of local collision cascade density; a high energy incident ion can induce greater local collision cascade density. The efficiency of transferring energy from incident ions to target electrons at the silicon surface is more than in silicon, and the recoil atoms dissipate most of their energy at the lattice sites where they are stopping. These results provide more insight into the radiation resistance of silicon-based devices.

Keywords: point defect; vacancy; interstitial atom; collision cascade; irradiation

1. Introduction

Silicon-based devices are commonly operated in many fields, such as biomedical engineering, particle physics experiments, and space [1]. Some of these fields present disadvantageous irradiation environments that may affect the operation of silicon-based devices. This is often due to the silicon substrate generating damage under energetic ions irradiation. When an energetic ion enters the silicon substrate, the energetic ion can collide with atoms in the substrate, which are then displaced and continue producing additional collisions. For instance, radiation-induced defects in the substrate often go through dynamic annealing processes, involving migration, clustering, and recombination of vacancies and interstitial atoms. Dynamic annealing often damages the lattice and reduces the radiation resistance in silicon. Hence, understanding the radiation-induced defects is critical to improve the radiation resistance of silicon-based devices.

Irradiation-induced defects in silicon and silicon compounds have been explored for decades [2–6]. The developments in studying effects in silicon and silicon compounds are often intertwined with experimental and computational developments [7–14]. During this period, the focus is on methods that either deal with the primary irradiation defects generation, or with such damage evolution that occurs due to irradiation [15,16]. Silicon surface can generate different morphology under irradiation, such as cleavage crack [17], nanocrystallite [18], crater [19], swelling [20], exfoliation [21], and so on. Our previous works show collision cascade generation and evolution in silicon under pulsed ion beam irradiation [22,23]. Such morphologies are induced generation by atomic movement in silicon. Therefore, investigating the mechanism of atomic movement in silicon under irradiation is necessary to understand the reasons for such morphology's formation. However, due to transient performance and microscopic property of atomic movement, it is difficult to carry out real-time observation of defect evolution. Nowadays, the computational simulation method is a powerful supplement to the experimental observation method. Computational simulations indicate collision cascade, which is one of the significant features of atomic movement, is mass fractal in silicon depending on ion mass, energy, and the depth from the sample surface [24]. Collision cascade density can be used to predict complicated radiation defect dynamics in silicon [25]. Recently, it has been difficult to study the influence of incident ion energy on collision cascade density using experimental methods. According to the above survey, we studied how the ion and silicon substrate affect collision cascade density and energy transfer process using computational simulations.

Ar ions irradiation as a useful method in material surface modification has been applied in scientific experiments and device processing [18,26,27]. In this paper, for a qualitative understanding of the radiation resistance of silicon-based devices under Ar ions irradiation, we investigated silicon substrate defects evolution in silicon-based devices under energetic ions irradiation, as Figure 1 shows. The vacancy probability density, local collision cascade density, ionization energy loss, and real-time defect evolution are illustrated.

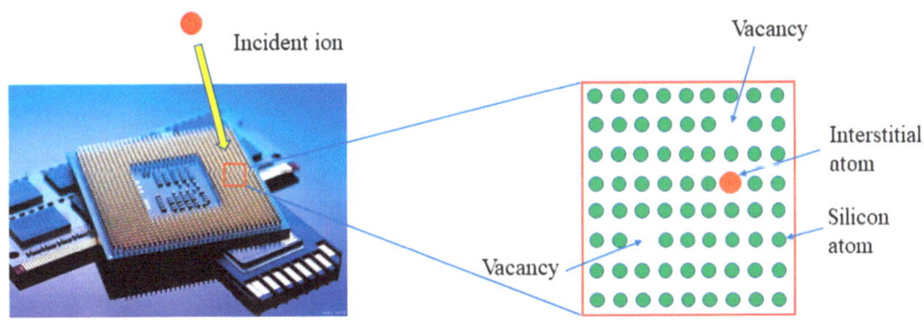

Figure 1. Schematic diagram of point defect generation in the silicon-based device under energetic ion irradiation. The left diagram shows a silicon-based device in the irradiation environment is irradiated under incident energetic ions. The right diagram shows point defect generation when incident ions collide with silicon atoms.

2. Modelling and Methods

2.1. Monte Carlo Simulation

In order to investigate the collision cascade in the silicon target during energetic Ar ions irradiation, the software package concerning the Stopping and Range of Ions in Matter (SRIM-2013) [28] was first performed in this work. We selected the method of detailed calculation with full damage cascade, which can follow every recoil atom until its energy drops below the displacement energy of any target atom, to analyze collision cascade in the target. We selected Ar ions as incident ions in the periodic table of the ion data menu, and 7 degrees of the angle of incidence was chosen. The silicon target was selected in the target layers menu and the width of target layer was 1 micrometer. The incident direction of Ar ion was defined with a positive x-axis being depth into the target. According to irradiation-induced defects often occurring near silicon surface, the impact of 100 keV, 300 keV, 500 keV, 700 keV, and 900 keV Ar ions on the silicon target surface were calculated, respectively. About ~6000 individual collision cascades were stored and analyzed.

2.2. Molecular Dynamics Simulation

Real-time observation of defect evolution is necessary for understanding collision cascade. Molecular dynamics (MD) as one of the methods for real-time observation at the nanoscale level has been used in energetic ions irradiation in recent decades. The classical molecular dynamics simulations of the impact of incident Ar ions on a silicon surface were performed by using the Large-scale Atomic/Molecular Massively Parallel Simulator (LAMMPS-23 Jun 2022) [29]. The three-dimensional MD model of incident ion impact on monocrystalline silicon target surface is constructed as shown in Figure 2. The cubic bulk represents the silicon target, which has a diamond lattice structure and contains about ~263,321 atoms. The lengths of the bulk edges are the same as 20a along the x and y-axis and 80a along the z-axis. The lattice constant a is given as 5.431 Å. The blue spot represents an incident Ar ion. Periodic boundary conditions are applied in the x and y-axis directions, free boundary conditions are imposed in the z-axis direction. The fixed layers that have 2a thickness at the bottom of the silicon target avoid incident Ar ions penetration target. The boundary layers that include 2a thickness above the fixed layers and 2a thickness at four lateral boundaries of the target are imitated to decrease energy dissipation with the Berendsen method.

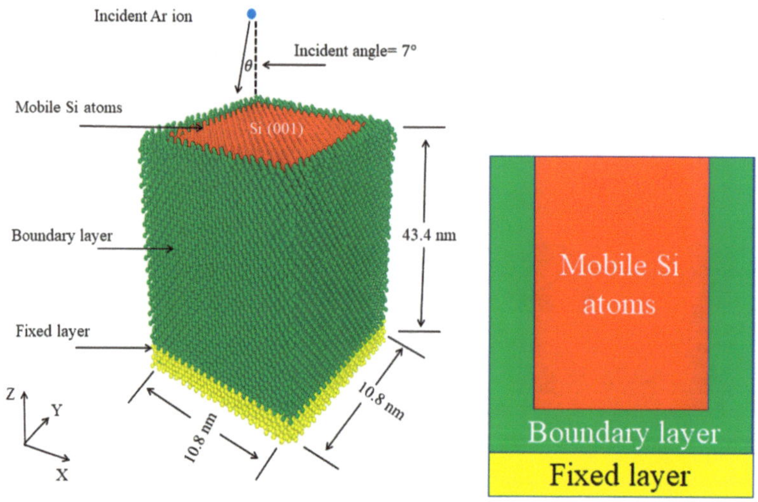

Figure 2. The three-dimensional MD model for incident Ar ion impact on monocrystalline silicon target (**left**). Blue spot, red spots, green spots, and yellow spots represent incident Ar ion, mobile silicon atoms, boundary silicon atoms, and fixed silicon atoms, respectively. The x–z cross-section of the three-dimensional model is shown in the (**right**) figure.

The interatomic interaction in materials during collision cascades can be briefly described by an empirical potential function. A short-range of repulsive Ziegler-Biersack-Littmark (ZBL) potential [30] is used to describe the binary collision between Ar ions and silicon atoms at short interatomic distances. The Tersoff potential [31] is used to describe the long-range interactions in covalent systems. According to the distinctive application of the Tersoff potential and the ZBL potential in covalent systems, the Tersoff/ZBL complex potential, which is the Tersoff potential splined smoothly to the ZBL potential by a transition function, is used to describe the interaction between silicon atoms in this paper.

$$E = \frac{1}{2}\sum_{i}\sum_{j\neq i} V_{ij} \tag{1}$$

$$V_{ij} = \left[1 - f_F(r_{ij})\right] V_{ij}^{ZBL} + f_F(r_{ij}) V_{ij}^{Tersoff} \tag{2}$$

$$f_F(r_{ij}) = \frac{1}{1 + e^{-A_F(r_{ij} - r_c)}} \tag{3}$$

where V_{ij}^{ZBL} and $V_{ij}^{Tersoff}$ indicate the ZBL interatomic potential portion and the Tersoff interatomic potential portion, respectively. Also, r_{ij} is the distance between the i'th atom and the j'th atom, r_c is the cutoff radius for the ZBL potential, and $f_F(r_{ij})$ is a transition function that controls the smoothness of the transition between the ZBL potential and the Tersoff potential. Finally, $f_F(r_{ij})$ is controlled by the parameter of A_F, the larger A_F is corresponding to the sharper transition function, and vice versa.

Before performing Ar ion irradiation, we relaxed the whole system by using a conjugate gradient energy minimization algorithm. If incident ions had moved along the crystal axis, they might have generated strong channel effect and quickly penetrated through the target. It is difficult to understand the interaction between incident ions and target atoms. Therefore, incident Ar ions bombarded the silicon target at 7 degrees off the normal direction of the target top surface to minimize the channel effect. Finally, the system was calculated at 10,000 timesteps during collision cascades between Ar ions and silicon atoms. The MD timestep was kept at a value of 0.1 fs, which was sufficient for Ar ion irradiation on the silicon target [32]. The open visualization tool (OVITO) [33] software was used to quantitatively measure the microscopic morphology in silicon.

3. Results and Discussion

3.1. Vacancy Probability Density

The impact of incident energetic ions on the target surface will result in primary knock-on atom (PKA) generation in the target. PKA often have enough kinetic energy to collide with other target atoms and induce a lot of target atom displacement; this process is so-called cascade collision. Vacancy and interstitial atoms generate in the process of target atoms displacement. Vacancy evolution was investigated during different energetic ion irradiation. Figure 3 shows vacancy probability density distribution with the depth of the target. Figure 3a indicates vacancy probability density function (VPDF) variation with the depth of the target. The VPDF firstly reaches its peak value at ~900 Å in target under 100 keV Ar ion irradiation in contrast to other Ar ions with higher kinetic energy. Compared with 300 keV–900 keV Ar ions, it is shown that Ar ions with higher kinetic energy can induce deeper vacancies formation while its VPDF peak is lower. It is suggested that the penetrating power of higher energy incident ions is deeper than that of lower energy incident ions, but the straggling power of lower energy incident ions is better than that of higher energy incident ions. The reason for this phenomenon can be explained based on our previous work [22]. Higher energetic ions can penetrate the silicon target quickly and generate fewer vacancies at their tracks. In contrast, lower energetic ions merely reach and stop at shallow depth, transfer their energies to neighbor target atoms and generate more vacancies.

Interstitial atoms immediately generate when recoil target atoms leave their initial lattice sites. Figure 3b shows the depth dependence of interstitial atoms distribution in the target. In the cases of 100 keV, 300 keV, and 500 keV Ar ions irradiation, the depth of the interstitial atom's peak is almost equal to the depth of vacancy's peak. As the dashed line guide, it is indicated that interstitial atoms mainly assemble nearby vacancies after they leave from their original lattice sites. But the depths of interstitial atom's peak under 700 keV and 900 keV Ar ions irradiation are shallower than the depth of vacancy's peak under the same Ar ion irradiation. We deduced that interstitial atoms may move backward opposite the Ar ion's incident direction and assemble at shallow lattice sites. This is because higher energy ions firstly reach the boundary layers, which can act as a source and emit interstitial atoms to the target [34], then go back to the target and recombine with more vacancies.

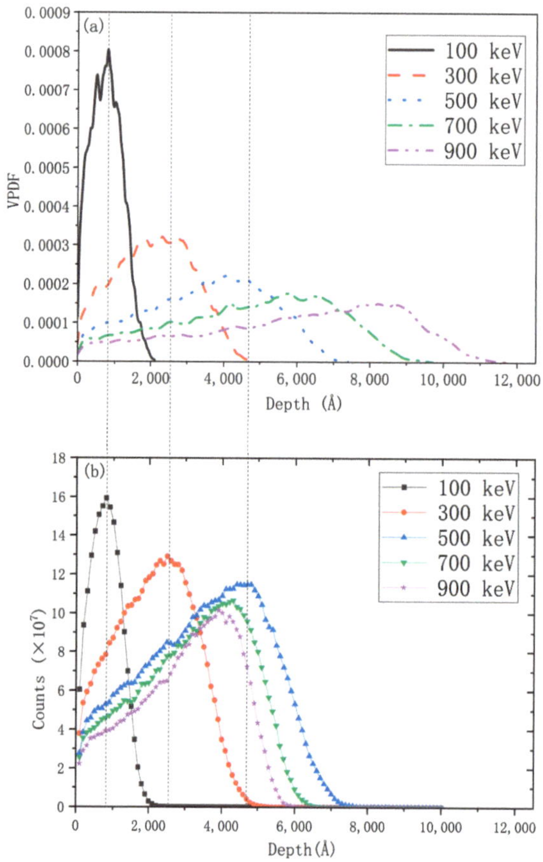

Figure 3. Depth dependence of vacancy probability density function (VPDF) (**a**) and interstitial atoms distribution (**b**) for 100 keV, 300 keV, 500 keV, 700 keV, and 900 keV Ar ions irradiation on the silicon target. Dashed lines are to guide the eye.

3.2. Local Collision Cascade Density

The complexity of the dependence of irradiation conditions on defect evolution can be reduced to a deterministic effect of a single parameter, the local collision cascade density (LCCD), calculated by taking into account the fractal nature of collision cascade [25]. The LCCD is defined as the local density of lattice vacancies within individual collision cascades with the same radius of 10 nm, which was chosen based on the effective diffusion length of the defect displacement in the silicon target [24].

Figure 4 shows the comparison of LCCD in the silicon target under 100 keV–900 keV Ar ions irradiation. The value of LCCD is lowest when 100 keV Ar ion impacts on the silicon target surface, and it is indicated that collision cascades under 100 keV Ar ion irradiation occur less than that under 300 keV–900 keV cases. This is due to shallower penetration for lower energy incident ions, which are in qualitative agreement with the results of our previous work [22]. For 300 keV–900 keV Ar ions, the LCCD quickly increases to the peak with increasing incident kinetic energy, and quickly declines after that. Compared with the five cases, it is seen that they all have the largest count of the LCCD at the same density of ~0.03 Vacancies/nm^3, although the counts of LCCD are different from each other. It is indicated that collision cascade shows similar local collision frequency under different energetic ions irradiation.

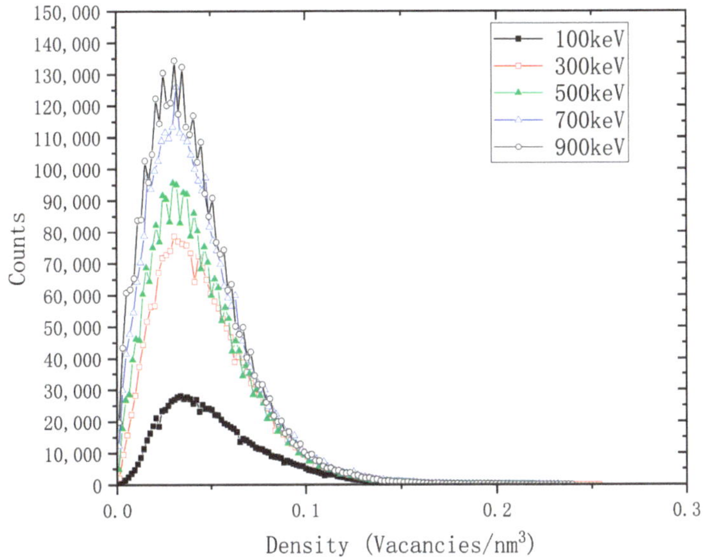

Figure 4. Histogram of local collision cascade density (LCCD) calculated in silicon target during 100 keV, 300 keV, 500 keV, 700 keV, and 900 keV Ar ions irradiation.

3.3. Ionization Energy Loss

The total energy loss of incident Ar ions can be divided into two parts: the energy transferred by the incident ions to the target electrons and to the target nuclei. The energy loss of the incident ions to the target electrons is often called ionization energy loss, which is much larger than the energy loss transferred to the target nuclei under hundreds of keV irradiation, therefore we chiefly investigated ionization energy loss during irradiation.

Figure 5 shows the depth dependence of ionization energy loss under 100 keV–900 keV Ar ions irradiation. For all of the incident Ar ions, the ionization energy loss is largest at the target surface and gradually decreases in the target; it is seen that the efficiency of transferring energy from incident ions to target electrons at the target surface is more than in the target. This is because the electrons at the target surface first make contact with the incident ions and gain more energy to ionize compared to the electrons at the inner target. The ionization energy loss of the recoil silicon atoms is lowest at the target surface and gradually increases to its peak at a depth in the target. The depth becomes deeper with the increasing kinetic energy of the incident ions, and after that the ionization energy loss of the recoil silicon atoms quickly decreases to zero. The recoil atoms do not transfer a lot of energy to target electrons during moving, and dissipate most of their energy at the lattice sites where they stop. This point can be more clearly understood in the next table.

In order to understand the energy loss of incident ions and recoil atoms, we calculated the percentage of transferring energy from incident ions and recoil atoms to ionization, vacancies, and phonons, which are three main destinations of energy loss as listed in Table 1. For incident Ar ions, ionization is the main energy loss compared to vacancies and phonons, ionization energy loss can be more significant with the increasing kinetic energy of incident ions. Meanwhile, the energy loss induced by vacancies and phonons gradually decreases with increasing the kinetic energy of incident ions. For recoil silicon atoms, ionization and phonon are two primary energy loss patterns, and they are gradually decreased by increasing the kinetic energy of incident Ar ions. Compared with incident ions, the energy loss induced by phonons from recoil atoms is more significant. Most of the energy of recoil atoms is dissipated by means of thermal vibration at their lattice sites after they have stopped moving.

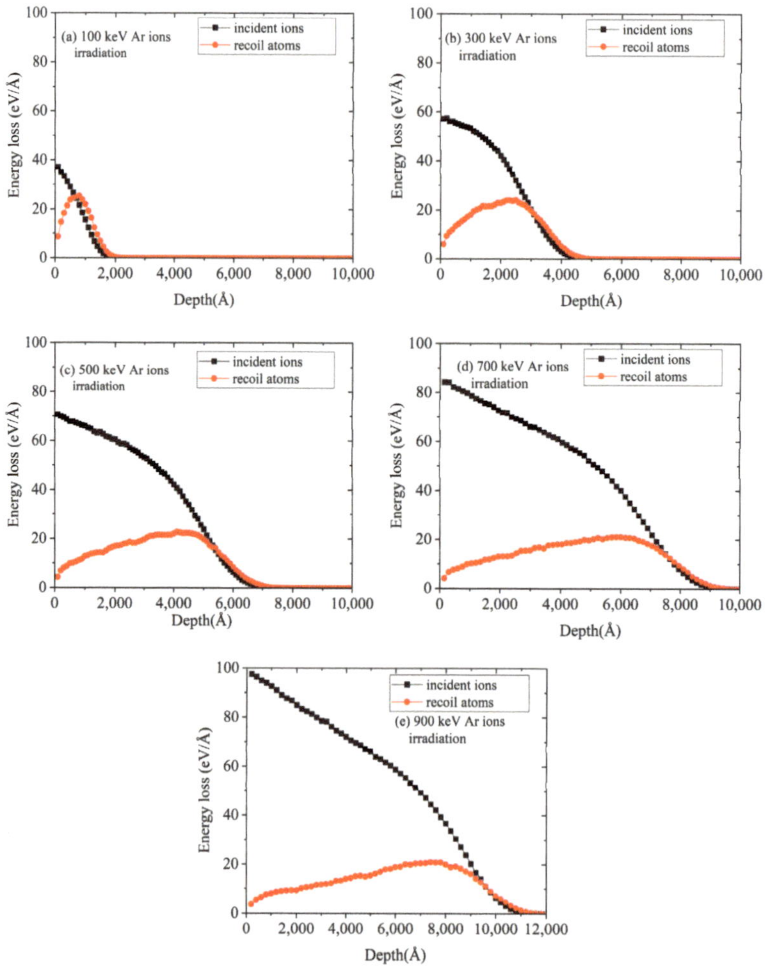

Figure 5. Ionization energy loss of incident ions and recoil atoms under 100 keV (**a**), 300 keV (**b**), 500 keV (**c**), 700 keV (**d**), and 900 keV (**e**) Ar ions irradiation.

Table 1. The energy loss percentage of incident ions and recoil atoms under 100 keV, 300 keV, 500 keV, 700 keV, and 900 keV Ar ions irradiation.

	Incident Ions (keV)					**Recoil Atoms (keV)**				
	100	**300**	**500**	**700**	**900**	**100**	**300**	**500**	**700**	**900**
Ionization	30.99%	47.83%	56.38%	62.16%	66.97%	28.02%	23.59%	20.58%	18.32%	16.14%
Vacancies	0.17%	0.10%	0.08%	0.07%	0.06%	3.23%	2.26%	1.82%	1.54%	1.33%
Phonons	0.52%	0.33%	0.27%	0.24%	0.22%	37.08%	25.88%	20.87%	17.67%	15.28%

3.4. Defect Evolution

Real-time observation of defect evolution with time during energetic ions irradiation is essential for understanding collision cascade in silicon. Figure 6 shows vacancies and interstitial atom evolution with time in silicon. As PKA have enough kinetic energy to collide with lots of target atoms and induce their displacement, the interstitial atom leaves from its original lattice site and generates a vacancy in this site as shown in Figure 6a. To avoid the channel effect, the incident direction of Ar ions is at 7 degrees off the normal direction of the target top surface, inducing the interstitial atoms' movement towards the target boundary. Figure 6b shows interstitial atoms reach the target boundary at 300 fs, then the target boundary can emit interstitial atoms to the inside of the target and generate vacancies, which is consistent with the result of our previous study [23]. After that, most of the interstitial atoms begin thermal vibration at the lattice sites where they stop because their kinetic energy is less than threshold displacement energy, as shown in Figure 6c,d.

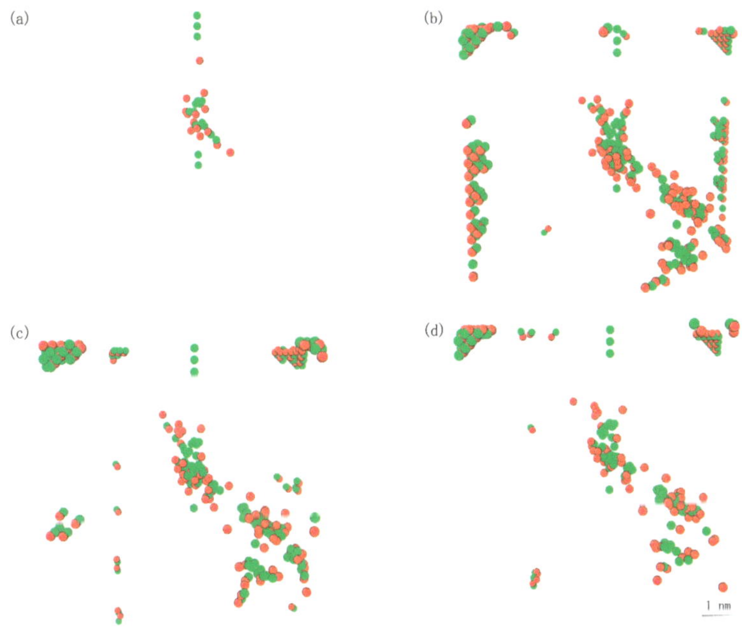

Figure 6. Vacancies and interstitial atoms evolution with time of (**a**) 80 fs, (**b**) 300 fs, (**c**) 500 fs, and (**d**) 700 fs under 1 keV Ar ions irradiation. Green spots represent vacancies and red spots represent interstitial atoms. The same scale (1 nm) is shown in (**d**).

Figure 6 can indicate vacancies and the interstitial atoms' evolution with time, but the interface between an amorphous region and perfect crystal is limited. In order to distinguish an amorphous region from perfect crystal region, we calculated and drew the interface of the two regions as shown in Figure 7. The size of the interface increases rapidly along depth in the first ~200 fs. It is illustrated that collision cascade in the target is very dense at the beginning of Ar irradiation. This is because PKA have enough kinetic energy to collide with a lot of target atoms and induce many collision cascade generation. The interface reaches the boundary at ~300 fs and continues to extend towards the deeper region due to the boundary emitting interstitial atoms to the inside of the target [34]. Finally, interstitial atoms stop in the target and the interface will not extend. Local recombination of interstitial atoms and vacancies induces a decrease in the size of the interface and generates cracks in the interface as shown in Figure 7d,e.

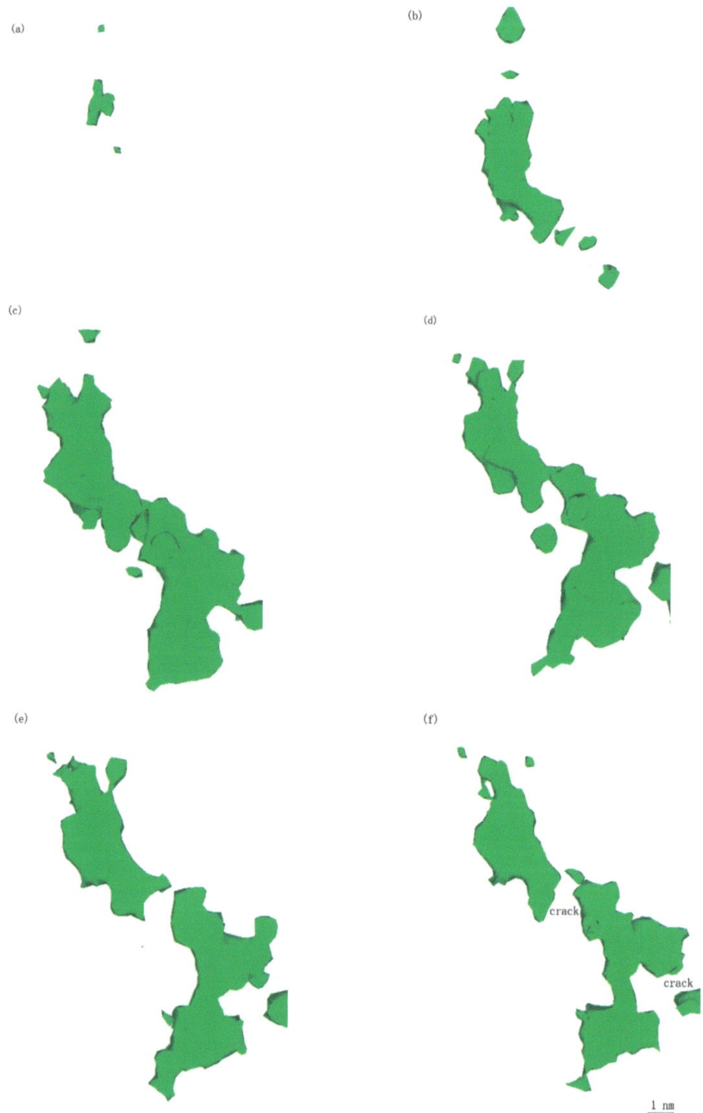

Figure 7. The evolution of interface between an amorphous region and perfect crystal with time of (**a**) 60 fs, (**b**) 100 fs, (**c**) 200 fs, (**d**) 300 fs, (**e**) 500 fs, and (**f**) 700 fs under 1 keV Ar ion irradiation. The same scale (1 nm) is shown in (**f**).

4. Conclusions

In this work, silicon substrate defects in silicon-based devices under energetic ions irradiation were studied by using Monte Carlo and molecular dynamics simulations. Vacancy probability density represents vacancy generating probability at a depth in silicon. The difference of vacancy probability density under different energetic incident ion irradiation is caused by the penetrating power and the straggling power of incident ions. The penetrating power of a higher energy incident ion is deeper than that of a lower energy incident ion, but the straggling power of a lower energy incident ion is better than that of a higher energy incident ion. Local collision cascade density represents the intensity of collision cascade in silicon. The kinetic energy of an incident ion determines the size of local collision cascade density. High energy incident ions can induce greater local collision cascade density. Based on ionization energy loss, it is seen that the efficiency of transferring energy from incident ions to target electrons at the target surface is more than that in the target, and the recoil atoms dissipate most of their energy at the lattice sites where they are stopping. Real-time observations of defects indicate defects such as vacancies and interstitial atoms have directional movement, and the shape and size of the interface between an amorphous region and perfect crystal are constantly changing during incident ions irradiation.

Author Contributions: Conceptualization, G.L.; methodology, G.L.; formal analysis, G.L. and B.X.; investigation, X.W.; software, G.L. and B.X.; writing—original draft preparation, G.L.; writing—review and editing, G.L.; visualization, X.W.; supervision, G.L.; funding acquisition, G.L. and B.X. All authors have read and agreed to the published version of the manuscript.

Funding: This research was funded by the National Natural Science Foundation of China, grant number 11705099 and the Dezhou University Project, grant number 2021xjrc313.

Institutional Review Board Statement: Not applicable.

Informed Consent Statement: Not applicable.

Data Availability Statement: All data are contained within the article.

Acknowledgments: The authors would like to thank Joseph Wallace of Lawrence Livermore National Laboratory for his technical assistance, Xiaoyun Le of Beihang University, Sha Yan of Peking University, and Gennady Efimovich Remnev of National Research Tomsk Polytechnic University for their data analysis assistance.

Conflicts of Interest: The authors declare no conflict of interest.

References

1. Leroy, C.; Rancoita, P.G. Particle Interaction and Displacement Damage in Silicon Devices Operated in Radiation Environments. *Rep. Prog. Phys.* **2007**, *70*, 493–625. [CrossRef]
2. Aji, L.B.B.; Wallace, J.B.; Kucheyev, S.O. Effects of Collision Cascade Density on Radiation Defect Dynamics in 3C SiC. *Sci. Rep.* **2017**, *7*, 44703. [CrossRef]
3. Wallace, J.B.; Aji, L.B.B.; Martin, A.A.; Shin, S.J.; Shao, L.; Kucheyev, S.O. The Role of Frenkel Defect Diffusion in Dynamic Annealing in Ion-Irradiated Si. *Sci. Rep.* **2017**, *7*, 39754. [CrossRef]
4. Bayu Aji, L.B.; Wallace, J.B.; Shao, L.; Kucheyev, S.O. Non-Monotonic Temperature Dependence of Radiation Defect Dynamics in Silicon Carbide. *Sci. Rep.* **2016**, *6*, 30931. [CrossRef] [PubMed]
5. Wallace, J.B.; Charnvanichborikarn, S.; Bayu Aji, L.B.; Myers, M.T.; Shao, L.; Kucheyev, S.O. Radiation Defect Dynamics in Si at Room Temperature Studied by Pulsed Ion Beams. *J. Appl. Phys.* **2015**, *118*, 135709. [CrossRef]
6. Azarov, A.Y.; Titov, A.I.; Karaseov, P.A.; Hallén, A. Effect of Collision Cascade Density on Radiation Damage in SiC. *Nucl. Instrum. Methods Phys. Res. Sect. B Beam Interact. Mater. Atoms* **2009**, *267*, 1247–1250. [CrossRef]
7. Pelenitsyn, V.; Korotaev, P. First-Principles Study of Radiation Defects in Silicon. *Comput. Mater. Sci.* **2022**, *207*, 111273. [CrossRef]
8. Mihai, M.D.; Iancu, D.; Zarkadoula, E.; Florin, R.A.; Tong, Y.; Zhang, Y.; Weber, W.J.; Velişa, G. Athermal Annealing of Pre-Existing Defects in Crystalline Silicon. *Acta Mater.* **2023**, *261*, 119379. [CrossRef]
9. Al-Sharif, A.; AlAderah, B.; Obeidat, A.; Talla, J. Influence of Transition Metal Defects on Electronic and Magnetic Properties of Bulk Silicon: Ab-Initio Simulation. *Mater. Today Commun.* **2023**, *34*, 105415. [CrossRef]
10. Cherkova, S.G.; Volodin, V.A.; Skuratov, V.A.; Stoffel, M.; Rinnert, H.; Vergnat, M. Formation of Light-Emitting Defects in Silicon by Swift Heavy Ion Irradiation and Subsequent Annealing. *Nucl. Instrum. Methods Phys. Res. Sect. B Beam Interact. Mater. Atoms* **2023**, *535*, 132–136. [CrossRef]
11. Huang, N.; Zhou, P.; Goel, S. Microscopic Stress Analysis of Nanoscratch Induced Sub-Surface Defects in a Single-Crystal Silicon Wafer. *Precis. Eng.* **2023**, *82*, 290–303. [CrossRef]
12. Tang, S.; Qi, X.; Chang, C.; Wang, Q.; Liu, L. Reduction of the Initial Defects Generated during Casting of Quasi-Single Crystalline Silicon by Reserving Gaps between Seed Crystals. *Sol. Energy Mater. Sol. Cells* **2023**, *263*, 112571. [CrossRef]
13. Li, Q.; Lin, X.; Lin, X.H.; Xing, Y.; Zhou, Z. Experimental Study on the Thermal Evolution of Silicon Defects with Zonal Characteristics Induced by Precisely Localized Irradiation of Focused Helium Ion Beams. *Vacuum* **2023**, *218*, 112615. [CrossRef]
14. Fan, Y.; Xu, Z.; Song, Y.; Sun, T. Molecular Dynamics Simulation of Silicon Vacancy Defects in Silicon Carbide by Hydrogen Ion Implantation and Subsequent Annealing. *Diam. Relat. Mater.* **2021**, *119*, 108595. [CrossRef]

15. Nordlund, K. Historical Review of Computer Simulation of Radiation Effects in Materials. *J. Nucl. Mater.* **2019**, *520*, 273–295. [CrossRef]
16. Vaitkus, J.V.; Rumbauskas, V.; Mockevicius, G.; Zasinas, E.; Mekys, A. An Evidence of Strong Electron-Phonon Interaction in the Neutron Irradiation Induced Defects in Silicon. *Nucl. Instrum. Methods Phys. Res. Sect. A Accel. Spectrometers Detect. Assoc. Equip.* **2015**, *796*, 114–117. [CrossRef]
17. Starinskiy, S.V.; Rodionov, A.A.; Shukhov, Y.G.; Safonov, A.I.; Maximovskiy, E.A.; Sulyaeva, V.S.; Bulgakov, A.V. Formation of Periodic Superhydrophilic Microstructures by Infrared Nanosecond Laser Processing of Single-Crystal Silicon. *Appl. Surf. Sci.* **2020**, *512*, 145753. [CrossRef]
18. Evseev, A.P.; Kozhemiako, A.V.; Kargina, Y.V.; Balakshin, Y.V.; Zvereva, E.A.; Chernysh, V.S.; Gongalsky, M.B.; Shemukhin, A.A. Radiation-Induced Paramagnetic Defects in Porous Silicon under He and Ar Ion Irradiation. *Radiat. Phys. Chem.* **2020**, *176*, 109061. [CrossRef]
19. Shen, J.; Yu, X.; Zhang, Y.; Zhong, H.; Zhang, J.; Qu, M.; Yan, S.; Zhang, G.; Zhang, X.; Le, X. Novel Microstructures on the Surfaces of Single Crystal Silicon Irradiated by Intense Pulsed Ion Beams. *Nucl. Instrum. Methods Phys. Res. Sect. B Beam Interact. Mater. Atoms* **2015**, *365*, 26–29. [CrossRef]
20. Shen, J.; Shahid, I.; Yu, X.; Zhang, J.; Zhong, H.; Cui, X.; Liang, G.; Yu, X.; Huang, W.; Yan, S.; et al. Fracture Analysis of Surface Exfoliation on Single Crystal Silicon Irradiated by Intense Pulsed Ion Beam. *Nucl. Instrum. Methods Phys. Res. Sect. B Beam Interact. Mater. Atoms* **2017**, *413*, 6–12. [CrossRef]
21. Shen, J.; Shahid, I.; Yu, X.; Zhang, J.; Zhong, H.W.; Yu, X.; Huang, W.Y.; Liang, G.Y.; Cui, X.J.; Yan, S.; et al. Surface Exfoliation Analysis on Single-Crystal Silicon under Compressed Plasma Flow Action. *Laser Part. Beams* **2018**, *36*, 129–135. [CrossRef]
22. Liang, G.; Zhong, H.; Zhang, S.; Xu, M.; Kuang, S.; Ren, J.; Zhang, N.; Yan, S.; Yu, X.; Remnev, G.E.; et al. Molecular Dynamics Study of Damage Nearby Silicon Surface Bombarded by Energetic Carbon Ions. *Surf. Coat. Technol.* **2020**, *385*, 125350. [CrossRef]
23. Liang, G.; Zhong, H.; Wang, Y.; Zhang, S.; Xu, M.; Kuang, S.; Ren, J.; Zhang, N.; Yan, S.; Yu, X.; et al. Molecular Dynamics Simulations of Vacancy Generation and Migration near a Monocrystalline Silicon Surface during Energetic Cluster Ion Implantation. *Coatings* **2020**, *10*, 146. [CrossRef]
24. Wallace, J.B.; Aji, L.B.B.; Shao, L.; Kucheyev, S.O. Fractal Analysis of Collision Cascades in Pulsed-Ion-Beam-Irradiated Solids. *Sci. Rep.* **2017**, *7*, 17574. [CrossRef] [PubMed]
25. Wallace, J.B.; Aji, L.B.B.; Shao, L.; Kucheyev, S.O. Deterministic Role of Collision Cascade Density in Radiation Defect Dynamics in Si. *Phys. Rev. Lett.* **2018**, *120*, 216101. [CrossRef]
26. Bayu Aji, L.B.; Wallace, J.; Kucheyev, S. Radiation Defect Dynamics in 3C-, 4H-, and 6H-SiC Studied by Pulsed Ion Beams. *Nucl. Instrum. Methods Phys. Res. Sect. B Beam Interact. Mater. Atoms* **2018**, *435*, 8–11. [CrossRef]
27. Myers, M.T.; Charnvanichborikarn, S.; Shao, L.; Kucheyev, S.O. Pulsed Ion Beam Measurement of the Time Constant of Dynamic Annealing in Si. *Phys. Rev. Lett.* **2012**, *109*, 2–5. [CrossRef]
28. Ziegler, J.F.; Ziegler, M.D.; Biersack, J.P. SRIM—The Stopping and Range of Ions in Matter (2010). *Nucl. Instrum. Methods Phys. Res. Sect. B Beam Interact. Mater. Atoms* **2010**, *268*, 1818–1823. [CrossRef]
29. Plimpton, S. Fast Parallel Algorithms for Short-Range Molecular Dynamics. *J. Comput. Phys.* **1995**, *117*, 1–19. [CrossRef]
30. Ziegler, J.; Biersack, J.; Littmark, U. *The Stopping and Range of Ions in Solids*; Pergamon Press: New York, NY, USA, 1985.
31. Tersoff, J. Modeling Solid-State Chemistry: Interatomic Potentials for Multicomponent Systems. *Phys. Rev. B* **1989**, *39*, 5566–5568. [CrossRef]
32. Xiao, Y.J.; Fang, F.Z.; Xu, Z.W.; Hu, X.T. Annealing Recovery of Nanoscale Silicon Surface Damage Caused by Ga Focused Ion Beam. *Appl. Surf. Sci.* **2015**, *343*, 56–69. [CrossRef]
33. Stukowski, A. Visualization and Analysis of Atomistic Simulation Data with OVITO-the Open Visualization Tool. *Model. Simul. Mater. Sci. Eng.* **2010**, *18*, 015012. [CrossRef]
34. Bai, X.M.; Voter, A.F.; Hoagland, R.G.; Nastasi, M.; Uberuaga, B.P. Efficient Annealing of Radiation Damage near Grain Boundaries via Interstitial Emission. *Science* **2010**, *327*, 1631–1634. [CrossRef] [PubMed]

coatings

MDPI

Review

Electrochemical Detection of Hormones Using Nanostructured Electrodes

Naila Haroon and Keith J. Stine *

Department of Chemistry and Biochemistry, University of Missouri-Saint Louis, St. Louis, MO 63121, USA; nhyym@umsl.edu
* Correspondence: kstine@umsl.edu

Abstract: Hormones regulate several physiological processes in living organisms, and their detection requires accuracy and sensitivity. Recent advances in nanostructured electrodes for the electrochemical detection of hormones are described. Nanostructured electrodes' high surface area, electrocatalytic activity, and sensitivity make them a strong hormone detection platform. This paper covers nanostructured electrode design and production using MOFs, zeolites, carbon nanotubes, metal nanoparticles, and 2D materials such as TMDs, Mxenes, graphene, and conducting polymers onto electrodes surfaces that have been used to confer distinct characteristics for the purpose of electrochemical hormone detection. The use of aptamers for hormone recognition is producing especially promising results, as is the use of carbon-based nanomaterials in composite electrodes. These materials are optimized for hormone detection, allowing trace-level quantification. Various electrochemical techniques such as SWV, CV, DPV, EIS, and amperometry are reviewed in depth for hormone detection, showing the ability for quick, selective, and quantitative evaluation. We also discuss hormone immobilization on nanostructured electrodes to improve detection stability and specificity. We focus on real-time monitoring and tailored healthcare with nanostructured electrode-based hormone detection in clinical diagnostics, wearable devices, and point-of-care testing. These nanostructured electrode-based assays are useful for endocrinology research and hormone-related disease diagnostics due to their sensitivity, selectivity, and repeatability. We conclude with nanotechnology–microfluidics integration and tiny portable hormone-detection devices. Nanostructured electrodes can improve hormone regulation and healthcare by facilitating early disease diagnosis and customized therapy.

Keywords: hormone; nanostructures; modified electrode; voltammetry; biosensor

1. Introduction

The term "hormone" is used in a comprehensive manner, encompassing molecules that are not conventionally classified as hormones, like eicosanoids, neurogenic amines, interleukins, and cytokines, along with the secretions of the principal endocrine glands, namely, the pancreas, adrenal glands, gonads, and thyroid [1]. Hormones are chemicals that are made and secreted by cells, and they can influence the behavior of neighboring cells. Both the actions that hormones perform and the molecular structures that they possess allow for their classification [2]. Hormones can be classified or categorized according to many criteria. The topics of interest include (a) region, (b) mechanism, (c) composition, (d) endocrine stimulation, and (e) effect. The initial classification of hormones consisted of three distinct categories, namely, steroid hormones, amino acid-related hormones, and protein hormones. Nevertheless, the classification of hormones into solely three categories grew increasingly insufficient and imprecise with the discovery of additional hormones. There are four distinct hormonal transmission systems, including neurotransmitters, systemic, paracrine, and autocrine [3]. Hormones can be broadly classified into two categories: endogenously synthesized hormones, which are naturally created by the human body, and exogenously synthesized hormones, which are artificially produced with industrial means [4]. The probability of developing breast cancer is related to increased exposure to elevated amounts of both exogenous and endogenous hormones [5].

Several glands in the body release hormones, which make it easier to control important biochemical processes. Hormone imbalances (either too much or not enough) can be caused by several conditions, many of which lead to serious illness and a wide range of systemic effects. So, measuring and monitoring hormone levels can help diagnose a person's condition. Also, hormones from outside the body can be used to treat many diseases, like diabetes, so being able to measure hormone concentrations is important for personalized and effective treatment. Since diseases that affect the endocrine system are becoming more common, the global endocrine testing market is growing quickly [6]. Hormones serve as signaling molecules within the human body and are synthesized and secreted by many endocrine glands, including the adrenal, thymus, pituitary, thyroid, pancreatic, and pineal glands. Male and female biological hormones in males and females are created by their respective reproductive organs, namely, the testicles in males and the ovaries in females. They are transported throughout the body via the circulatory system and ultimately reach their intended destinations within diverse tissues and organs. Hormones are recognized for their multifaceted functions within the human body, encompassing several aspects such as metabolism, sexual function, development, reproduction, and growth. Moreover, they play a crucial role in regulating circadian rhythms and the innate sleep–wake patterns of individuals. Furthermore, there have been documented instances of their interaction affecting the immune response. They perform a fundamental function in the regulation of metabolism within the human body. In addition, hormones are known to fulfill crucial functions in maintaining homeostasis, such as regulating blood pressure and glucose levels. The functionality of these hormones is typically facilitated with the utilization of positive and negative feedback mechanisms.

Maintaining hormonal equilibrium is essential for promoting overall well-being and reducing the risk of cancer, irrespective of an individual's religious convictions. The natural occurrence of hormone fluctuations is observed in various conditions, including premenopause, puberty, menopause, and detrimental lifestyle choices. These fluctuations can disrupt hormone levels, resulting in a range of symptoms such as hair damage, hair loss, infertility, irregular periods, unwanted weight gain or loss (not attributable to intentional dietary changes), insomnia, anxiety, depression, fatigue, reduced libido, digestive problems, changes in appetite, and numerous other manifestations [7]. Hormonal disorders, diseases of the cardiovascular system, eating disorders, and osteoporosis are just some of the serious conditions that can develop as a result of hormonal imbalances or an inability to produce normal amounts of hormones [8]. The ability to detect and analyze hormones in both invasive and non-invasive ways using bodily fluids is a powerful diagnostic tool that has been used for years to identify or confirm diseases and anomalies and to distinguish between distinct physiological states or changes. Doping in sports can also be checked using hormone detection methods. Over the years, other approaches have emerged, each one more refined and precise than the last. Bioassays, immunoassays, and receptor assays are the big three conventional methods for detecting and analyzing hormones [3]. Certain cells, such as Leydig cells, mostly found in glands, are responsible for the production of testosterone hormones, which can subsequently reach their desired target cells either via the process of simple diffusion or by the circulation of blood. The identification of hormones is essential for the disease diagnosis process since hormones often circulate at low concentrations (less than or equal to 1 nM) [9]. Hormones, being complex molecules, have a crucial function in animal growth and several biochemical processes. Both synthetic and natural hormones are extensively used in the agricultural and dairy industries, despite their significant health risks. They are responsible for overseeing substantial modifications in organisms that impact several biological processes, including glucose metabolism, stress response, pigmentation, and reproductive well-being. Given the limited presence of hormones in organisms, it is imperative to use efficient techniques for hormone detection. Nevertheless, traditional analytical techniques used for the recognition of animal hormones exhibit many restrictions. These downsides encompass prolonged analysis duration, high costs, frequent occurrence of inaccurate outcomes, and inherent complexity in their application [10].

The ability to frequently monitor hormones is constrained by the limitations of existing conventional centralized analytical instruments, which lack the capability to provide a rapid response. The utilization of electrochemical sensing has been widely regarded as an optimal method for hormone detection due to its several advantages, including a rapid response time, ease of use, affordability, and potential applicability in point-of-care environments [11]. Scientists have placed a premium on plant hormone detection for several reasons, including the identification of novel hormones and metabolites and tracking the presence of individual hormones within a cell. Science has progressed to the point where we can categorize the analytical tools used to assess plant hormones and other elements as either accurate, fast, or sensitive. Existing analytical methods are tested using traditional detection procedures. Extensive processes are required for sample preparation prior to running a bioassay, immunoassay, or chromatographic method. In addition, current methods are not advanced enough to isolate and quantify compounds deep within plant tissues. Wearable devices, electrochemical sensors, biosensors, and spectroscopic methods have replaced traditional laboratory approaches because of their improved quality and sensitivity. These have allowed for sensitive and selective recognitions at the specific tissue level, greatly enhancing the detection limits and sample recoveries [12].

Biomolecules play a crucial role in the human body since they participate in numerous metabolic pathways [13]. Endocrine cells are responsible for the secretion of hormones, which are subsequently transported to target tissues by the circulatory system. Due to their minimal presence, a significant level of sensitivity is required for their detection. Endocrine cells, located within endocrine glands, are responsible for the synthesis and release of hormones [4]. For hormones to reach their targets, a specific cell type—typically located in a gland—must first secrete them. Without entering the target cell, certain hormones trigger reactions on the inside by binding to certain receptors on the plasma membrane. Some hormones, however, penetrate the cell membrane and bind to specific receptors inside the target cell. Hormones are often difficult to detect because they are secreted at such low amounts (1 nM or less) [9]. Hormones possess significant physiological roles encompassing metabolic regulation, developmental processes, somatic expansion, and reproductive functions. They also facilitate a range of environmental signals, injuries, and stress factors. The agricultural industry extensively utilizes both natural and synthetic hormones despite the potential for serious health risks associated with their use. Therefore, the expeditious and precise identification of hormones holds significant significance. Nevertheless, conventional chromatographic methods are characterized by their extended duration, high cost, frequent lack of precision, and challenging implementation. Biosensors are devices that have three essential constituents: a transducer, a biorecognition element, and an output mechanism. The utilization of biosensors for hormone detection is becoming increasingly significant due to their rapidity, convenience, and affordability, as well as their notable selectivity and sensitivity. Therefore, the selection of biosensors as a means for hormone detection offers numerous advantages and has numerous applications when compared with conventional chromatographic approaches. Figure 1 shows the broad application of hormone biosensors in various fields covering medicine, healthcare, and agriculture [14].

1.1. Classification of Hormones

Hormones can be categorized based on their chemical composition, mode of operation, functional characteristics, physiological consequences, and the stimulus they provide to endocrine glands as shown in Figure 2 [7]. Hormones can originate from lipid or peptide sources [14]. The structure of hormones discussed in this review are shown in Table 1, along with an example and reference of how their electrochemical detection is related to the hormone structure undergoing a redox change. Not all electrochemical hormone detection relies on a hormone undergoing oxidation or reduction but may be associated with a change in the amount or activity of a redox-active marker.

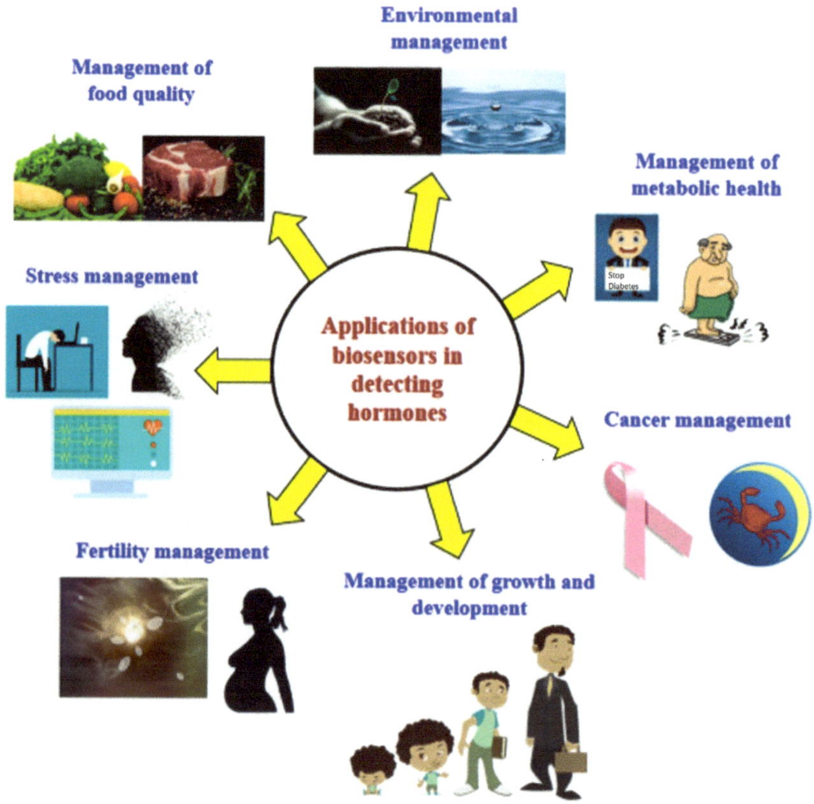

Figure 1. A broad range of applications of hormone biosensors. Reproduced from reference [14] with permission from Elsevier, copyright.

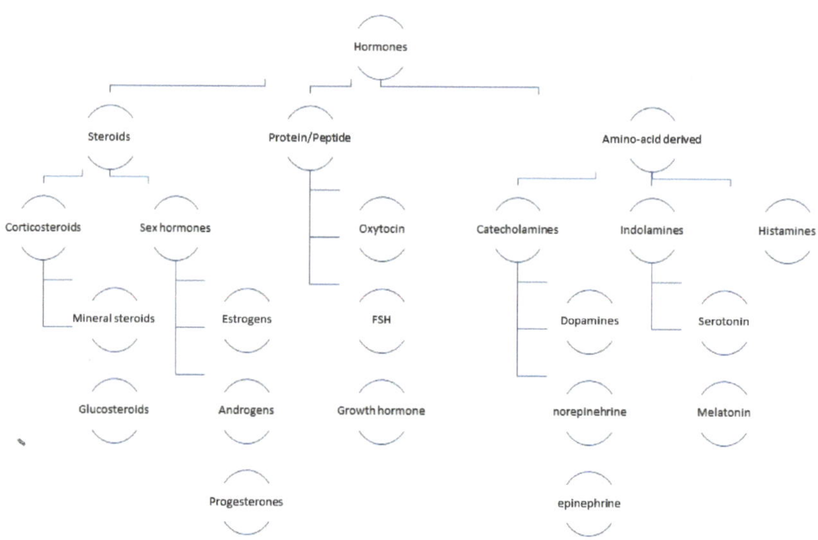

Figure 2. Classification of hormones based on their chemical structures.

1.1.1. Steroid Hormones

Steroid hormones are a vital group of chemical compounds that play a crucial role in regulating numerous physiological activities. Thorough detection of steroid hormones can provide valuable insights into the physiopathologic mechanisms behind illnesses associated with these hormones. Identifying steroid hormones in biological samples poses significant challenges because of their limited endogenous levels and inadequate ionization efficiency. Steroid hormones play a crucial role as signaling molecules in various physiological activities, including transcriptional regulation, maintenance of secondary sexual characteristics, and coordination of the immunological and endocrine systems. Consequently, their significance to the human body is exceptional. Recent studies have demonstrated a direct correlation between irregularities in steroid hormone metabolism in humans and several ailments, such as endocrine disorders, breast cancer, prostate cancer, endometrial cancer, and cardiovascular disease. To comprehend the pathophysiology of different illnesses, it is imperative to use an analytical approach that is highly sensitive, stable, and trustworthy, enabling quantitative profiling of steroid hormones. The utilization of a research methodology that specifically targets the identification of a limited quantity of steroid hormones has the potential to introduce inaccuracies in the process of pathological characterization and the diagnosis of illnesses. Quantitative metabolic profiling of a broader range of steroid hormones not only enhances the comprehensive characterization of the steroid hormone metabolism network but also facilitates the provision of more precise and reliable diagnostic outcomes, hence reducing the occurrence of false positives or false negatives. Researchers encounter significant challenges and obstacles while investigating the metabolic network of steroid hormones due to the predominantly low concentrations of these hormones within the human body [15]. The amount of steroid hormones is an important part of investigating endocrinological disorders that affect how the adrenal glands or gonads work [16].

1.1.2. Peptide Hormones

Peptide hormones are a class of hormones composed of amino acid chains, which exert their primary physiological effects on the endocrine system. Hormones can be categorized into two systems, namely, amino acid-based or steroid-based, depending on their building units. Peptide hormones can exert their effects on target cells with the utilization of secondary messengers, owing to the inclusion of amino acids within their composition. There is a distinction between steroid hormones, which possess lipid solubility, enabling them to traverse the plasma membranes of target cells and exert their effects within the nuclei [17]. Peptide hormones play a crucial role as messengers within the signaling network that connects neurological coordination, endocrine glands, the gastrointestinal tract, and energy storage, hence governing the regulation of metabolism and feeding behavior [18]. They are known to play a crucial role in mediating endocrine transmission between the gonads and brain in vertebrates, hence regulating reproductive development. However, the impact of these molecules on reproductive growth in invertebrates remains relatively understudied [19].

1.1.3. Amino Acid-Derived Hormones

Regular hormone release is crucial for sustaining optimal health. The hypothalamus's intrinsic pacemaker functions as the principal regulator of circadian periodicity, whereas several hormones exhibit oscillatory patterns with varying frequencies and amplitudes. The aforementioned rhythms are required to exhibit responsiveness to external stimuli, retain their resilience when confronted with significant disruptions, and convey pertinent data regarding the state of sound physiological operation [20].

Table 1. Summary of hormone structures and their electrochemically detectable groups.

Hormones	Structures of Hormones	Purpose/Function
Cortisol	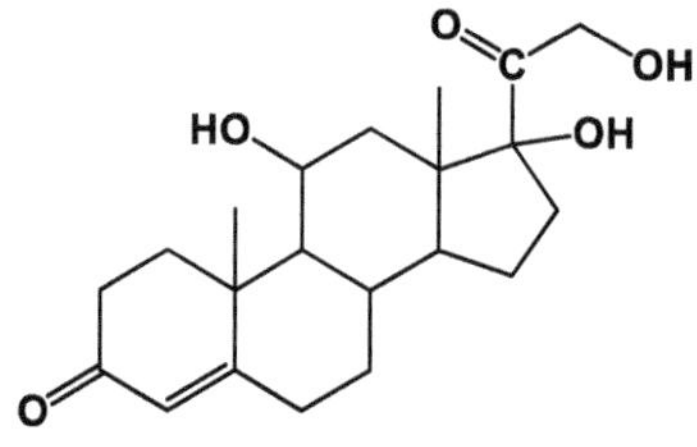	The single-bonded keto group becomes oxidized during electrochemical detection [21].
17β-Estradiol	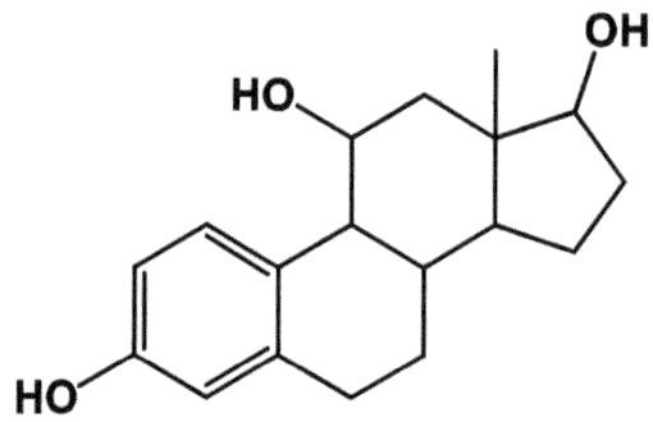	The irreversible oxidation of the hydroxyl group in the aromatic ring of the 17β-estradiol molecule is ascribed to a singular oxidation mechanism, resulting in the formation of its associated ketone derivative. The cause of this irreversible behavior is thought to be an electron transfer control mechanism used by CV and DPV to regulate the 17β-estradiol reaction on the electrode surface [22,23].
Estriol	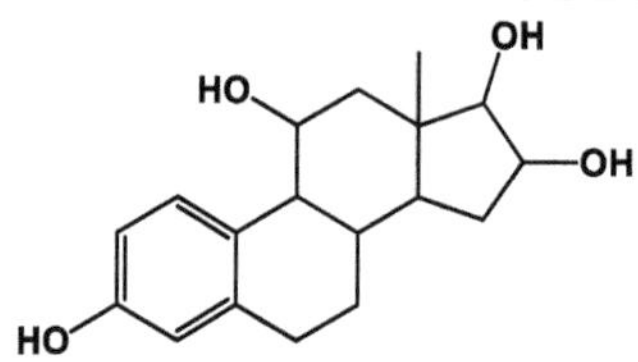	During the oxidation of estriol, a highly reactive phenoxy radical and a C=O were formed, corresponding to anodic peaks seen at LSV [24].
Testosterone	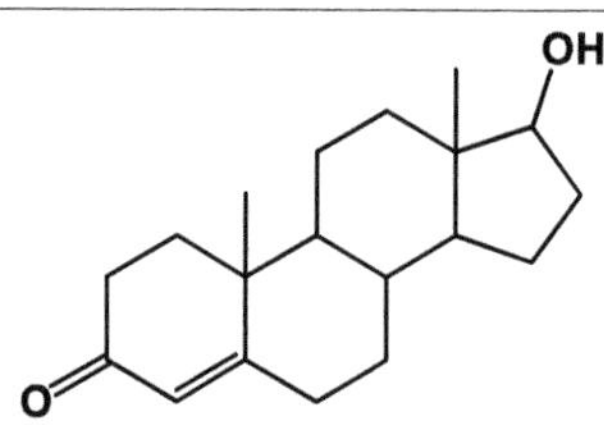	The C-3 keto group of testosterone is first subjected to one-electron reduction, forming an unprotonated radical; then, those above radicals undergo protonation and then participate in a reaction with another radical, ultimately leading to the creation of a dimeric configuration [25].
Progesterone	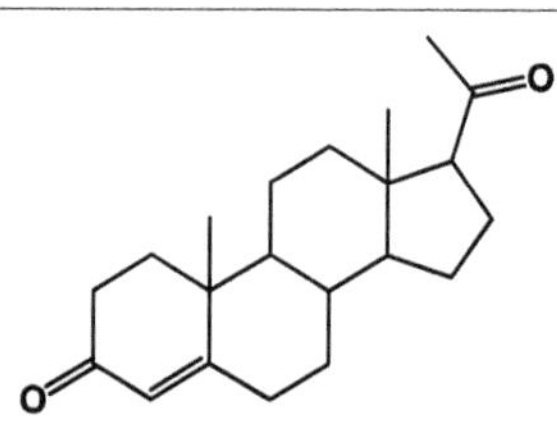	The P4 molecule undergoes a single electron reaction that reduces the C-3 keto group and has a larger positive electron density than the C-20 ketone group, making it simpler to acquire electrons for reduction [26].
Melatonin	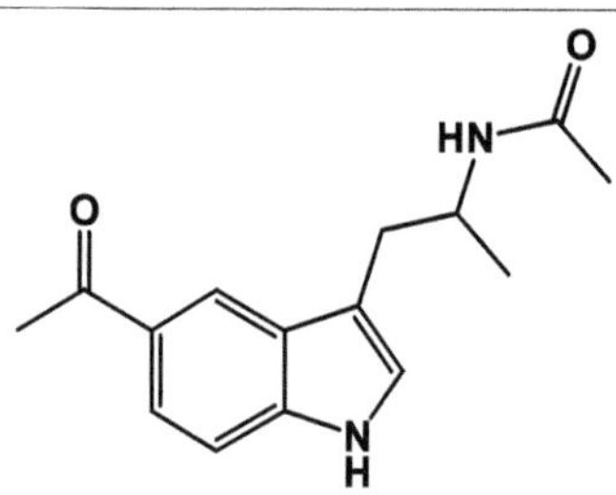	The process of electrooxidation of MT involves the loss of two electrons and a proton, resulting in the formation of an intermediate compound. This intermediate compound is susceptible to nucleophilic attack, leading to the formation of a derivative known as 4,7-dihydroxy indole. This derivative exhibits a pair of quasi-reversible redox peaks when compared with its quinone counterpart [27].

Table 1. *Cont.*

Hormones	Structures of Hormones	Purpose/Function
Thyroxine	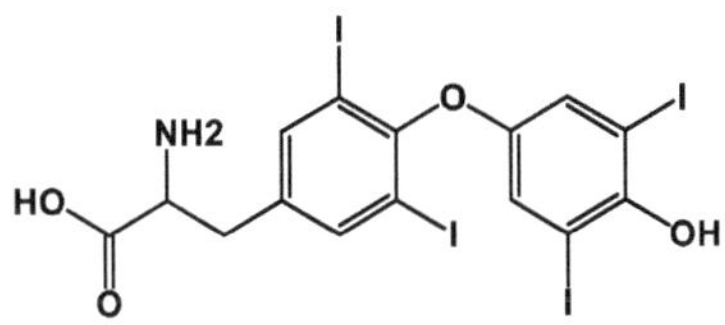	The first anodic peak attributed to the oxidation of hydroxyl (OH) groups present on the phenol of T4 and the first cathodic peak attributed to the reduction of iodine atoms, since the iodine atoms present on the phenol group of T4 exhibited significant reactivity. Additionally, the subsequent peaks observed to the oxidation and reduction products resulting from the first step [28].
Oxytocin	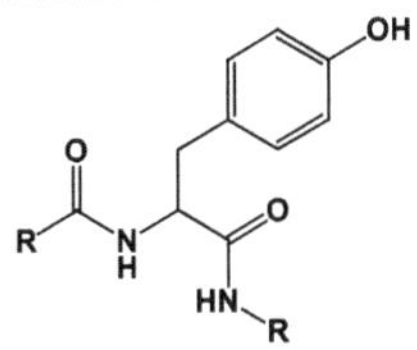	The redox reaction attributed to the reaction involving a quinoic structure, which is one of the products resulting from the oxidation of phenolic OH of oxytocin [29].
Insulin	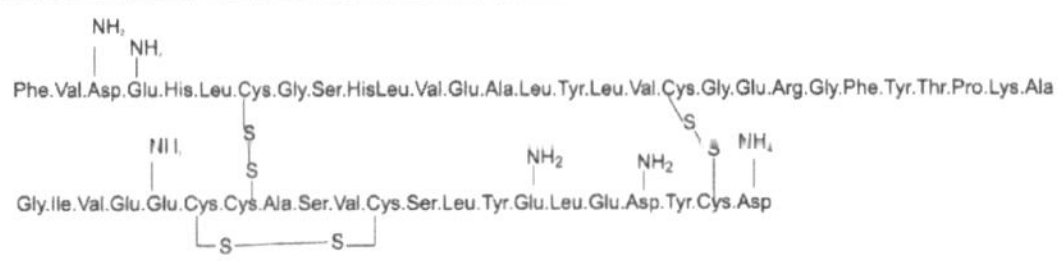	Oxidation of amino acids in insulin [30].

Several methods have been used to determine the identity of hormones. These include high-performance liquid chromatography (HPLC), gas chromatography/mass spectrometry (GC-MS), and high-performance liquid chromatography/mass spectrometry (HPLC-MS) [9]. Tests for estrogens are often conducted using tried-and-true analytical methods such as gas chromatography, HPLC, MS, and immunoassays. HPLC and GC methods are typically highly automated, exceedingly sensitive, and extremely specific, making them ideal for the quantification and identification of a wide range of species at trace levels. These methods offer sensitive and accurate detection, but are impractical due to their high cost, requirement of specially trained operators, extensive time required between sample collection and analysis, and inability to be used in the field [31]. Electrochemical sensing techniques present a comparatively more streamlined and expeditious approach for hormone analysis in comparison with methodologies such as high-performance liquid chromatography/tandem mass spectrometry (LC-MS/MS) [4] because of their low cost, rapid response, high sensitivity, and user-friendliness. These techniques also do not require specialized tools or expert technicians. Electrochemical techniques possess notable benefits over alternative approaches in hormone detection, including cost-effectiveness, heightened sensitivity, rapid response, and operational simplicity. Moreover, these methodologies do not necessitate the use of costly apparatuses or highly skilled personnel [9]. Enzymes or antibodies are often used as biorecognition elements in electrochemical biosensors to detect hormones. Electrochemical biosensors work on the basic idea that specific molecules undergo reactions with target substances and produce a product that can be oxidized or reduced on an electrode surface. There are a few problems that biosensing systems have to deal with: enzymes have a limited lifespan, biosensors sometimes need extra treatments before each use, and some of the sensors we have now do not give a steady response over time [4]. Electroanalytical techniques offer several advantages in the field of analytical chemistry. These techniques demonstrate high selectivity and sensitivity toward electroactive species, allowing for the precise detection of trace amounts of analytes. Additionally, they possess a wide linear range, enabling the quantification of analytes across a broad concentration range. Furthermore, electroanalytical techniques exhibit very low detection limits, ensuring the detection of even minute quantities of analytes. Moreover, the portability and affordability of instruments used in these techniques make them a practical and cost-effective option for routine analysis of various analytes in diverse sample media [30].

2. Electrochemical Methods for the Detection of Hormones

High-performance liquid chromatography, fluorescence and calorimetric tests, spectrometry, enzyme-linked immunosorbent assay, surface plasmon resonance (SPR), and electrochemical methods are among the analytical techniques commonly used for the assessment of purity and efficacy. Except for electrochemical methods, all treatments require demanding and time-consuming steps, specialized equipment, and skilled operators. Many pharmaceutical compounds have electrochemical activity, rendering electrochemical methods capable of detecting and differentiating them, even within biological fluids. Electrochemical technologies possess several notable advantages, including their compact dimensions, rapid response time, heightened sensitivity, cost-effectiveness, extended linear detection range, and exceptional selectivity; these options are favored over other choices, as deemed suitable [32]. The main advantages of electrochemical detection methods are their low detection threshold, high accuracy, fast reaction times, and easy system incorporation [33]. Due to their sensitivity, affordability, and intrinsic compactness, electrochemical detection technologies have drawn a lot of interest. Simple devices and affordable electrodes can be combined to make quick measurements in portable devices that are easy to use and small. Electrochemical biosensors' relatively straightforward nature is one of their significant benefits. One of the best things about electrochemical biosensors is how easy they are to use. Simple electronics can be easily put together with cheap electrodes to make portable systems that are small and easy to use [34]. Figure 3 illustrates the structure and components of an electrochemical biosensor and different electrochemical methods [9].

Potentiometry is a method used to determine the electrical potential of a two-electrode system when there is no current flowing between the electrodes. The solution composition remains unaltered. Ion-selective electrodes (ISEs) are highly favored potentiometric sensors that have found extensive application in the assessment of ion activity in several types of samples, for example, those of biological, chemical, and environmental nature. These electrodes are recognized for their simplicity and effectiveness in this regard. Ion-selective electrodes (ISEs) provide the capability to selectively identify ions in the presence of other chemicals, making them a cost-effective platform. Potentiometry is widely recognized for its utility in the determination of pH levels in various liquids. Moreover, the utilization of potentiometric methods for trace analysis of metals has been widely used in environmental studies [35]. Amperometry is a technique used to quantify the current variations versus time due to oxidation or reduction of an indicator while maintaining a continuous potential relative to a reference electrode. Chemical compounds may undergo either oxidation or reduction processes when exposed to inert electrodes under conditions of constant potential in a suitable range. Amperometry typically offers enhanced sensitivity in terms of detection limits, albeit with the limitation of being applicable just to electroactive species, and systems are commonly used in enzyme-based sensors, wherein the enzyme facilitates a redox process [35].

Electrodeposition for forming nanomaterials on the surface of a working electrode can be provided using chronoamperometry (CA) [36]. The electrochemical technique of chronoamperometry also finds widespread application in enzymatic biosensors. Electrochemical analyses are simplified by the constant potential configuration because enzymes catalyze only certain redox reactions. The sensor activity can be measured using CA [37]. The effective explanation of a chronoamperogram requires a comprehensive depiction of the concentration profiles of all species present in the solution relative to the working electrode surface. The Cottrell equation predicts the occurrence of a diffusion-only line, which eventually reaches a steady-state current as time progresses. The two sets of graphs exhibit significant differences, particularly in terms of the involvement of both Faradic and non-Faradic processes. It is evident that the bulk of the current may be attributed to non-Faradaic effects. The Faradaic current's behavior exhibits a notable level of detail, which may be ascribed to the small delay in the electric field's reaction near the electrode when there is a modification in the applied potential. The observed substantial rise in Faradaic current may be attributed to the enhanced favorability of the potential driving force for electrolysis, which occurs when a more negatively charged double layer is formed. At significantly extended durations, the rate of electron transfer becomes rapid enough to the extent that the current is only constrained by the transport of the electroactive species. In the context of this limit, the current exhibits a diminishing trend over time due to the expansion of the depletion layer away from the electrode. In the scenario when the electroactive species is in an uncharged state, the observed behavior of current over time aligns precisely with the predictions made using a diffusion-only model. Nevertheless, in the case where the species carries a negative charge, there is an observed increase in current when compared with the diffusion-only model over extended periods of time. This may be attributed to the flow of an anion down the positive potential gradient toward the electrode, which provides an extra contribution to the overall mass transport [38].

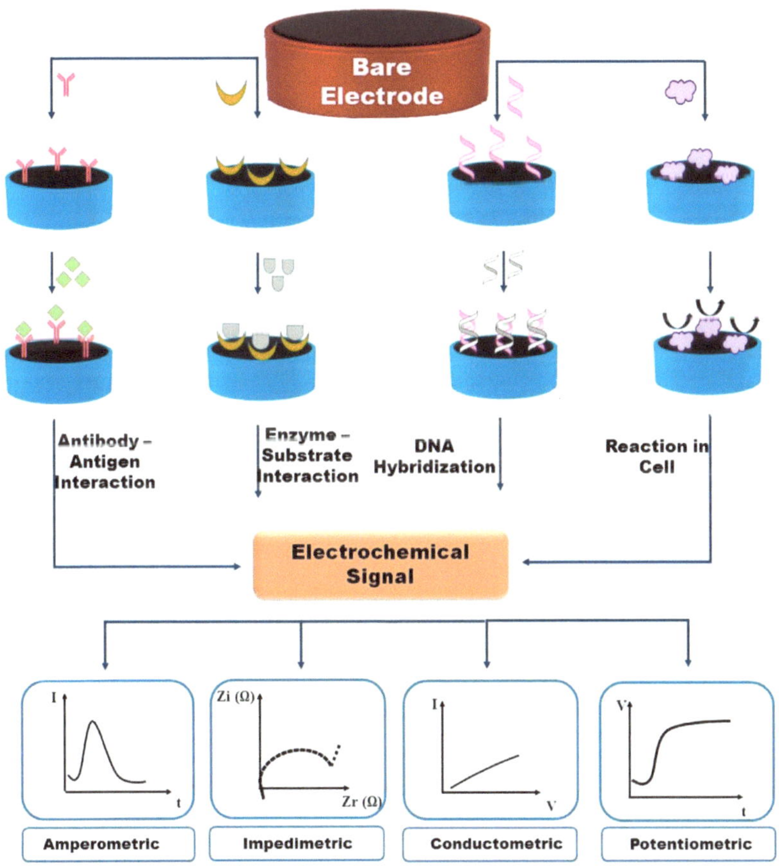

Figure 3. Schematic representation of an electrochemical biosensor. Reproduced from reference [9] with permission from Elsevier, copyright 2015.

Voltammetry is a technique used to determine the current generated by an electrochemical cell, which is then measured in relation to a potential that varies with time. In the field of voltammetry, it is possible to manipulate the applied potential and afterward measure the resulting current over a specified duration. In the context of voltammetric measurements, it is common practice to use a three-electrode system. Voltammetric methods can be categorized into several subtypes, including cyclic voltammetry (CV), alternating current voltammetry (ACV), differential pulse voltammetry (DPV), linear sweep voltammetry (LSV), square wave voltammetry (SWV), anodic stripping voltammetry (ASW), and cathodic stripping voltammetry (CSW). Voltammetric measures primarily yield qualitative and quantitative information regarding both direct and indirect redox reactions [35]. Voltammetry is widely recognized as a prominent technology used in the field of electrochemical immunosensors, and such approaches are used to manipulate the decay rates of both the charging and Faradaic currents using various waveforms to scan the potential. The enhanced proportion of Faradaic current relative to non-Faradaic current facilitates a reduction in the limit of detection and an increase in sensitivity for both reversible and irreversible events. In comparison with the impedance method, voltammetric methods require less expensive equipment [39].

The potentiostat, which is the most widely used instrument in electrochemical analysis, is frequently utilized for the acquisition of CV data. In this method, a composite working electrode, integrating both a reference electrode and a counter electrode, is used. The application of the source of voltage for the potential scan occurs in the region between the counter electrode and the working electrode. To maintain the appropriate potential at the working electrode relative to the reference electrode, the overall voltage is adjusted after the measurement of the potential between the reference electrode and the working electrode utilizing a voltmeter. The term "scan rate" refers to the linear rate at which the potential is scanned in this particular scenario [40]. For example, 3-thiophene acetic acid and palladium nanoparticles (Pd NPs) formed the basis of a new type of electrochemical sensor for progesterone detection, which was then characterized with CV. Palladium nanoparticles (Pd NPs) were used as a cross-linking agent to fabricate a three-dimensional network film exhibiting remarkable conductivity. This enhanced conductivity was due to improved interparticle charge transfer [41].

Differential pulse voltammetry (DPV) is another electroanalytical technique used in biosensor development. DPV is the utilization of linear sweep voltammetry, wherein a linear potential sweep is accompanied by a sequence of regular voltage step pulses. The measurement of the current is conducted over a short time window immediately prior to the upward (i1) and downward (i2) steps of a pulse and then the difference between these two currents (i2–i1) is recorded. Therefore, by minimizing the contribution of the charging current, it is possible to attain a heightened level of sensitivity [42]. DPV responses are assessed using varying doses of hormone according to the ideal experimental circumstances. It is evident that the peak currents of DPV displayed an upward trend in conjunction with the rise in thyroid-stimulating hormone concentration. A strong linear correlation may exist between the peak currents observed in DPV and the logarithmic values of the concentrations of the hormone under investigation, as observed for a sensor detecting thyroid stimulating hormone [43]. The findings of one study show that DPV has greater sensitivity than CV in determining the concentration of parathyroid hormone with an LOD of 0.17 pM and 0.33 pM for DPV and CV, respectively [35]. A progressive decline in the peak current was seen as the amount of hormone increased in both the DPV and CV methods. The observed phenomenon involves a reduction in the number of binding sites present on the molecular layer that recognizes the PTH molecule. This drop may be attributed to an increase in steric hindrance on the surface, generated by the binding of parathyroid hormone (PTH). Consequently, electron transport is suppressed because of these combined effects. A correlation was established between the peak current and the concentration for PTH, indicating a dynamic linear relationship [34,44].

Electrochemical impedance spectroscopy (EIS) is a technique used to examine the response of an electrochemical system to an alternate current (AC) as it varies with frequency. The analysis of EIS data involves examining the variations in current response to an oscillating potential at a specific series of frequencies over a wide range, enabling the determination of the electrochemical characteristics of the system. EIS, a technique known for its responsiveness and non-destructive nature, has found extensive application in several domains including corrosion, protective coatings, conductors, fuel cells, batteries, electrocatalytic reactions, and the study of interfacial parameters of biosensors. The elimination of the requirement for labeled electroactive groups or indicators is the primary distinguishing characteristic of the EIS approach. However, it is important to note that EIS should be used in conjunction with other analytical approaches to fully comprehend interfacial properties [35]. The utilization of surface-modified electrodes presents several advantages when using the EIS approach. The impedance spectra represented as a Nyquist plot usually exhibit two distinct regions, namely, a semicircular segment and a linear segment. The linear portion of the graph represents the diffusion-limited phase of the electrochemical process, while the semicircular region corresponds to the electron transfer resistance (R_{ct}). The rate at which electron transfer occurs for an added redox probe is governed by the charge transfer resistance (R_{ct}) at the electrode surface [45]. Impedance measurements are often conducted under open circuit potential conditions. The acquisition of each impedance spectrum typically requires a duration of around 3 to 4 min [46]. The utilization of EIS proves to be quite advantageous in the evaluation and analysis of interfacial properties pertaining to surface-modified electrodes. Impedance data are usually treated using two distinct sorts of graphs, which are the Bode and Nyquist plots [32]. Nyquist plots were obtained with EIS to investigate the interfacial characteristics of an electrode, including electrocatalytic and conductivity capabilities, both before and after changes. The $[Fe(CN)_6]^{3-/4-}$ redox probe was used for this purpose [34]. There is a correlation between the diffusion-limited process and the semicircular region of the EIS curve [47].

Square-wave voltammetry is a pulse-based electrochemical technique that possesses the notable advantages of rapid analysis and high sensitivity for both reversible and irreversible electrochemical reactions [48]. The electrode potential is modified by SWV with the utilization of a distinct waveform composed of a square wave combined with staircase potential, serving to regulate the decay rates of both the charging and Faradaic currents. The enhanced proportion of Faradaic to non-Faradaic current results in a reduced limit of detection and heightened sensitivity in both reversible and irreversible events. The primary variables associated with SWV are prepotential, start potential, final potential, time at pre-potential, pulse period, pulse amplitude, pulse frequency, and potential step. SWV is considered to possess a superior sensitivity [49]. The utilization of SWV is deemed highly suitable for the electroanalysis of species that have been adsorbed onto the electrode surface [50].

3. Electrodes for Hormone Detection

The selection of an appropriate electrode is of paramount importance in analyte detection, notwithstanding the numerous advantages offered by electrochemical methods. For example, the utilization of bare electrodes is commonly linked to electrode fouling, low selectivity, inadequate repeatability, elevated overpotential, and slow electrode kinetics. To mitigate these limitations, it is common practice to improve the catalytic activity of electrodes by including noble metals such as palladium (Pd), platinum (Pt), gold (Au), and ruthenium (Ru), which not only improve conductivity but also augment catalytic performance. The implementation of modifications serves to mitigate the development of fouling layers [32]. Microelectrodes are also useful for electrochemical sensors because of their fast mass transfer, low electroactive area, low interfacial resistance, and low ohmic drop at the surface of the electrode due to radial diffusion. The total enhanced surface area is useful to obtain improved ranges of detection, a better signal-to-noise ratio, and a wider dynamic range. However, due to a weak response from the analyte, these microelectrodes can only be used to test for a certain range of hormones. Strategies for changing electrode platforms are very important for improving our ability to analyze and identify hormones. Nanostructured electrodes with a high surface area have often been shown to be amongst the best platforms for electrochemical biosensors [34].

3.1. Interdigitated Array Electrodes

The interdigitated electrode array (IDA) as shown in Figure 4 was made by using a standard photolithographic method to make gold designs on a silicon (Si) chip that was treated with boron to give it a resistance of about 1 to 30 Ω cm. Before depositing the gold, an oxide layer 100 nm thick was formed on the Si base to stop any leaky current from going through the Si chip. The oxidized Si wafer was then treated with LOR™ (lift-off resist) and positive photoresist that was a few hundred nanometers thick. It was then soft baked, exposed to UV light via a chrome mask with negative IDA electrode patterns that were lined up, and developed in MIF solution. The area of photoresist that was exposed to UV light became easy to dissolve during development, making it possible for the negative pattern in the IDA electrode structure to be seen on the Si chip base after the lithographic process. A binding layer of 5 nm thick Cr on top of the printed Si base and a layer of 50 nm thick platinum or gold was then placed on top of these. The IDA electrode on the Si substrate was completed using the standard lift-off process to get rid of the last few resists [51]. Since electrochemical assay measurements are homogeneous, they are not impacted by dispersion in bulk solution, and the basic equipment can be downsized to make detection in tiny volumes more accessible. Ino et al. used two-band microelectrodes for the detection of an enzyme produced by cells; however, IDA electrodes and a reversible redox couple can provide additional amplification. The metal fingers of an IDA electrode are flat, parallel, and arranged in two interdigitated comb arrays. Molecules oxidized at an electrode finger of one array can be reduced at the nearby fingers of the other array, making the molecule available for a subsequent reoxidation, thanks to the independent regulation of the potential of each array. The electrochemical responses are improved by the redox cycling across the arrays. To maximize the signal strength, the electrode finger width and the spacing between them are critical. Electrochemistry coupled with spectroscopy, dielectrophoretic cell separation, and the use of IDA electrodes with biological substances are all examples of applications. Electrochemical monitoring has been used to track the activities of enzymes such as glucose oxidase and glutamate oxidase immobilized in hydrogels, urease immobilized in a sol-gel, and alkaline phosphatase. A homogeneous competitive test for hormones was connected with an IDA electrode to dual-electrode voltammetry [52].

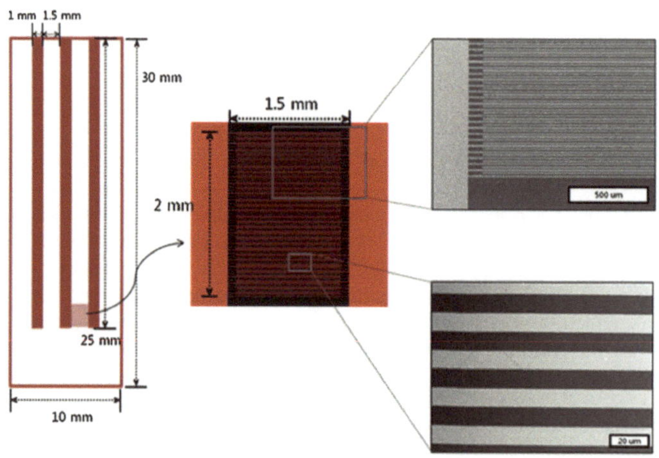

Figure 4. Schematic illustration of IDA. Reproduced from reference [51] with permission from Elsevier, copyright 2012.

3.2. Screen-Printed Electrodes

The usage of screen-printed electrodes (SPEs) has allowed for the mass manufacture of disposable sensors that are both cheap, remarkably repeatable, and dependable. Different inks can be printed on various substrates including circuit boards, plastic, or ceramic to create SPEs. The analytical procedure's needed selectivity and sensitivity are based on the chemical makeup of the various printing inks. Carbonaceous materials (carbon black (CB), graphene (GR), carbon nanotubes (CNTs), etc., are just a few examples of the types of nanomaterials that have been successfully used in SPE surface modifications to enhance analytical features and performance. Graphite working/counter/reference electrodes and an Ag/AgCl reference electrode make up the SPEs in use. Graphite is a working electrode surface with an external counter and reference electrodes used for uniformity [53]. The utilization of screen-printed carbon electrodes (SPCEs) is crucial in the progress of recyclable biosensors and electrochemical sensors because of their advantageous surface features, straightforward manufacturing procedure, cost-effectiveness, environmentally friendly nature, and production suitability for large-scale manufacturing [26].

3.3. Gold Electrodes

Gold's versatility as a substrate for thiol conjugation chemistry has led to its extensive adoption in sensor development, where it is used to tailor surface functionality for a wide range of chemical and biosensing applications. Fabricating and characterizing gold or gold nanostructure-modified electrodes is a crucial first step in creating reproducible and resilient thiol conjugated surfaces, which are necessary for accurate sensor performance. The most common metals for solid electrodes in the fields of electrochemistry and electroanalysis are gold and platinum. This is due to the metals' high quality, ease of use in manufacturing, and inertness in the presence of almost any chemical. Two major applications are driving the rise in demand for gold electrodes: stripping analysis and research into surface changes via self-assembly. A gold working electrode is used for numerous voltammetric procedures, while a platinum counter electrode and silver/silver chloride (Ag/AgCl) reference electrode complete the standard three-electrode electrochemical setup [54].

3.4. Glassy Carbon Electrode

Glassy carbon electrodes (GCEs) have been extensively used in numerous electrochemical investigations due to their affordability, wide potential range, and ease of surface modification. There exist many techniques for preparing and enhancing a GCE's surface prior to utilization, including mechanical polishing, ultrasonication, and electrochemical treatments. The electrochemical treatment approach is widely utilized due to its straightforward operation, resulting in the formation of a nanoporous coating on the GCE with a significantly increased effective surface area. The proposed technique exhibits potential for utilization as a surface modification method for GCE due to its remarkable reproducibility and efficiency [45]. Glassy carbon is extensively utilized in the field of electrochemistry owing to its exceptional characteristics, including excellent tolerance to elevated temperatures, low electrical resistance, low density, and hardness. Glassy carbon (GC) is a variety of non-graphitic carbon that is produced with the process of pyrolysis of specific polymeric precursors at a temperature exceeding 2000 °C. The microstructures of graphene composite consist of separate pieces characterized by curved carbon planes, resembling nanoparticles with imperfections like those found in fullerene structures. Under these conditions, the resulting graphene structure manifests as a network comprising stacked ribbon-like molecules that resemble graphite. The utilization of GC as an electrode material for electroanalysis is prevalent due to its characteristics of high hardness, low reactivity, impermeability, and superior electrical conductivity and impermeability [55].

3.5. BDD Electrodes

Boron-doped diamond, a novel and highly promising solid material has many applications in the field of electrochemistry. The application of this material is prevalent in the examination of pharmaceuticals and their metabolites, bioactive compounds, metallic ions, and organic contaminants [56]. Researchers can produce BDD electrodes using either home-built systems or commercial growth systems such as hot filament (HF), plasma chemical vapor deposition (CVD) reactors, and microwave (MW) reactors. Key factors that are important when evaluating the electrochemical behavior include the boron content, surface morphology, surface termination, surface polish, and non-diamond-carbon (NDC) presence despite the source [57]. It has been used as a viable substitute for high-performance electrodes in electrochemical investigations due to its inert nature, hardness, elevated thermal conductivity, and electrical conductivity. Furthermore, the BDD electrode is often favored in this domain because of its notable characteristics, including a decreased background current, a wide electrochemical potential range, and exceptional resilience and longevity in both alkaline and acidic environments. Additionally, the BDD electrode has several advantages in comparison with traditional solid electrodes, including enhanced hardness, durability, and optical characteristics, and improved response parameters, such as time, stability, and precision. In the realm of electrochemical investigations, it is imperative to ensure the cleanliness or activation of the exterior of the BDD electrode. This is necessary to achieve consistent and highly responsive outcomes on the electrode's outermost layer, a requirement shared by other solid electrodes. The presence of impurities on the electrode surface might result in either a reduction or enhancement in the strength of the signal used for analysis. Consequently, this leads to inaccuracies in the outcomes of the analysis. Because of the high sensitivity of the BDD electrode surface, it is imperative to exercise caution and use a gentle mechanical cleaning procedure. The cleaning procedure described in this study involves the activation of an electrode that has been rendered free from surface contamination. This activation process occurs in various solutions, with the specific potential applied to the electrode being dependent on the prevailing working conditions. In the existing body of literature, various studies have been conducted to investigate the cleaning and activation of the BDD electrode. These studies used the application of potentials in both the anodic and cathodic directions, albeit in different solutions [58]. Boron-doped diamond (BDD) electrodes, when subjected to varying levels of boron doping, exhibit distinct variations in binding energies and carrier densities, which have significant implications for their utilization in semiconductor applications. The aforementioned features have a notable impact on the present state and oxidation peak potential of the substance, hence influencing the overall effectiveness of analyte detection [59].

3.6. Fused Deposition Modeling

To fabricate electrodes using fused deposition modeling (FDM), a combination of carbon-based conductive particles and thermoplastic materials is used to create conductive filaments. In the recent literature, there have been reports on various conductive polymeric filaments, including polylactic acid (PLA)–graphene filament, carbon nanofiber–graphite–polystyrene, PLA-carbon black (PLA-CB), polybutylene terephthalate–carbon nanotube–graphene, polypropylene-CB, and acrylonitrile butadiene styrene (ABS)-CB. The PLA-CB filament has received significant attention in recent research endeavors due to its use in the practical fabrication of printable electrodes that can be utilized in electroanalytical applications. The utilization of printed electrodes has demonstrated encouraging outcomes in the identification and quantification of biomolecular species. Up to the present time, there has been an absence of exploration of the utilization of 3D-printed electrodes for the purpose of hormone analysis [13]. For electrochemical hormone determination, a wide variety of carbon electrodes, including glassy carbon (GCE), pencil graphite (PGE), edge plane pyrolytic graphite (EPPGE), carbon paste (CPE), and screen-printed carbon (SPCE), have been utilized widely [4].

4. Nanostructures for the Detection of Hormones

The scientific community has increasingly shown interest in the advantages of various nanomaterials for a broad range of biosensor functions [60]. Nanomaterials have garnered considerable attention for their capacity to augment catalytic processes and their substantial surface area, rendering them promising candidates for deployment in sensors and several other technological devices [24]. Numerous nanomaterials have been produced, modified, created, and investigated within the realm of electrochemical biosensors in order to leverage their exceptional intrinsic benefits and achieve heightened sensitivity and selectivity [61].

Figure 5 shows the usage of various nanomaterials along with biorecognition elements for detecting steroidal hormones [62].

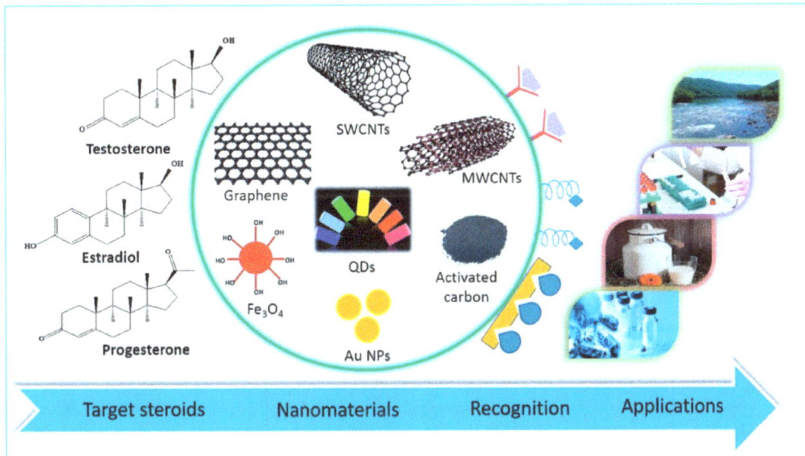

Figure 5. Schematic representation of nanomaterials for detecting various hormones. Reproduced from ref. [62] with permission from Elsevier, copyright 2022.

4.1. Two-Dimensional (2D) Nanomaterials

Of all the nanomaterials, two-dimensional (2D) nanomaterials are among the newest and most interesting. Since Andre Geim and Konstantin Novoselov found graphene in 2004, other materials like boron nitride (BN), graphite carbon nitride ($g\text{-}C_3N_4$), transition metal dichalcogenides (TMDs, like MoS_2 and WS_2), transition metal oxides (like MoO_3, WO_3, and MnO_2), MXenes, silicene, germanene (2D germanium), hexagonal boron nitride, borophene, and black phosphorus have been found as depicted in Figure 6. So far, 2D nanoparticles and their nanocomposites have shown great chemical, physical, electronic, and optical characteristics. These qualities make them useful for many applications, like catalysis, drug delivery, antibacterial, therapy, and bioimaging [63–65].

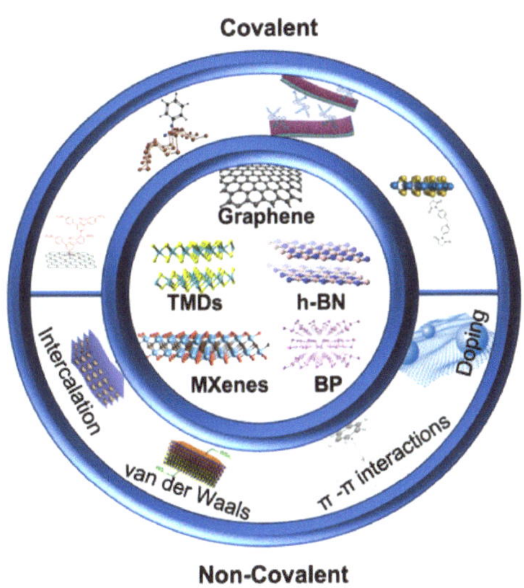

Figure 6. Covalent and non-covalent materials used in biosensors. Reproduced from ref. [65] with permission from Elsevier, copyright 2022.

4.1.1. Graphene

Since its initial identification in 2004, graphene has sparked a significant surge in attention within the realm of electrochemistry owing to its remarkable electronic transport characteristics, electrocatalytic capabilities, and expansive surface area. Graphene, akin to carbon nanotubes (CNTs), can also serve as a substrate that can be subjected to various modifications with distinct species [66]. It is a special two-dimensional (2D) material made of carbon atom monolayers and has a honeycomb-like structure. Due to its novel physical and chemical properties, it has attracted attention since its introduction in 2004. The oxidized form graphene oxide (GO) has a useful surface chemistry because of the existence of oxygen groups that permit functionalization [24]. SEM images of graphene oxide and the reduced form of graphene oxide are should in Figure 7. Research has been conducted on the utilization of graphene as an electrode material in the realm of high-sensitivity detection of biological molecules. This interest stems from graphene's remarkable catalytic activity, electrical conductivity, and intrinsic physicochemical features. Graphene nano-sheets experience restacking due to robust π–π interactions and van der Waals attractive forces among them. Consequently, the layered structure of graphite is partially restored, resulting in a reduced specific surface area of graphene when compared with its theoretical value [67]. Graphene oxide magnetic nanocomposites were used as electrodes for the loading of bioreceptors. The sensitivity of these sensors was improved with the large increase in electrically conductive surface area [68].

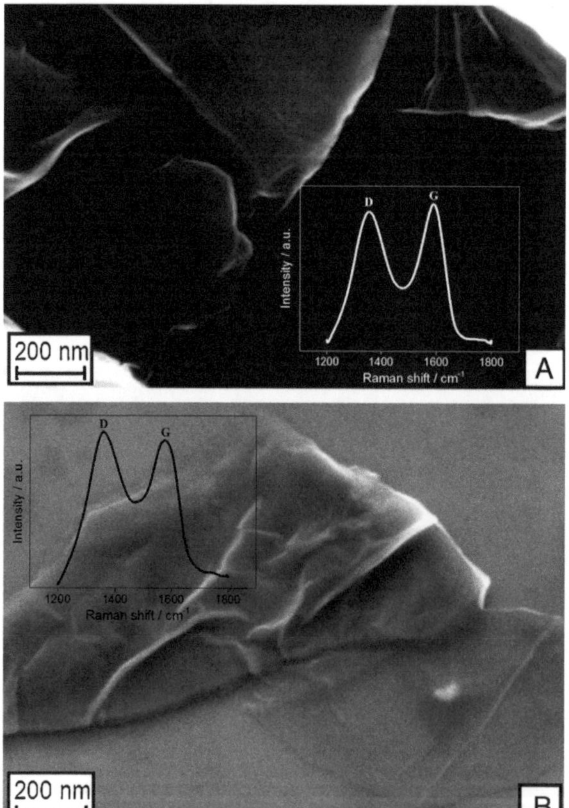

Figure 7. FEG-SEM (field emission scanning electron microscopy) micrographs showing (**A**) GO and (**B**) reduced GO. The G band arises from C-C bond stretching and the D band arises from disorder in the structure of graphene and would not be present in perfectly ordered graphene. Reproduced from ref. [66] with permission from Elsevier, copyright 2015.

4.1.2. Transition Metal Dichalcogenides (TMDs)

Since the discovery of graphene in 2004, research on two-dimensional (2D) materials has increased exponentially. Transition metal dichalcogenides (TMDs) as shown in Figure 8 belong to a category of 2D materials in which the bonding between individual covalently bound X–M–X layers (where M represents the transition metal and X represents the chalcogen) is facilitated by weak van der Waals forces. This characteristic enables the synthesis of these materials with precise control over the number of layers. Single-layer TMDs exhibit a noteworthy shift from indirect to direct band gaps, accompanied by intriguing optical and electrical characteristics. The presence of limited layer-dependent opto-electronic characteristics and band gaps render single-layer TMDs more advantageous compared with graphene, hence facilitating its use in many domains [69]. The commendable characteristics of TMDs include their remarkable capacity for charge transfer, their substantial surface-to-volume ratio (S/V), the tunability of their energy band gap based on the number of layers, their significant interaction with light, and their mechanical resilience. Additionally, these resources are both economically efficient and readily available. The distinctive characteristics, structural properties, and current use of nanostructured TMDs render them very suitable for producing electrochemical biosensors. The commonly recognized 2D TMDs include tungsten diselenide (WSe_2), molybdenum disulfide (MoS_2), tungsten disulfide (WS_2), and molybdenum diselenide ($MoSe_2$) [64,70]. For example, molybdenum diselenide ($MoSe_2$) was used to make a sensor for 17β-estradiol. $MoSe_2$ nanosheets mixed with carbon aerogel spheres ($MoSe_2$-CA sensor) using a chemical method worked very well with 17β estradiol, detecting as little as 0.2 pM in a concentration range of 5 pM to 5 nM. The RSDs of 1.7% and 3.1% found in two sets of six detection tests showed that the $MoSe_2$-CA sensor was very reliable. Its stability was also shown by the fact that it retained 94.8% of its initial current response after being stored at a low temperature. It was found that carbon aerogel not only kept $MoSe_2$ from sticking together and improved electron transfer, but it also made the efficiency of composites more stable, all of which were important for the 17β-estradiol detection application [71].

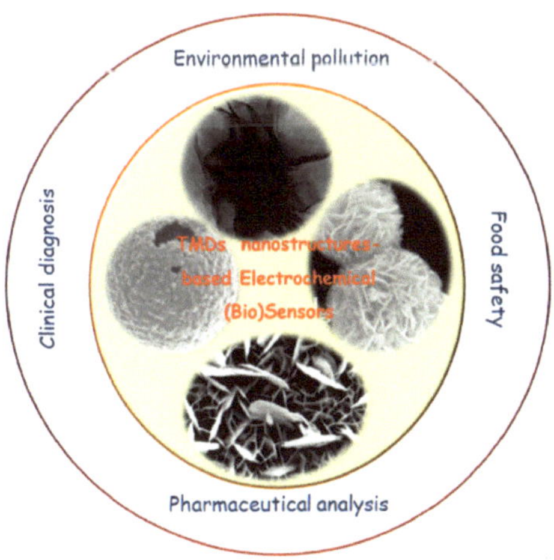

Figure 8. Nanostructures of TMDs enable electrochemical biosensing. Reproduced from ref. [70] with permission from Elsevier, copyright 2022.

4.1.3. Mxenes

MXene (transition metal carbides and nitrides) have been the subject of intensive study because they are 2D materials. $M_{n+1}X_nT_x$ is the chemical formula used for MXene. M can be any early transition metal, like Mo, V, Nb, or Ti. X can be N or C, and n can be any value from 1 to 3. Tx stands for different surface endings, which can be F, O, or OH groups. MXene has special physical and chemical features, including the ability to carry large amounts of charge and conduct electricity, and have good mechanical qualities. In addition to these traits, one of the attractive features of MXene is that it is biocompatible. This makes it a promising material for designing advanced biosensor systems [72]. SEM, TEM and EDS data for a MXene-MWCNT composite are shown in Figure 9. Due to its unique layered structure, high conductivity, high surface area, high thermal stability, and low environmental impact, MXenes have recently attracted a lot of scientific interest for use in sensing, catalysis, electronics, and energy storage. It is well-suited for the development of rapid-performance electrochemical biosensors due to its great hydrophilicity from surface functional groups, high electrical conductivity, and exceptional ion intercalation behavior. Recent years have seen the development of high-performance materials in the form of ultrathin MXenes for stretchy and flexible conductive coverings [73,74]. A new reusable electrochemical impedimetric immunosensor using a $Ti_3C_2T_x$ MXene-loaded laser-burned graphene (LBG) flakes 3D electrode network and PDMS was developed to detect cortisol in human sweat without touching the person. The sensor has a microfluidic path and chamber. The sensor was put on the skin to collect sweat and sweat moved through the tube to the detection chamber with its own weight. The $Ti_3C_2T_x$ MXene/LBG/PDMS-based patch cortisol immunosensor worked well with linearity and a LOD of 0.01–100 nM and 88 pM, respectively. It also worked well for the detection of cortisol at the point of care [75].

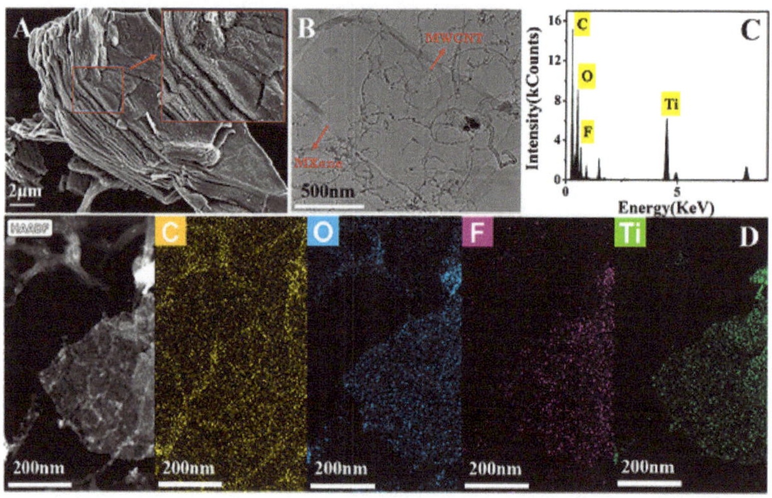

Figure 9. (**A**) SEM, (**B**) TEM, and (**C**,**D**) EDS of an MXene-MWCNT composite. Reproduced from ref. [72] with permission from Elsevier, copyright 2023.

4.1.4. Black Phosphorous

BP is a semiconductor material characterized by a layered structure; in which discrete atomic layers are assembled using van der Waals forces. The formation of a puckered honeycomb structure occurs because each phosphorus atom is covalently bonded to three neighboring phosphorus atoms. BP has a direct bandgap, notable wrinkled structure, elevated hole mobility, and good mechanical, electrical, and optical characteristics, making it a material with considerable potential for several applications. However, the limited use of this substance in many industries is mostly attributed to its inherent instability and rapid chemical deterioration when exposed to typical environmental conditions. In recent times, BP has been used in several domains including sensing, energy storage devices, photocatalysis, and medicinal applications. The commonly used method for synthesizing few-layer BP involves liquid-phase exfoliation followed by chemical vapor deposition (CVD) [65]. BP has garnered significant attention as a novel nanomaterial that presents the establishment of a unique immunofiltration strip approach that utilizes temperature as the readout signal, leveraging the photothermal impact of BP nanosheets. Using an indirect competitive approach, this method offered a straightforward, expeditious, highly responsive, and cost-effective framework for the identification of 17β-estradiol, an endocrine-disrupting chemical often found in ambient water or food samples. A negative correlation was shown between the content of 17β-estradiol in the sample and the binding of BP nanosheets to the strip surface. Additionally, a decrease in temperature variation was observed when the sample was subjected to intense laser irradiation. A detection limit of 0.104 ng mL^{-1} was attained under optimal circumstances. The test's practicality was evaluated using a standard addition procedure in samples of water and milk. The results demonstrated satisfactory performance and suggested that the assay has potential usefulness for convenient, cost-effective, and direct monitoring of 17β-estradiol [76].

4.1.5. Two-Dimensional Metal–Organic Frameworks

Most of the current research in the field of synthesizing metal–organic frameworks (MOFs) in two-dimensional (2D) structures has been primarily concerned with augmenting the advantages of porous materials with the amplification of the exposed surface area. There have been reports on the synthesis procedures and applications of 2D MOFs in energy-related research, which has generated significant interest. The secondary building units (SBUs) consist of metal-containing nodes and organic linkers that serve as bridges. These components combine to produce ordered network structures with large pore volumes and surface areas in crystal formations, which may be one, two, or three-dimensional in nature. There are several techniques available for the integration of 2D conductive MOFs into devices. One such approach involves the use of drop casting, where an MOF is suspended in a solvent and then placed as a coating onto the desired substrate. The framework architectures, pore layouts, and sizes of MOFs may be modified using various metal centers and organic linkers. With the chemical modification of linkers and other alterations, it is possible to modify their chemical properties. The dimensionality of the emerging framework structure is influenced significantly by the coordination geometry of the structural components. Metal–ligand coordination bonds have been widely used for the purpose of arranging molecular building blocks into diverse supramolecular structures, which give rise to one-dimensional (1D), two-dimensional (2D), and three-dimensional (3D) networks, usually known as coordination polymers (CPs) or metal–organic frameworks (MOFs) [77]. In the process of immune recognition, a Cd^{2+}/Au/polydopamine/Ti_3C_2 (Cd^{2+}/Au/pDA/Ti_3C_2) composite-modified electrode was utilized for immobilizing the E2 antibody (E2-Ab). Subsequently, the E2-conjugated bovine serum albumin (E2-BSA) was labeled with a Cu-MOF and competed with E2 for binding to the E2-Ab. The Cu-MOF exhibited both electroactivity and effective electrocatalytic performance toward H_2O_2. Therefore, the quantification of E2 was determined by analyzing the peak current change in the Cu-MOF on the differential pulse voltammetry (DPV) curve or by measuring the variation in the current of H_2O_2 reduction.

4.1.6. Two-Dimensional-Doped Materials

The incorporation of metal into 2D materials or the addition of metal decorations to 2D materials has shown to be a useful approach for modifying the electrical and chemical characteristics of layered materials. This modification often leads to improved applications, particularly in the field of sensors. The addition of metal atoms or the incorporation of metal decorations has been shown to be beneficial in several cases for enhancing catalytic activity, improving the optical characteristics of layered materials, and enhancing the sensitivity and selectivity of sensors [78].

The electronic properties of TMDs can be modified by introducing metal dopants or functionalizing them with metal nanostructures. This process involves the transfer of electrons or holes, which reduces the activation energy required for reactions with gaseous molecules. Consequently, the sensitivity of TMDs toward nonpolar gas molecules is enhanced, leading to improved selectivity [78]. N-doped MoS_2 nanoflowers were better at photocatalysis because they had a larger surface area, which meant more active sites for surface adsorption. This was because of their shape, the narrow band gap that came from N doping, which made the material respond better to visible light, and the absorption edge moving farther out in the visible light region [64].

A novel approach for the synthesis of polydopamine-functionalized black phosphorus (PDA/BP) with enhanced biocompatibility and chemical stability indicated that the application of polydopamine as a coating on black phosphorus surfaces has the dual effect of preventing oxidation and creating a biocompatible matrix. This matrix offers a favorable microenvironment for protein immobilization, allowing proteins to maintain their biological activity. Consequently, polydopamine-coated black phosphorus shows great potential as a versatile component in the development of innovative biological probes and biosensing platforms. Subsequently, PDA/BP (polydopamine-coated black phosphorus) facilitated the activation of an ultrasensitive fluorometric immunoassay, where the PDA/BP complex, which was coupled with antibodies, served as a quencher. The outcome yielded a detection limit of 83 (pg mL^{-1}) for diethylstilbestrol, which showed that the immunosensor had exceptional sensitivity, specificity, and stability, making it suitable for a wide range of applications. Its versatility allowed for the quantitative detection of many targets in both food and medical specimens [79].

The stable configurations, electrical characteristics, and interactions among pristine and Al- and Si-doped boron nitride nanosheets (BNNSs) with methimazole medication (MM) were studied using density functional theory (DFT) computations. The results suggested that MM exhibited physical interactions with pristine BNNSs, whereas it demonstrated chemical interactions with Al- and Si-doped BNNSs. Due to the poor association and small change in the energy gap (E_g) seen between BNNSs and the sensing material, it seemed that this system was not well-suited for possible sensing applications. However, the introduction of silicon-doped BNNS demonstrated a more favorable interaction and a significant shift in the energy gap upon the absorption of the sensing material. This suggests that silicon-doped BNNSs might serve as a promising option for the development of a sensing device. Aluminum-doped BNNSs exhibited a pronounced affinity toward MM, resulting in a rapid recovery period. This observation suggests that this particular system has favorable characteristics for the degradation of MM [80].

In this situation, the measurement of DPV was conducted by including Cd^{2+} as an internal reference, resulting in the acquisition of a ratio readout that exhibited a high level of reproducibility. The dual-mode E2 electrochemical immunosensor, as produced, exhibited a satisfactory linear correlation throughout the concentration ranges of 1 pg mL^{-1}–10 ng mL^{-1} (DPV) and 10 pg mL^{-1}–10 ng mL^{-1} (chronoamperometry technique). The respective detection limits were determined to be 0.47 and 5.4 pg mL^{-1}. Moreover, the dual-mode electrochemical immunosensor showed favorable practicality in the examination of actual samples [81].

4.2. Carbon-Based Nanostructures

Various carbonaceous materials, such as fullerenes, graphene, graphene oxide, reduced graphene oxide, carbon dots, carbon black, carbon nanofibers, and carbon nanotubes have garnered significant interest and have been extensively studied in the field of electrochemical sensing for a wide range of metabolites. The rapid flow of electrons between electroactive metabolites and the electrode surface is accelerated by the large surface area and the high degree of electrical conductivity of these materials and their utilization around sensors and biosensors. Furthermore, carbon-based materials have the potential to improve the functioning of biosensors, owing to their broad potential window, exceptional electrochemical stability, significant mechanical resilience, and compatibility with biological systems. Nevertheless, the high cost of their manufacture and the considerable challenges associated with immobilizing them on electrode surfaces present an inherent drawback. These limitations may account for the current lack of large-scale production of carbon nanostructures intended for utilization in sensing applications [82,83]. Carbon-based materials, particularly the ones at the nanoscale, have exhibited numerous benefits compared with traditional electrode materials. This is primarily attributed to their distinctive properties and functionalities, including but not limited to a functionalized structure, elevated electrical conductivity, and large surface area. These characteristics can be used efficiently in chemically modified electrochemical devices [84].

4.2.1. Carbon Nanotubes (CNTs)

Carbon nanotubes (CNTs) possess exceptional strength and rigidity as an allotropic form of carbon, which are characterized by their cylindrical nanostructure at the nanoscale. Carbon nanotubes (CNTs) possess unique qualities that make them appropriate for various applications, specifically in the field of electrochemical determination of electrically active moieties. These properties include great mechanical strength, improved electrical conductivity, chemical stability, increased surface area, and distinctive electronic properties. In addition to carbon nanotubes (CNTs), graphite paste-based electrodes have also been found to exhibit similar capabilities. Consequently, both of these materials are used as crucial sensing materials were used in a previous study, albeit in composite form [85]. CNTs make biosensors more stable, biocompatible, and electrocatalytic and provide improved electron transport. Enhanced electron transfer capacities for biomolecule electrochemical activity can be provided by CNT-based modified electrodes, which make biosensors more stable, biocompatible, electrocatalytic, and better at moving electrons [86]. Multi-walled carbon nanotubes (MWCNTs) are interesting one-dimensional nanomaterials that are used to make biosensors because they have special properties like the ability to transfer electrons easily, electrocatalytic effects, a large specific surface area, and the ability to adsorb molecular species strongly. When MWCNTs are added to surfaces, they not only change how electrochemically reactive analytes are, but they also lower electrolyte resistance. Because of these features, MWCNTs are very appealing as surface enhancers in numerous electrochemical devices [87].

4.2.2. Carbon-Based Quantum Dots

Since their discovery in the 1980s, quantum dots (QDs) have emerged as versatile tools in various fields including the development of optical biosensors, bio-sensing, gene delivery, glucose monitoring, drug delivery, pharmaceutical purposes, and cellular imaging. These applications encompass a broad range of areas, such as the examination of HeLa cells, human lung cancer cells, and bio-imaging, among others. The diverse range of applications of QDs renders them very versatile materials with several uses [88]. Many types of biomedical applications have been established using quantum dots (QDs), including in vivo bioimaging, cellular labeling, targeted medication administration, and disease detection. The primary reason behind this is the high quality of their optics. In addition to their usage in medicine, QDs are also poised for rapid expansion in the fields of solar cells, sensors, and LEDs (light-emitting diodes) [89]. Carbon-based quantum dots (CQDs) have the potential to serve as a novel platform for the electrochemical detection of insulin at physiological pH due to their appropriate antifouling features, stability, and enhanced insulin oxidation current. Furthermore, the straightforward and cost-effective technique for synthesizing colloidal quantum dots (CQDs) on graphene-based substrates, such as reduced graphene oxide, and graphene oxide presents a promising avenue for using CQDs as an attractive substitute substance for numerous electrochemical applications [90]. Graphene quantum dots (GQDs) have been extensively used in the development of electrochemical bio-sensing techniques within the realm of nanomaterials. This is mostly attributed to their intrinsic properties, including biological compatibility, stability, excellent conductivity, quick electron transfer, and a significant surface area [91].

4.2.3. Laser-Induced Graphitized Surfaces

Recently, laser-induced graphitized (LIG) surfaces have received consideration as a favorable novel material for electrochemical sensing applications due to their unique features. For instance, graphene generated with laser ablation has been formed on the surface of polyimide (PI) films using a thermal procedure. This technology is characterized by its high speed and cost-effectiveness, making it particularly suitable for applications in versatile electronics and energy storage devices, such as supercapacitors. In addition to their fascinating features, porous carbonaceous materials are inexpensive and readily available, making them ideal for a broad range of ablation applications. Electronics, motion sensors, electrochemical sensors, ultraviolet (UV) photodetectors, and supercapacitors are just some of the areas where LIG structures have been put to good use. Heat transmission from a laser to the PI or another target substrate is reported to be the mechanism underlying LIG porosity creation. Other methods, such as controlling the laser head using a 3D printer, have been reported more recently using diode lasers operating at distinctive wavelengths. Three-dimensional printing is a cutting-edge technology that offers exciting prospects for improving the production of sensors and analytical instruments. Research and industrial institutions are increasingly using 3D printing setups due to the adaptability of the manufacturing approach; this can be taken even further by integrating low-cost diode lasers into the production process. A 3D printer/laser hybrid system was used to make electrochemical sensor platforms. The effect of varying processing factors was investigated on the final product in terms of electrical conductivity and morphology. These parameters included substrate distance, scan speed, and laser power to focus. The flexible 3D printer/laser system was also used to demonstrate the primary use of LIG for the electrochemical detection of estradiol at a low cost [92,93]. A multimodal analysis device that incorporates machine learning (ML) and utilizes LIG electrodeposited with polysulfide molybdenum (eMoSx) as the sensing material facilitated the concurrent identification of uric acid and tyrosine in sweat and oral fluid. The sensor's analytical capabilities make it suitable for many applications in medicinal products, environmental toxin detection, and research in the life sciences [94]. Yoon et al. effectively developed an acetic acid-treated LIG electrochemical glucose sensor. An electrochemical and physical investigation revealed that the use of the acetic acid treatment improved electrical properties and allowed for stable, efficient electrodeposition of PtNPs onto LIG without accumulation, resulting in a homogeneous distribution. The reduced electron-transfer rate and hydrophobicity of the acetic acid-treated LIG facilitated a minimal background current in PBS, owing to fewer interactions with species like sweat and PBS. The newly developed glucose sensor using acetic acid-treated LIG showed an excellent sensitivity of 4.622 μA/mM, an extremely low LOD of 300 nM, a linear dynamic range of up to 2.1 mM, and an R-square of above 0.99 after a linear regression [95].

4.3. Zeolites

In the present day, there is much emphasis placed on the application of zeolite-modified electrodes (referred to as ZMEs) in the domain of electroanalysis. The interest garnered by zeolites can be mostly attributed to their unique capabilities in the areas of cation exchange and electrocatalysis. Noteworthy energies have been dedicated to the advancement of ZMEs to facilitate the creation of biosensors utilizing voltammetric techniques. Notably, these techniques have been useful in the detection and analysis of several analytes, including NADH, cytochrome c, and glucose. In this context, a notable use of ZMEs pertains to the integration of metal ions and metal nanoparticles. This prospect arises from the complicated three-dimensional architectures of ZMEs, which provide diverse sites for accommodating these metallic species. Significant improvements in detection sensitivity have been observed in several instances, such as the analysis of ascorbic acid (AA) and dopamine (DA). Notably, the utilization of zeolite-modified electrodes included with Fe^{3+} and Cu^{2+} ions has resulted in heightened levels of detection sensitivity [96]. Zeolitic imidazolate frameworks (ZIFs) are made up of transition metal particles (like Zn, Co, Fe, and Cu) that are connected by imidazolate or its derivatives. ZIFs are different from other MOFs in that they have great properties like high temperature and hydrothermal stability, resistance to chemical changes, hydrophobic properties, and low cytotoxicity. ZIFs have a lot of promise for use in many different fields, such as storing energy, separating gases, sensing chemicals, delivering drugs, and catalyzing reactions. In the field of zeolitic imidazolate frameworks (ZIFs), zeolitic imidazolate framework-8 (ZIF-8) has received a lot of attention because of its well-defined pores and high temperature and chemical stability. Even though ZIF-8 is being used more and more to make sensors, its low conductivity makes it difficult for electrons to move between the sensor surface and the analyte contact. But this problem with conduction can be solved by adding nanomaterials like metal nanoparticles and carbon-based nanomaterials, which make ZIF-8 much better at conducting electricity [97].

4.4. Metal–Organic Frameworks (MOFs)

Metal–organic frameworks possessing adjustable active functional sites, porosity, and fluorescence properties have acquired substantial interest as capable materials for the advancement of opto-electrochemical sensors. MOFs possess a surface that is readily amenable to functionalization, allowing for facile modification. Additionally, their pore size may be adjusted to meet specific requirements. MOFs also exhibit inherent luminescence and possess a favorable adsorption capacity. These characteristics collectively contribute to an enhancement in MOF-target analyte interactions and enable the conversion of these interactions into quantifiable optical responses [98]. MOFs have garnered significant interest because of their potential utilization across diverse domains, including but not limited to adsorption, environmental applications, storage, separation processes, and sensing technologies. The field of biomedical applications experienced increased attention in the previous decade, primarily due to the remarkable structural characteristics exhibited by these materials, such as enduring porosity, exceptional specific surface areas, customizable pore size and structure, adaptable modifications, and compatibility with biological systems. These characteristics exemplify a broad range of MOF applications as biosensing platforms. These platforms are designed to facilitate the rapid discovery of various diseases such as cancer or diabetes, the identification of pathogens, the quantification of drugs and their metabolites, and the detection of analytes in biological samples. Consequently, these MOFs enable the expedited diagnosis of diseases with the use of rapid testing methods [99].

4.5. Nanoparticles

In the field of electrochemical sensing, different metal nanoparticles have been used to find chemical and biochemical species with a very high level of sensitivity [100]. Electroanalyses use a lot of different metallic nanoparticles (MNPs), like palladium, antimony, silver, and gold, because they are good conductors, have a large surface area, are chemically stable, can have their surface chemistry changed, and are cheap.

4.5.1. Metallic Nanoparticles

Silver nanoparticles (AgNPs) are one of these MNPs, and they have antibacterial, catalytic, optical, electrical, and surface-enhanced Raman activity. AgNPs are used in spectroscopy, antibacterial materials, and sensors and are used to enhance the functioning of solar materials because of these qualities [101]. The utilization of gold nanoparticles (AuNPs) in the development of drug delivery systems and biosensors has garnered significant and ongoing interest in the last decade. The use of AuNPs is motivated by their distinctive physicochemical characteristics, which facilitate advancements in biomarker identification and the development of drug delivery systems [102]. Electrochemical biosensors, immunoassays, vaccine development, biosensorics, drug transport, and detection are only some of the more standard uses of gold nanoparticles in the biomedical industry. The distinctive chemical and physical features of these particles form the basis for their extensive range of applications. However, the detection of biological and chemical materials in environmental and biomedical settings is also critically important, making features like cost-effectiveness, high specificity and sensitivity, convenience, and low detection of limits (LOD) crucial. Nanoparticles, especially gold nanoparticles, can be used as biological and chemical sensors in various material and analyte detection methods because of their unique chemical and physical properties [103]. Nanomaterials composed of palladium (Pd) have notable electro-catalytic efficacy toward specific objectives, such as hydrogen evolution, fuel cells, and electrochemical sensors [67].

4.5.2. Metal Oxide Nanoparticles (MO-NPs)

Metal oxide nanoparticles possess several advantageous characteristics, including a significant surface area-to-volume ratio, exceptional endurance, precise control over morphology, and remarkable chemical stability [104]. Cobalt-based nanoparticles have garnered significant interest within the realm of nanomaterials. Due to their notable reactivity, CoO nanoparticles (NPs) have been widely used in diverse fields including energy storage devices, sensing, biosensing, field-effect transistors, heterogeneous catalysis, Li-ion battery anode materials, photocatalysis, solar energy absorption, pigments, and various biomedical applications. Cobalt oxide (CoO) nanoparticles (NPs) have been used in a diverse range of electrochemical sensors. An experimental setup involving the utilization of a glassy carbon electrode (GCE) that was coated with cobalt oxide nanoparticles (CoO NPs) was used for the purpose of detecting arsenic (III) ions. Cobalt oxide nanoparticles (CoO NPs) were manufactured using a surfactant-assisted hydrothermal process and were utilized for the detection of glucose [104].

4.6. Plasmonic Nanostructures

Plasmonic nanostructures possessing specific textures and topologies have been found to boost the sensitivity of surface-enhanced Raman spectroscopy (SERS)-based molecular analysis. Furthermore, these nanostructures have been observed to improve the intensity of Raman scattering. The distinctive optical characteristics exhibited by gold (Au) and silver (Ag) have resulted in their extensive utilization as nanoparticles (NPs) in plasmonic resonance methodologies. The intense scattering of light by plasmon nanoparticles (NPs) results in the localization, measurement, and molecular imaging of analytes. Various types of nanostructures, such as nanospheres, nanowires, hollow nanocages, and top-down nanochips, with large aspect ratio nanorods (NR) are subjected to modifications in plasmonic NP structures based on specific requirements. Examples of methods that enhance the chemical catalytic performances of modified plasmonic nanoparticles (NPs) while also possessing notable biochemical sensing capabilities include nanohybrids based on molybdenum disulfide (MoS_2) and a revolutionary green technique involving photochemical reduction using femtosecond laser pulses [105].

4.7. PEDOT Nanostructures

PEDOT NPs, or poly(3,4-ethylenedioxythiophene) nanoparticles, have been utilized frequently because they have a low bandgap and are conductive, stable, and doped transparent [88]. Poly(3,4-ethylenedioxythiophene), or PEDOT, is a PTh family derivative that is one of the most stable and hopeful CPs on the market today. It has been found to be interesting because it has a high and stable conductivity, relatively few structural defects caused by wrong linking, and good qualities for making films. Because of these favorable qualities, there has been a recent focus on PEDOT, and it can be used in a very broad range of fields. Specifically, PEDOT has been used to make chemo/bio-sensors, nerve probes (prostheses), and drug-releasing vehicles because it is biocompatible and stable in water and electrolytes [106]. The structure of imprinted PEDOT films showed more compactness than that of non-imprinted PEDOT fields [107].

4.8. Co_3O_4/g-C_3N_4 Heterojunctions

The utilization of heterojunctions comprising g-C_3N_4 and Co_3O_4 nanoparticles exhibits potential as an electrocatalyst for the detection of phenolic hormones in the environment. The fabrication of Co_3O_4/g-C_3N_4 heterojunctions was achieved using a straightforward one-pot thermal decomposition process. The Co_3O_4/g-C_3N_4 heterojunctions were thoroughly examined using a range of advanced analytical techniques. When compared with g-C_3N_4, the Co_3O_4/g-C_3N_4 heterojunctions exhibited superior characteristics such as a higher donor density, a narrower band gap, and a greater surface area, as determined with Brunauer–Emmett–Teller (BET) isotherm analysis, which provides information about the porosity of a material. These attributes are beneficial for enhancing the electrocatalytic activity of the heterojunctions in the degradation of environmental phenolic hormones. Furthermore, the Co_3O_4/g-C_3N_4 heterojunctions were effectively used for the electrochemical detection of phenolic hormones in the environment [108].

5. Applications of Nanostructures for Hormone Detection

Materials with nanoscale characteristics have garnered significant attention due to their unique surface qualities, which make them extremely desirable for the development of sensors that exhibit rapid response times, as well as providing excellent sensitivity and selectivity in detecting chemical groups and biological molecules [109].

5.1. Applications in Cortisol Detection

Cortisol is a steroid hormone that has been used as a biomarker for assessing stress exposure due to its status as a physiological consequence of the hypothalamic–pituitary–adrenal (HPA) axis [110,111]. The impact of stress extends to both the physical and emotional well-being of individuals. The aforementioned issue is widely acknowledged as a significant risk element and an inconspicuous contributor to a range of health-related ailments [112]. Cortisol, the primary human glucocorticoid, helps keep blood pressure stable, but too much of it, either systemically or locally, causes hypertension [113]. In contemporary society, stress has emerged as a significant contributing factor to a range of pathological conditions in the human population. The implementation of real-time monitoring of stress-related biomarker levels in bodily fluids can considerably improve the efficacy of stress management strategies [114].

Cortisol is an adrenal hormone crucial for the preservation of homeostasis that plays a vital role in supporting life. It is secreted by the hypothalamic–pituitary–adrenal system and is synthesized as a component of the body's physiological reaction to stress. The cortisol hormone, commonly known as the "stress hormone", plays a significant role in influencing and regulating a range of physiological processes, including but not limited to blood pressure, sugar levels, immunological responses, cardiac contractions, field-effect transistors, and central nervous system activation. The diurnal variation in cortisol levels has been well-documented, exhibiting a circadian rhythm characterized by peak levels during the early morning hours and a gradual decline throughout the day, ultimately reaching its nadir during the nighttime. Therefore, it is crucial to comprehend the implications of deviant cortisol levels. Aberrations in cortisol levels serve as diagnostic markers for chronic ailments such as Cushing's disease, which is characterized by excessive cortisol production, and Addison's disease, which is characterized by inadequate cortisol production and adrenal insufficiencies. The measurement of cortisol at point-of-care (POC) has become crucial in comprehending human behavioral patterns due to the influence of changes in behavioral and environmental factors on cortisol secretion. Recently, electrochemical immunosensing has been recognized as a potentially useful method for point-of-care (POC) detection of cortisol in bio-fluids because of its simplicity, inexpensive price, and lack of a need for a tag [115]. Human cortisol is widely assessed using urine, saliva, interstitial fluid (ISF), sweat, and blood samples (including feces and milk from animals). Despite their reliability, these methods can only detect cortisol levels that have been rapidly circulating (saliva or plasma) or absorbed for the course of the examined sampling period (urine), which is usually a little more than 24 h [116].

The precise determination of cortisol in saliva was proved with the development of an electrochemical-based aptasensor. Because of the direct covalent connection that can be achieved with thiol bonding, gold nanoparticles have received significant attention as a potential immobilization substrate for thiol-terminated aptamers. In a previous study, a truncated 14-mer aptamer with a thiol on the 5′ end that is specific to cortisol was immobilized onto gold nanoparticles that were electrodeposited onto screen-printed gold electrodes. The unmodified sites on the Au nanoparticle surfaces were blocked with 6-mercapto-1-hexanol. The binding of cortisol resulted in a change in the aptamer conformation that resulted in a decrease in the current seen for the oxidation of the $[Fe(CN)_6]^{3-/4-}$ redox probe as measured with DPV. With a detection limit of 0.25 pg/mL in buffer and 0.28 pg/mL in artificial saliva matrix, the aptasensor had a large dynamic range from 0.1 pg/mL to 100 ng/mL. In PBS buffer, the sensor had a sensitivity of 1.53 μA $(pg/mL)^{-1}$, while the sensitivity of the sensor in the artificial saliva matrix was 1.33 μA $(pg/mL)^{-1}$. The suggested aptasensor was shown to have a high selectivity for cortisol and did not show any change in the signal when exposed to structural analogs of cortisol [117].

Continuous stress monitoring using biological indicators like cortisol helps spot diseases earlier. Continuous cortisol measurement was made possible using an electrochemical aptamer sensor with a conjugated redox probe that changed conformation after the binding of cortisol. In the study, a 40-mer cortisol-specific aptamer with a thiol on the 3′-end was used to modify gold disc electrodes. Methylene blue was used as a redox probe and was covalently conjugated to the 5′-NH_2 ends of the aptamer with an amide linkage. The binding of cortisol caused a conformational change in the aptamer that resulted in a decrease in log(current) from methylene blue oxidation, as measured with SWV, which was linearly related to log([cortisol]). The sensor could detect clinically significant cortisol concentrations from 0.05 to 100 ng/mL, and a detection limit of 0.25 ng/mL was found for a 100-fold diluted serum sample. The measurements without reagent were also shown using unadulterated human blood serum. The novel sensor can measure cortisol to help diagnose and treat cortisol-related diseases [118].

The efficient detection of the steroid hormone cortisol holds potential advantages for the diagnosis of medical conditions associated with adrenal gland problems and chronic stress. Yeasmin et al. presented a new electrochemical sensor that uses a poly o-phenylenediamine (poly-o-PD) molecularly imprinted polymer (MIP) doped with Au nanoparticles formed with electropolymerization in the presence of simultaneous reduction of $HAuCl_4$. This sensor selectively detected low cortisol concentrations with improved sensitivity, and the measurement of cortisol levels was accomplished using the sensor by determining the decrease in the current from the oxidation of the redox-active probe $[Fe(CN)^6]^{3-/4-}$. This decrease in current was a result of the binding of imprinted sites existing in the polymer with the target cortisol. The Au@MIP sensor was fabricated with a simple one-step process including in situ electropolymerization and gold reduction. This method effectively dispersed gold nanoparticles with a high density near the binding sites. The process of the polymerization reaction was facilitated with in situ gold, hence increasing the sensor's effective surface area. The incorporation of doped nano gold additionally promoted charge transfer upon exposure to redox reagents. The effective blockage of charge transfer was made possible with the occupation of the cavities by cortisol, leading to an increased sensitivity and an improved detection response. The Au@MIP sensor had a strong binding affinity for cortisol, as indicated by a dissociation constant (K_d) of around 0.47 nM. It also displayed a linear detection range spanning from 1 pM to 500 nM, with a detection limit of approximately 200 fM. Furthermore, the sensor exhibited specificity toward cortisol against other steroid hormones that possess structurally comparable characteristics. The effective demonstration of the sensor involved the determination of cortisol concentrations spiked into saliva within both normal and high limits. The efficacy of this cortisol-detection approach, which does not require the use of antibodies, was demonstrated to possess a high level of sensitivity and selectivity. Consequently, it was deemed appropriate for utilization in point-of-care testing scenarios. Figure 10 depicts a simplified version of this strategy and the corresponding CV response [119].

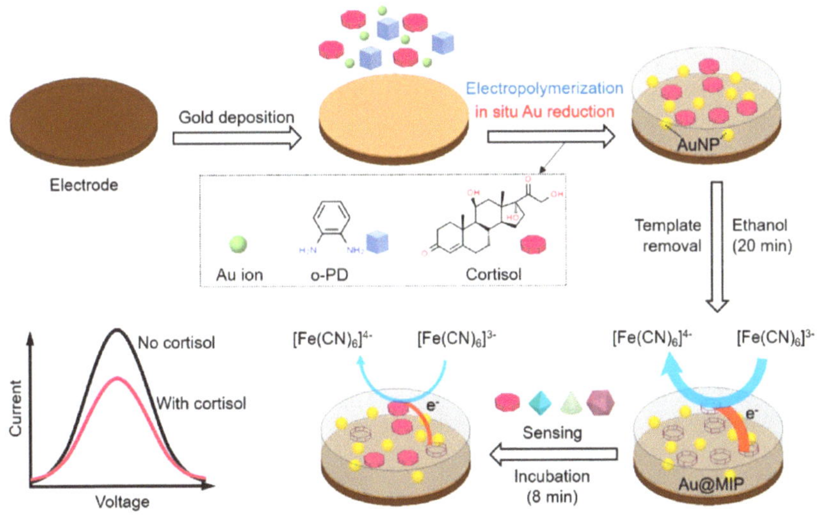

Figure 10. The fabrication method and the mechanism for sensing of a cortisol sensor constructed on a gold-doped molecularly imprinted polymer (Au@MIP) that is selective for cortisol. Reproduced from ref. [119] with permission from Elsevier, copyright 2022.

A silver nanoparticle-doped, molecularly imprinted polymer electrochemical sensor showed sensitive cortisol detection. Shama et al. developed molecularly imprinted poly (hydroxyethyl methacrylate-N-methacryloyl-(l)-histidine methyl ester)-coated pencil graphite electrodes doped with silver nanoparticles (AgNPs). DPV was used to detect cortisol in aqueous solution and biological samples. The cortisol-imprinted pencil graphite electrode (PGE) had an enhanced surface area due to doped AgNPs that also provided improved electroactivity. Cortisol-imprinted polymer-coated PGEs (MIP), MIP@AgNPs, and NIP@AgNPs were tested for selective and sensitive detection of cortisol in aqueous solution. The MIP@AgNPs were exposed to five cortisol doses (0.395, 0.791, 1.32, 2.64, and 3.96 nM), and a DPV response due to cortisol oxidation was detected. The modified electrode had a low limit of detection of 0.214 nM and better cortisol electrocatalytic activity from 0.395 to 3.96 nM. The MIP@AgNPs sensor showed promise for selective and sensitive cortisol measurement in biological trials [120].

Biosensing techniques based on textiles and clothing make it possible to put diagnostic devices in a worn format to create non-invasive ways to measure biomarkers in bodily fluids. The modification of textiles with nanostructured materials, onto which bioreceptors can be immobilized, is a new and attractive strategy. In one study, conductive carbon fibers (CCY) were modified with the deposition of SnO_2 nanoflakes (SnO_2/CCY) and used to sense cortisol. Anti-cortisol antibodies were noncovalently assembled onto SnO_2/CCY, likely with hydrogen bonding. In the presence of cortisol, the current observed with DPV for the oxidation of the redox probe [$Fe(CN)_6{}^{3/4-}$] decreased linearly with log([cortisol]). The electrochemical immunosensor reacted to a wide range of cortisol concentrations, from 10 fg/mL to 1 g/mL, with a detection limit of 1.6 fg/mL for the detection of cortisol. Under optimum circumstances, the immunosensor was tested for measuring the amount of cortisol spiked into human sweat, and an excellent recovery of 99.02 to 102.93% was achieved [121].

A thread-based electrochemical immunosensor was developed for non-invasive cortisol detection in human sweat, which was achieved by immobilizing anti-cortisol antibodies on a conductive thread electrode that was modified with L-cys/AuNPs/MXene. The utilization of AuNPs and MXene in the context of the conductive thread electrode served to augment the surface area, hence facilitating the immobilization of anti-cortisol antibodies and consequently allowing for an improved sensitivity of the sensor. The immobilization of anti-cortisol antibodies onto the electrode was achieved with the utilization of N-hydroxysulfosuccinimide and 1-ethyl-3-(3-dimethylaminopropyl) carbodiimide coupling agents, in conjunction with L-cys/AuNPs/MXene modification. The electrochemical detection method for cortisol relied on the reduction of oxidation current for the redox probe [$Fe(CN)_6^{3/4-}$] due to the hindrance of electron transport caused by the binding contact between the antigen and antibody. This immunosensor had excellent sensitivity, a broad linearity range of 5–180 ng/mL, and a low detection limit of 0.54 ng mL^{-1} while being less affected by interferences. In addition, this immunosensor exhibited a notable degree of reproducibility and demonstrated a remarkable capacity for long-term storage stability, with a minimum duration of six weeks. In conclusion, this approach was effectively utilized for the identification of cortisol in synthetic perspiration, with good outcomes. Figure 11 shows that this particular technology exhibits potential suitability for utilization as a wearable electrochemical sensor for the detection of cortisol in human sweat when integrated into a watchstrap [122].

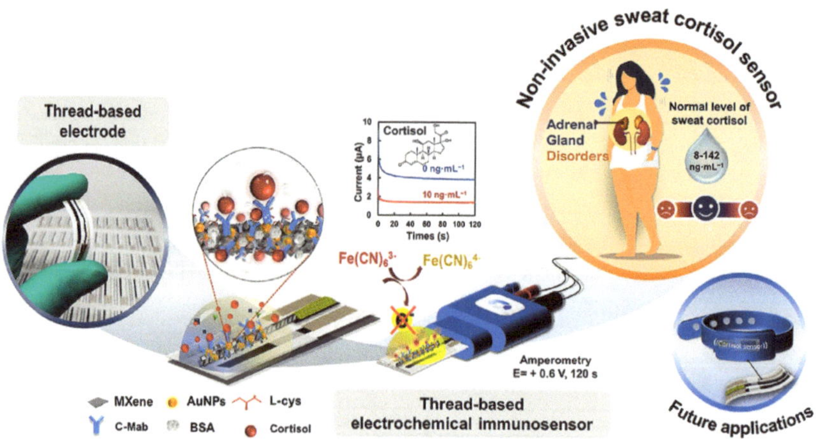

Figure 11. Fabrication of a cortisol-detecting electrochemical immunosensor using thread. Reproduced with permission from ref. [122]. Copyright 2022, Elsevier.

Zubarev et al. investigated the sensing characteristics of multi-layer graphene in conjunction with pyrrole to develop an affordable and highly sensitive material for the detection of cortisol [123]. Pyrrole and graphene nanoplatelets were dispersed in 1 M HNO_3 with the utilization of an intense ultrasound probe for a duration of 10 min. Subsequently, centrifugation was carried out for a period of 30 min at a rotational speed of 4000 revolutions per minute. Polymerization was subsequently conducted using cyclic voltammetry. The pyrrole–graphene composite was subjected to testing for ultra-low concentrations of cortisol in synthetic saliva, which closely matched the quantities found in human oral samples. Raman spectroscopy was used to conduct a more in-depth analysis of the structure of the composite. Additionally, the interactions between cortisol and the layer were simulated using Marvin Beans® software (https://chemaxon.com/marvin, accessed on 30 October 2023). The results demonstrated a notable degree of sensitivity in the measurement of salivary cortisol levels using cyclic voltammetry, with the capability to detect cortisol concentrations as low as 0.5 ng/mL [123].

The synthesis of nitrogen-doped carbon nanotubes with a bamboo-like structure, incorporating nickel nanoclusters, was achieved using a single-pot pyrolysis technique. The study focused on the investigation of a molecularly imprinted electrochemical sensor (MIECS) that possessed recognition sites for cortisol. Molecularly imprinted polymer films were fabricated with a straightforward and manageable electropolymerization technique, using o-phenylenediamine as the functional monomer and cortisone as the template. Following a sequence of optimized experimental settings, the MIECS (microfluidic immunoassay electrochemical sensor) demonstrated a linear range spanning from 10^{-14} M to 10^{-9} M for the detection of cortisol. Moreover, this sensor exhibited a low detection limit of 2.37×10^{-15} M. Furthermore, this sensing system was successfully utilized for identifying the presence of cortisol in saliva samples, yielding a good recovery [124]. Table 2 summarizes studies reported using nanostructure-modified electrodes to detect cortisol.

Table 2. Summary of electrochemical studies using nanostructure-modified electrodes to detect cortisol with the linear range and limit of detection.

Hormone	Electrochemical Method	Type of Nanostructures	Linear Range	LOD	Reference
Cortisol	CV	L-cys/AuNPs/MXene modified thread electrode	5–180 ng/mL	0.54 ng/mL	[122]
Cortisol	DPV	Apt/AuNP/SPE	0.1 pg/mL–100 ng/mL	0.25 pg/mL	[117]
Cortisol	CA, DPV	BSA/AuNW/GA/Au	1–1000 nM	0.51–0.68 nM	[112]
Cortisol	CV, DPV	SGCN/CFP	-	15.8×10^{-8} M	[32]
Cortisol	CV	GQDs/SPE	0.1 pg/mL–100 ng/mL	0.1 pg/mL	[114]
Cortisol	CV	Graphene–pyrrole composite	0.5–5 ng/mL	0.5 ng/mL	[123]
Cortisol	EIS, CV	NiNCs-N-CNTs/GCE	10^{-14}–10^{-9} M	2.37×10^{-15} M	[124]
Cortisol	CV, SWV, CA	AuNC-integrated MB-aptamer with hybrid hydrogel	0.1–50 ng/mL	0.1 ng/mL	[111]

5.2. Applications in Melatonin Detection

Melatonin (MLT) is a hormone with electroactive properties that plays a crucial role in regulating and controlling the circadian rhythms of human beings. The indoleamine hormone is situated within the epithalamus of the vertebrate pineal gland and synthesized during periods of darkness. However, it is also produced by certain meninges, glial cells, and neurons. Melatonin (MT), often known as the hormone responsible for regulating sleep, plays a crucial role in governing a range of physiological processes within the brain. Additionally, it exhibits properties such as neuroprotective, anti-apoptotic, antioxidant, and anti-inflammatory actions. Moreover, music therapy has been found to be advantageous in the treatment of brain damage due to its potential to efficiently permeate the blood–brain barrier. Over the course of the previous decade, a substantial number of studies demonstrated the restoration of tissue after MT treatment in numerous conditions associated with oxidative stress. Additionally, these studies indicated encouraging therapeutic outcomes of MT in diverse sleep and neurological disorders, encompassing depression, Parkinson's disease, epilepsy, Alzheimer's disease, and alcoholism [125,126]. Because of its great sensitivity, potential for miniaturization, and extreme compatibility with micro- and nanotechnologies, electrochemical detection of melatonin has been demonstrated to be an appropriate and effective analytical approach [127].

Melatonin is therapeutically significant for immune system, circadian cycle, blood pressure, and cortisol maintenance. Thus, real-time detection is essential for body function monitoring. Electrochemical sensors can detect melatonin with high sensitivity in point-of-care analyses with a rapid reaction time, ease of use, and low cost. For biosensing applications, natural polymer-based biocomposites such chitosan, gum acacia, xanthan gum, and chitin are commonly used due to their availability, low cost, biocompatibility, and high surface area. In one study, tungsten oxide (WO_3) nanospheres coated with functionalized chitosan (FCH) were used for melatonin detection. Chitosan functionalization created abundant amine groups and the inter-hydrogen bonding needed to generate a WO_3/FCH biocomposite. The nanocomposite's conductivity and melatonin detection were enhanced by the high density of amine groups, which bound WO_3 efficiently. An electroanalysis showed the biocomposite's good electrocatalytic efficacy against melatonin with a 4.9 nM detection limit. The proposed nanocomposite was selective, reproducible, and stable. Its actual applicability was tested using real samples, which proved its effectiveness in detecting therapeutically significant biomolecules [128].

An electrochemical sensor for melatonin was developed using sonogel–carbon electrode material (SNGCE) and gold nanoparticles. The ultrasound-assisted sonogel technology produced an inexpensive and eco-friendly SNGCE material. Chemically produced AuNPs were 12.3 ± 3.9 nm in size. The SNGCE/AuNP sensor was characterized using electrochemical impedance spectroscopy, and it was found that the Au nanoparticles reduced the charge transfer resistance. The SNGCE and SNGCE/AuNP sensor's linear response range, accuracies, selectivities, and detection limits were examined using chronoamperometry. The SNGCE sensor displayed a linear current response to melatonin oxidation over the range of 0.5–20 μM with a detection limit of 100.2 nM. The optimized SNGCE/AuNP sensor displayed a second linear current response over the range of 0.02–0.3 μM and a detection limit of 8.4 nM. The recovery of melatonin spiked into human peripheral blood serum showed good accuracy for the suggested sensor's real sample analysis and anti-matrix performance [129].

An electrochemical impedance monitor was developed for melatonin. Polydopamine formed on glassy carbon electrodes was coated with Au nanoparticles (AuNPs/PDA/GCE). Melatonin was detected using a method that combined the signal-off strategy and electrochemical impedance spectroscopy (EIS). When the AuNPs/PDA/GCE electrode was put into a buffered solution with melatonin and the oxidation current of AuNPs was measured, it was seen that melatonin molecules made the oxidation current of AuNPs decrease. When -0.2 V was applied for 150 s in an acetate buffer solution (ABS) with a pH of 5, melatonin molecules adsorbed to the electrode surface and slowed the electron transfer process, which is important for oxidizing AuNPs. This resulted in a decrease in the electrical current. EIS was used to detect melatonin over the range of 1 to 18 pM. The sensor was able to reach an outstanding detection limit of 0.32 pM. Importantly, these new sensors showed that they had very good sensitivity, specificity, stability, and repeatability. The use of the sensors was proposed for a wide range of fields, such as medical diagnosis, drug evaluation, environmental surveillance, and production control [130].

The voltammetric signals for melatonin and tryptophan overlap, making the detection of melatonin in the presence of tryptophan difficult. A new method to detect melatonin was developed using SWV measurement of an electroactive product of melatonin. This product was generated by applying a positive prepotential of +0.8 V (vs. Ag/AgCl) for 10 s to a carbon paste. Using this approach, the electrochemical signal of melatonin was successfully separated from that of tryptophan by generating an electroactive product that can be oxidized at +0.4 V. The optimal response was found at pH 2 with a prepotential of +0.8 V applied for 10 s. Over the range of 5.00×10^{-7}–8.00×10^{-5} mol L^{-1}, a linear relation was found between the peak current and melatonin concentration. This method gave a detection limit of 8.30×10^{-8} mol L^{-1} (S/N = 3). The method was very selective because the signal from melatonin did not change when there were more substances that could interfere, such as ascorbic acid, tryptophan, glucose, etc. The sensor was used to determine melatonin concentration in samples of drugs and food [131].

The precise identification of melatonin, a potent antioxidant with broad-ranging effects and a crucial hormone produced by the pineal gland, holds considerable importance. A new ternary nanocomposite α-Fe_2O_3/CeFe-gC_3N_4 with exceptional electrocatalytic activity for the nanomolar-level electrochemical detection of melatonin was synthesized by embedding porous flower-like hematite (α-Fe_2O_3) nanoparticles into carbon and iron-codoped graphitic carbon nitride (C-Fe-gC_3N_4) using a sonochemical method. The manufacture of CeFe-gC_3N_4 involved the use of a thermal polymerization strategy, whereas the synthesis of α-Fe_2O_3 nanoparticles was achieved with a hydrothermal process. A comprehensive analysis of the synthesized materials was conducted with a thorough materials characterization. Additionally, electrochemical investigations were carried out using CV and DPV techniques to assess the electrochemical performance of the prepared nanocomposite in quantifying melatonin. The melatonin sensor exhibited a remarkably low detection limit of 4 nM, along with exceptional selectivity, stability, reproducibility, and repeatability. The study presented an investigation into the development of a screen-printed carbon electrode incorporating a composite of α-Fe_2O_3/CeFe-gC_3N_4 for the purpose of detecting melatonin, as shown in Figure 12 [132].

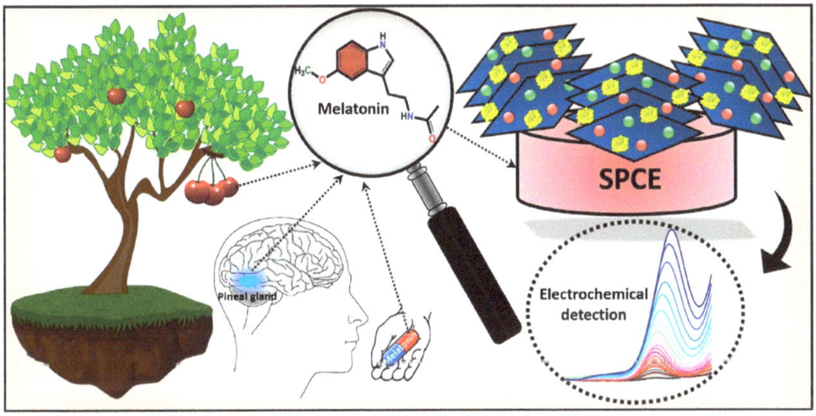

Figure 12. A melatonin-detecting electrode composed of α-Fe_2O_3/CeFe-gC_3N_4/screen printed-carbon. Reproduced from ref. [132] with permission from Elsevier, copyright 2023.

The efficient fabrication of an electrochemical sensing platform was developed using a probe sonication approach to self-assemble gold nanoparticles (Au NPs) on MoS_2 nanoflakes (NFs). The investigation focused on an initial electrochemical sensor (Au-MoS_2) utilized for the simultaneous detection of uric acid and melatonin using DPV. The Au-MoS_2 sensing platform was applied onto a glassy carbon electrode (GCE) using a drop-casting method. The utilization of Au-MoS_2 resulted in an increased electrochemical active surface area (EASA) of the GCE, reaching a value of 0.980 cm^2 due to its high surface-to-volume ratio. The electron transport at the interface of the gold–molybdenum disulfide (Au-MoS_2)-modified glassy carbon electrode (GCE) is accelerated due to the low charge transfer resistance (197.7 Ω), the high conductivity of the Au-MoS_2 nanocomposite, and the synergistic effect between the gold nanoparticles and molybdenum disulfide nanofibers (MoS_2 NFs). The simultaneous electrochemical detection of uric acid and melatonin was facilitated with the significant electrocatalytic activity and extensive surface area present at the Au-MoS2 interface. The limits of detection for uric acid and melatonin were determined to be 18.2 and 15.7 nM, respectively, within a linear range of 0.033–10.0 μM. The sensor that was created had a high degree of selectivity, exact reproducibility, and outstanding repeatability, and it demonstrated good stability in both operational and storage conditions. The experiment involved conducting the spiking of uric acid and melatonin into a urine sample under optimal conditions. The calculated recoveries obtained from the experiment fell within the range of 92.0% to 105.6%. The potential for parallel electrochemical detection of several electrochemical active species may be achievable by adjusting the sensing platform to accommodate varied conditions, thanks to the advantageous features of the simple Au-MoS_2 sensing platform [133].

Zhou et al. presented the concurrent quantification of serotonin and melatonin with a composite substrate consisting of highly ordered and vertically oriented mesoporous silica nanochannel films (VMSF) deposited on a highly electrochemically reduced graphene oxide–carbon nanotube (HErGO-CNT) platform [134]. A compact 2D-2D layered structure consisting of VMSF/HErGO-CNT was produced on indium tin oxide (ITO) electrodes using a two-step electrochemical process. The first step involved the electrodeposition of an ErGO-CNT film onto the ITO surface with CV. In the second step, VMSF is grown on the ErGO-CNT/ITO surface using an electrochemically assisted self-assembly method. This process involved subjecting graphene oxide (GO) to two rounds of electrochemical reduction to form HErGO. The carbon nanotubes (CNT) enclosed within the graphene oxide (GO) sheets functioned as conductive pathways for electron transfer. This not only enhanced the electrochemical reduction of GO but also imparted a certain level of water repellency, thereby facilitating the controlled electrodeposition of composite materials consisting of reduced graphene oxide (ErGO) and CNT onto indium tin oxide (ITO) substrates within a safe electrochemical potential range. Furthermore, the introduction of doping carbon nanotubes (CNTs) had the capability to increase the interlayer spacing between graphene sheets, thus enhancing the electroactive surface area and mass transfer properties of the nanocarbon composite substrate. In addition, the HErGO-CNT layer served as an effective electroactive component, while the outer VMSF layer exhibited electrostatic preconcentration and anti-fouling properties. The separation of the peaks observed in DPV was about 0.3 V, and for serotonin, the peak appeared near 0.45 V, and that for melatonin appeared near 0.76 V (vs. Ag/AgCl), on the ErGO-CNT/ITO, which were negatively shifted relative to the peaks on the bare ITO electrode. For serotonin, two linear ranges in the peak current response of 0.1–10 µM and 10–30 µM were observed with a detection limit of 5.4 nM, and for melatonin, two linear ranges of 1–20 µM and 20–100 µM were observed with a detection limit of 14.3 nM. This composite electrode enabled the direct electrochemical analysis of serotonin and melatonin spiked into two challenging biological fluids, namely, human whole blood and artificial cerebrospinal fluid [134].

A new melatonin sensor that utilized unfunctionalized macroporous graphene networks adorned with gold nanoparticles was designed for the purpose of detecting melatonin in pharmaceutical products using DPV. Metallic template-assisted chemical vapor deposition was used to create graphene structures with a high degree of porosity. Subsequently, they enhanced the active surface area and electrocatalytic activity of these structures by electrochemically depositing gold nanoparticles (ranging from 50 to 250 nm) onto their struts. The electrodes composed of graphene and gold demonstrated a remarkable level of sensitivity and performance for the electro-oxidation of melatonin. This sensitivity was characterized by a linear range spanning from 0.05 to 50 µM, a detection limit of 0.0082 µM (calculated as 3 times the standard deviation divided by the slope of the calibration curve), and a notable sensitivity value of 16.219 $\mu A\ \mu M^{-1}\ cm^{-2}$. Hence, the sensor's performance in terms of the achieved figures of merit surpassed that of numerous other electrodes used for melatonin detection. The electrochemical active surface area of the glassy carbon electrode was increased by a factor of 18. Additionally, the gold–graphene composites exhibited good conductivity and a synergistic catalytic effect, resulting in a reduction in electron transport resistance by 87%. Furthermore, the study demonstrated the long-term stability of the signal for a duration of around 14 days. The reproducibility of the results was deemed satisfactory. The electrodes used in the study exhibited selectivity against various interfering substances. The study demonstrated the possible application of the sensors for the accurate measurement of melatonin in medicinal samples of dissolved tablets [135].

Another study presented a melatonin sensor platform that exhibits good performance, utilizing carbon nanofiber arrays that are adorned with FeCo alloy nanoparticles. A carbon nanofiber variant, derived from the utilization of FeCo bimetallic alloy, was synthesized with the electrospinning technique. The electrocatalytic capability and electron transfer potential of the materials were increased with the synergistic effect of FeCo. Furthermore, the authors investigated the electrooxidation activity of melatonin and the corresponding reaction mechanism based on the composite material. Subsequently, they developed a highly efficient electrochemical sensor for the quantitative detection of melatonin. The sensor exhibited a wide detection range of 0.08–400 µM and a detection limit of 2.7 nM. In contrast to earlier studies, the melatonin sensor under consideration exhibited exceptional characteristics and a broad range of detectability [27]. Table 3 summarizes studies reported using nanostructure-modified electrodes to detect melatonin.

Table 3. Summary of electrochemical studies using nanostructure-modified electrodes to detect melatonin with the linear range and limit of detection.

Hormone	Electrochemical Method	Type of Nanostructures	Linear Range	LOD	Reference
Melatonin	CV, DPV	$Cu_3P_2O_8$/NbC/GCE	0.014–1517.68 μM	0.0083 μM	[126]
Melatonin	CV, DPV	α-Fe_2O_3/C–Fe-gC_3N_4/SPCE	0.1–233.1 μM	4 nM	[132]
Melatonin	Hydrodynamic voltammetry, electrophorometry	ME–SW-PTEs	50–500 μM	4 μM	[127]
Melatonin	DPV	AuNPs/MoS_2/GCE	0.033–10.0 μM	15.7 nM	[133]
Melatonin	CV, DPV	VMSF/HErGO-CNT/ITO electrode	1–100 μM	14.4 nM	[134]
Melatonin	EIS, CV, DPV, CA	FeCo@CNFs/GCE	0.05–250 μM	2.7 nM	[27]
Melatonin	EIS, DPV, SWV, CV	BDD electrode	0.4–600 μM	0.003 μM	[59]
Melatonin	SWV	BDD electrode	5.0×10^7 M to 4.0×10^6 M	1.1×10^{-7} M (0.025 μg/mL)	[48]
Melatonin	DPV, CV, EIS	Porous graphene–gold composites	0.05–50 μM	0.0082 μM	[135]
Melatonin	CV, EIS	SNGCE/AuNPs	0.02 to 0.3 μM, 0.5 to 20 μM	8.4 nM	[129]

5.3. Applications in Androgens and Testosterone Detection

Androgens play a crucial role in the development and maintenance of sexually mature hair and sebaceous glands. Research findings have provided evidence that androgens have the capacity to augment the diameter of hair fibers, enlarge hair follicles, and extend the duration of the anagen phase for terminal hairs. Acne vulgaris, a dermatological condition affecting the sebaceous gland, has been shown to be influenced by the increase in androgen levels throughout puberty. Consequently, androgens play a crucial role in the growth and differentiation of sebaceous glands [136]. Testosterone is an androgen that functions as the principal sex hormone for men. Testosterone is a type of anabolic steroid. Both free testosterone (testosterone that is not bound) and bound testosterone (testosterone that is attached to albumin or sex hormone-bound globulin) are possible forms of testosterone. Testosterone levels, including total testosterone and free testosterone, are intimately connected to a large variety of physiological events that occur within the body, and having low testosterone levels can cause a wide variety of serious problems [8].

A novel aptasensor utilizing graphene oxide (GO) adorned with gold nanoparticles (AuNPs) was developed to detect the androgen receptor (AR) secreted by prostate cells. This innovative approach aimed to enhance our understanding of the correlation between AR and prostate cancer, ultimately leading to improved diagnostic accuracy for cancer detection. The characterization of the surface-modified graphene electrode (GE) was conducted with the utilization of cyclic voltammetry and electrochemical impedance spectroscopy techniques. Ferrocene was used as a redox probe in these analyses. The electrochemical biosensor used in the study demonstrated a detection limit of 0.5 ng/mL (3σ/slope) and a linear range spanning from 0 to 110 ng/mL, under the optimum experimental conditions. Due to its exceptional sensitivity, specificity, and reliability, the current electrochemical biosensor exhibited promising potential for utilization in clinical diagnostics [137].

For medical diagnostics, therapy, pharmaceutical quality control, and doping detection, sensitive, easy, and rapid testosterone (TST) blood monitoring is essential. Human athletic abilities, protein synthesis, and muscular growth depend on TST, the principal male sex hormone. Thus, sportsmen abuse TST and its equivalents to increase muscle growth and performance. Steroid use is prohibited for fair play. One study measured TST using a pencil graphite electrode electrochemically modified with CuO nanoparticles (CuONPs). Electrodeposition of CuONPs onto the PGE surface at −0.6 V for 200 s resulted in a four-fold increase in electrode responsiveness relative to the original pencil graphite electrode (PGE). A morphological investigation with scanning electron microscopy and energy-dispersive X-ray spectroscopy verified the success of the modification. A square wave adsorptive stripping voltammetry analysis in Britton–Robinson buffer at pH 6.0 showed that the proposed sensor could detect TST within a linear range of 5–200 nM. The sensor has a detection limit of 4.6 nM (1.32 ng/mL). The sensor platform for precise, sensitive, and specific TST determination has great potential for point-of-care devices and lab-on-a-chip research [25].

Moura et al. investigated the voltammetric oxidation of testosterone (4-androsten-17-ol-3-one) using a cobalt oxide-modified edge plane glassy carbon electrode [138]. A distinct and precise peak was observed at a peak potential of 0.50 V vs. SCE, which was related to the reduction reaction of CoIV/CoIII. This reduction process is responsible for the oxidation of the steroid. A calibration curve was generated for testosterone by conducting measurements in triplicate within the concentration range of 0.33 to 2.00 M. The resulting correlation coefficient exceeded 0.99, indicating a strong relationship between the measured values. The detection limit for testosterone was determined to be 0.16 M. The relative standard deviations (R.S.Ds) ranged from 1.33×10^{-5} to 4.00×10^{-4}% with a 95% significance level ($p < 0.05$). The electrode that underwent modifications exhibited a consistent and repeatable reaction when used to detect this steroid. Consequently, the findings indicated that the electrode that underwent modification with cobalt oxide exhibited potential as a viable substitute for conducting straightforward, cost-effective, and expeditious analysis of testosterone [138].

A biosensor for the determination of androsterone was developed by immobilizing the enzyme 3a-hydroxysteroid dehydrogenase (3a-HSD) on a composite electrode platform. The platform consisted of a mixture of multi-walled carbon nanotubes (MWCNTs), octylpyridinium hexafluorophosphate ($OPPF_6$) ionic liquid, and NAD^+ cofactor. The arrangement facilitated the rapid, highly responsive, and consistent electrochemical identification of NADH produced during enzymatic activity. The optimization of all experimental factors pertaining to the making and functioning of the enzyme biosensor was conducted. Amperometry was conducted in agitated solutions at a potential of +400 mV, resulting in a linear calibration graph for androsterone within the concentration range of 0.5–10 mM. The slope value obtained from the study was found to be more than 200 times greater than the value reported in prior research. The achieved detection limit was 0.15 mM, and a low value of the apparent Michaelis–Menten constant (${K^M}_{app}$) of 36.0 mM, which was comparable to the value recorded for the enzyme in solution, was computed. The biosensor consisting of 3a-HSD/MWCNTs/$OPPF_6$/NAD^+ showed favorable outcomes when utilized for the quantification of androsterone in human serum samples that were artificially enriched with the compound [139].

A very sensitive electrochemical method for detecting the androgen receptor (AR) was devised, utilizing the safeguarding of DNA duplex with AR against restriction endonuclease-mediated digestion, followed by a hybridization chain reaction (HCR). Two DNA probes, namely, P1 and P2, were strategically chosen to create a hybridized probe known as the androgen receptor binding probe (ARBP). The ARBP molecule is characterized by the presence of a duplex structure at one terminus and two single-stranded tails at the opposite terminus. The section of the duplex that includes the recognition sites for the AR and NspI restriction endonucleases was affixed to the gold electrode. In contrast, the single-stranded portions were utilized as the capture probes to initiate a hybridization chain reaction (HCR). If AR is not present, the enzyme NspI can cleave the double-stranded DNA and subsequently release the capture probes. Consequently, the hybridization chain reaction (HCR) does not take place. Nevertheless, it was observed that AR can interact with ARBP, safeguarding the duplex from undergoing cleavage. Consequently, the capture probes can initiate the hybridization chain reaction (HCR) involving four meticulously engineered G-quadruplex generating hairpin probes and the capture probes. This process leads to the generation of a substantial number of G-quadruplex structures. The quantification of AR was performed using differential pulse voltammetry (DPV). The experimental analysis demonstrated a detection limit of 7.64 femtomolar (fM). The excellent specificity and practicality of this method in analyzing serum samples suggest its potential utility in clinical evaluation and illness diagnosis [140]. Table 4 summarizes studies reported using nanostructure-modified electrodes to detect androgens and testosterone.

Table 4. Summary of electrochemical studies using nanostructure-modified electrodes to detect androgens and testosterone with the linear range and limit of detection.

Hormone	Electrochemical Method	Type of Nanostructures	Linear Range	LOD	Reference
Androgens	CV, EIS	GC-rGO-AuNPs electrode	0–110 ng/mL	0.5 ng/mL	[137]
Androgens	DPV, CV	ARBP/Au	200 fM to 500 pM	7.64 fM	[140]
Testosterone	CV	GCE/CoOx electrode	0.33 to 2.00 μM	0.16 μM	[138]
Androsterone	CV	3a-HSD/MWCNTs/OPPF6/NAD$^+$	0.5–10 μM	0.15 μM	[139]
Testosterone	CV	PoPD-MIPs/SPE	1–25 ng/dL	1 ng/dL	[8]

5.4. Applications in Estrogens and the Detection of Its Related Categories

Estrogen is a powerful steroid hormone that is found in elevated concentrations in females over the period from adolescence to menopause, while males typically have lower amounts of this hormone. The notion that estrogen supplementation after menopause has a beneficial effect on reducing cardiovascular disease has been substantiated with a substantial body of anecdotal evidence gathered over an extended period. There exist two well-established estrogen receptors, namely, α and β. The modulation of the tissue response to estrogen is believed to be influenced by variations in the distribution of these receptors across different tissues. However, the confirmation of this hypothesis is still pending [141]. Estrogen, a type of sex steroid hormone, demonstrates a wide range of physiological actions that encompass the management of the menstrual cycle and reproductive processes, as well as the modulation of bone density, brain function, and cholesterol mobilization. Although endogenous estrogen has normal and advantageous physiological effects in women, unusually elevated levels of estrogen are linked to a higher occurrence of specific types of cancer, particularly breast and endometrial cancer [142]. Estrogens belong to a category of steroid hormones encompassing estrone (E1), estradiol (E2), and estriol (E3) [143].

For the examination of contraceptive pills and products used in hormone replacement therapy, a simple electroanalytical approach for simultaneous detection of estrogens, namely, synthetic 17-ethinylestradiol (EE2) or 17-estradiol (E2), was devised. The methodology used in the study utilized SWV to investigate the anodic oxidation on carbonaceous electrode materials in a pure acetonitrile solvent. The investigation focused on examining the electrochemical properties of these hormones when subjected to a glassy carbon electrode (GCE) in a nonaqueous solvent. CV was used as a technique to validate and understand the underlying reaction mechanisms involved. During the optimization process, it was observed that commercially available glassy carbon electrode (GCE) and boron-doped diamond electrodes had comparable analytical performance. This performance was characterized by a large linear range exceeding 100 μmol/L, with limits of quantification of around 1.6 μmol L^{-1} for each hormone. Mixtures of E2 and progestin hormone dienogest (DGN) were found to have clearly separated current peaks with SWV. The electroanalytical method that was developed was subsequently utilized for the quantification of female hormones found in pharmaceutical preparations. The obtained results were found to be similar to those obtained using a reference method that involved reversed-phase high-performance liquid chromatography coupled with a diode array detector. This correspondence between the two methods validates the practical applicability of the developed electroanalytical method for routine pharmaceutical analysis [144].

In another study, a magnetic molecularly imprinted sensing film (MMISF) was developed to detect estradiol (E2). The fabrication process involved the use of a magnetic glassy carbon electrode (MGCE) and magnetic molecularly imprinted polymers (MMIPs). The synthesis of the molecularly imprinted polymers (MMIPs) involved the in situ polymerization of glutathione (GSH)-functionalized gold (Au)-coated Fe_3O_4 (Fe_3O_4@Au-GSH) nanocomposites and aniline. The construction of the MMISF involved the utilization of MMIPs with a process known as "soft modification". This process entailed the assembly and immobilization of MMIPs onto the surface of MGCE, which could be achieved by either inserting a magnet into MGCE or leaving it magnet-free, hence allowing for the removal of MMIPs from the surface. The E2-MMIPs were synthesized using molecularly imprinted polymers (MMIPs) to selectively recognize E2 in the sample. Subsequently, an electrochemical detection method was used by including a modified glassy carbon electrode (MGCE) into the suspension liquid of E2-MMIPs, generating a "soft modification" sensing film. Subsequently, the "soft modification" magnetic microparticle immobilization system (MMISF) was detached from the electrode with the removal of the magnet from the magnetically guided carbon electrode (MGCE). The electrode contact had the potential to be rapidly rejuvenated with a straightforward treatment for subsequent detection. The investigation of the structures and morphologies of Fe_3O_4@Au-GSH, MMIPs, and MMISF was conducted using FTIR, UV-VIS spectroscopy, scanning electron microscopy, and atomic force microscopy. Furthermore, the MMISF (magnetic molecularly imprinted solid-phase extraction) technique demonstrated effective detection of E2 (estradiol) in milk powder, exhibiting favorable attributes such as high sensitivity, selectivity, repeatability, and efficiency. The linear range of the MMISF (microfluidic molecularly imprinted solid-phase extraction) method for detecting E2 (estradiol) was found to be 0.025–10.0 μmol/L. The limit of detection for E2 using this method was shown to be 2.76 nM, using a signal-to-noise ratio of 3 [145].

Arvand et al. established an electroanalytical technique for the concurrent quantification of steroid hormones, marking the first instance of such an approach. The selection of appropriate electrode materials is a crucial determinant in electrochemical methodologies. To achieve the desired objective, the immobilization of graphene quantum dot (GQD)-doped poly (sulfosalicylic acid) (PSSA) was carried out on a glassy carbon electrode (GCE). In addition to exhibiting a robust and consistent electrocatalytic response toward estradiol (E2) and progesterone (P4), the sensor under consideration demonstrated the capability to effectively differentiate between the two hormones' oxidation peaks. Under ideal circumstances, a study was conducted to selectively determine the concentration of E2 and P4. The results showed that there were strong linear associations between the peak current in linear sweep voltammetry and the concentrations of E2 and P4 in the range of 0.001–6.0 μmol L^{-1}. The detection limit for E2 was found to be 0.23 nmol L^{-1}, while the detection limit for P4 was 0.31 nmol L^{-1}. The sensor that was constructed demonstrated precise and prompt detection capabilities toward E2 and P4, exhibiting enhanced stability, selectivity, and repeatability. Furthermore, the electrochemical construction strategy presented in the study holds significant importance due to its simplicity and environmentally friendly nature. This approach has the potential to offer a cost-effective means of establishing nanocomposites or nanohybrid-based sensing platforms. Consequently, it expands the potential applications of electrochemical sensors for the efficient and sensitive analysis of electroactive compounds in biological systems and drugs while also promoting sustainability and ease of use [146].

Li et al. effectively synthesized Pd-decorated N-doped reduced graphene oxide (Pd/N-RGO) utilizing a straightforward wet-chemical approach, devoid of capping agents or stabilizers. The Pd/N-RGO composite was thoroughly characterized using a variety of techniques including XRD, XPS, and TEM. The efficient integration of nitrogen (N) into nitrogen-reduced graphene oxide (N-RGO) sheets was supported by the XPS spectra. Additionally, the presence of palladium (Pd) nanoparticles, ranging in size from 5 to 20 nm, on the N-RGO sheets was confirmed using TEM. The electrocatalytic activity of the Pd/N-RGO-based sensor was enhanced due to the combination of the wide surface area of N-RGO sheets and the efficient electron transfer capability of Pd nanoparticles. This integration resulted in the improved oxidation of estradiol. Hence, the Pd/N-RGO composite exhibits potential as a proficient sensing platform for the non-enzymatic identification of estradiol. Additionally, it was evident that the artificially created sensor exhibited two distinct linear concentration ranges, specifically 0.1–2 μM and 2–400 μM, while also possessing a notably low detection limit of 1.8 nM for the detection of estradiol. Ultimately, the sensor that was developed proved to be effective in detecting estradiol in actual samples, yielding good recoveries ranging from 96.0% to 103.9%. The Pd/N-RGO composite has several notable advantages, such as its straightforward fabrication process, favorable repeatability, and long-term stability. Consequently, it holds great potential as a viable option for implementing improved electrode materials in sensing applications [147].

The levels of estradiol in women have been found to be correlated with the occurrence of lung, uterine (endometrial), ovarian, and breast cancers. However, the precise mechanism by which these malignancies form is not yet comprehensively known. A novel, economically efficient sensor was developed by integrating conductive $NiFe_2O_4$ metal oxide into the electrochemical sensor platform of mesoporous carbon (MC) generated from cotton. This sensor was designed for the electrochemical detection of estradiol. A novel electrode, denoted as GCE/$NiFe_2O_4$-MC, was developed with the application of the drop casting technique. This modified electrode demonstrated remarkable electrocatalytic properties and was specifically designed for the detection of the electroactive estradiol molecule. The electrochemical determination of estradiol was conducted using SWV. The detection limit achieved was 6.88 nM, within a linear range of 20.0–566 nM in a phosphate buffer solution at pH 7.4. The study utilized a unique modified glassy carbon electrode (GCE) with a $NiFe_2O_4$-MC modification, which demonstrated the highest electron transfer rate among the tested electrodes. The study successfully conducted the simultaneous detection of estradiol in the presence of testosterone using a 0.1 M tetrabutylammonium tetrafluoroborate supporting electrolyte solution produced in acetonitrile, an anhydrous medium. Empirical experiments were conducted to demonstrate the practicality and accuracy of the newly developed electrode. The quantification of estradiol from pills was successfully achieved using the standard addition method, resulting in good recovery values [148].

A unique microfluidic electrochemical array device (μFED) that is entirely disposable was used for the detection of the biomarker estrogen receptor alpha (ERα). The microfluidic electrophoresis device (μFED) was fabricated utilizing affordable materials and a cost-effective home cutter printer, allowing for the production of several μFEDs within a short span of 2 h. The manufacturing cost per device was less than USD 0.20 in terms of material expenses. The microfluidic electrochemical device (μFED) includes counter and reference electrodes, as well as eight working electrodes made of carbon. These working electrodes were modified with DNA sequences called estrogen response elements (DNA-ERE), which have a specific binding affinity for ERα. In the study, paramagnetic particles that were extensively coated with anti-ERα antibody and horseradish peroxidase (MP-Ab-HRP) were used for the effective capture of ERα from the solution containing the sample. The bioconjugate ERα-MP-Ab-HRP was introduced into the microfluidic electrochemical detector (μFED) and allowed to interact with the DNA-ERE-modified electrodes. Subsequently, amperometric detection was performed by applying a potential of −0.2 V vs. Ag | AgCl. Simultaneously, a combination of H_2O_2 and hydroquinone was introduced into the microfluidic device. The proposed approach achieved an ultralow limit of detection of 10.0 femtograms per milliliter (fg mL^{-1}). The assay's performance, specifically its sensitivity and repeatability, was evaluated using undiluted calf serum. Notably, the detection of ERα in MCF-7 cell lysate yielded exceptional recoveries ranging from 94.7% to 108%. The μFED system possesses the capability to be readily assembled and utilized for the purpose of multiplex biomarker detection. This characteristic renders the device a highly commendable and economically viable substitute for cancer diagnosis, particularly in regions with limited resources and infrastructure [149].

A new electrochemical biosensor was developed for the quantitative analysis of natural estrogens. The biosensor was constructed using a carbon paste electrode modified with a composite of electrodeposited graphene and ordered mesoporous carbon (GR/OMC/CPE). The measurement of natural estrogens was carried out using SWV as the analytical technique. A modified electrode consisting of a composite of graphene oxide (GO) and ordered mesoporous carbon (OMC) was fabricated using electrodeposition. The fabrication process involved cyclically sweeping the electrode potential in the range of −0.8 to +2.4 V at a scan rate of 100 mV s^{-1} in solutions containing 0.1 mg/mL of OMC and GO, respectively. SEM was used to characterize the morphology of the GR/OMC composite. The study aimed to explore the electrochemical behaviors of estrogens at the graphene oxide-modified carbon paste electrode (GR/OMC/CPE) using CV. The obtained results demonstrated that the GR/OMC/CPE exhibited remarkable electrocatalytic activity for the oxidation of natural estrogens, specifically estrone, estradiol, and estriol. The solid-phase extraction (SPE) with vacuum technique was used for the quantitative analysis of estrogens. Under ideal circumstances, it was observed that the current response exhibited a linear relationship with the concentration of estrogens throughout the range of 5.0×10^{-9} to 2.0×10^{-6} mol/L. Furthermore, the detection limit was determined to be 2.0×10^{-9} mol/L using a signal-to-noise ratio of 3. The sensor that was suggested was effectively utilized to assess the concentration of estrogens in serum samples obtained from females [150]. Table 5 summarizes studies reported using nanostructure-modified electrodes to detect estrogens.

Table 5. Summary of electrochemical studies using nanostructure-modified electrodes to detect estrogens with linear range and limit of detection.

Hormone	Electrochemical Method	Type of Nanostructures	Linear Range	LOD	Reference
Estradiol	CV, DPV	Fe_3O_4@Au-GSH/MGCE	0.025 to 10.0 μM	2.76 nM	[145]
Estradiol	DPV, CV	Pd/N-rGO/GCE	0.1–2 and 2–400 μM	1.8 nM	[147]
Estradiol	CV, SWV, DPV	$NiFe_2O_4$-MC/GCE	20.0–566 nM	6.88 nM	[148]
Estradiol	LSV, CV, CA	Pt/Se/GCE	0.05 to 85.5 M	11.12 nM	[151]
Estrogen	CV	PDDA/GSH-AuNPs	16.6–513.3 fg mL^{-1}	10.0 fg mL^{-1}	[149]
Estrogen	SWV, CV	GR/OMC/CPE	5.0×10^{-9} to 2.0×10^{-6} M	2.0 nM	[150]
Estradiol	CV, DPV	GQDs/PSSA/GCE	0.001–6.0 μmol L^{-1}	0.23 nM	[146]

5.5. Applications in Progesterone Detection

Progesterone (a 12-carbon steroid hormone) is secreted mainly by the corpus luteum [152] and is synthesized from cholesterol. It has a crucial function in ensuring the stability and upkeep of pregnancy in mammalian species [153]; has a significant role in the female reproductive system, specifically in regulating the menstrual cycle, facilitating pregnancy, and supporting embryogenesis in both humans and animals [154]; and also has a significant function in other non-reproductive tissues, including the mammary gland in the anticipation of lactation, the circulatory system, the central nervous system, and skeletal structure. Progesterone (P4), synthesized by the gonads, is primarily transported through the bloodstream to exercise its physiological effects. It originates from the adrenal glands and undergoes significant conversion into glucocorticoids and androgens [155]. It is present in the bloodstream, where it is bound to cortisol-binding globulin (about 10%) and serum albumin [155,156], and exhibits a comparatively brief half-life of about five minutes within the human body. The liver primarily synthesizes sulfates and glucuronides of reduced derivatives as metabolites, which are then eliminated through urinary excretion [155].

A hydrogel conjugated aptamer-based electrochemical detector for detecting progesterone (P4) was synthesized with thiol-modified aptamers specific for P4 that were assembled onto gold nanocubes (AuNCs). Chitosan (CS) and hydroxyethyl cellulose (HEC) were mixed to make CS-gHEC hybrid hydrogels that were then mixed with the aptamer-modified AuNCs, as shown in Figure 13. SWV was used measure P4 aptamer binding of P4. When P4 bound to the P4 aptamer–hydrogel-modified electrodes, the electrochemical response of the redox probe $Fe(CN)_6^{3-/4-}$ decreased. The aptamer-AuNC-hydrogel-modified gold electrode was found to have a sensitivity of 0.78 μA ng mL^{-1}. The lowest amount of P4 that the aptamer–hydrogel sensor could detect was 1 ng mL^{-1}. When tested with analytes with similar structures, like testosterone (T) and cortisol (C), the aptamer based showed that it was very selective for P4. The aptasensor was successfully applied to detect P4 added in different amounts to a blood sample [157].

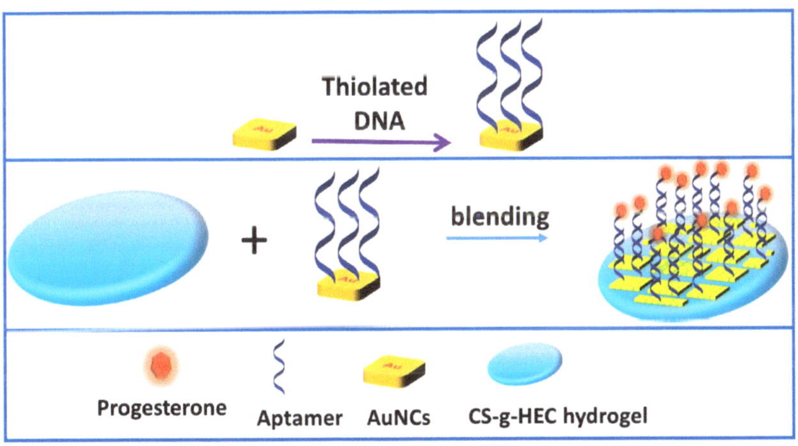

Figure 13. Schematic showing how P4 aptasensors are made. Reproduced from ref. [157] with permission from Elsevier, copyright 2021.

Progesterone serves as a precursor to several steroidal hormones and plays a crucial role in the regulation and maintenance of hormonal equilibrium during the reproductive cycle. The application of reduced graphene oxide (rGO), 5-Amino-2-mercaptobenzimidazole (AMBI), and gold nanoparticles (AuNPs) on a screen-printed carbon electrode (SPCE) for electrochemical sensing of progesterone was reported. This modified electrode demonstrated its potential for voltammetric determination of progesterone (P4), marking the first instance of such utilization. The characterization of the sample was conducted using various analytical techniques, including SEM, TEM, XPS, XRD, and Raman spectroscopy. The characteristics of the electrode were assessed with the utilization of CV and EIS techniques. The assessment was conducted using SWV at an operating voltage of −1.64 V. The figures of merit, including the linear dynamic response for the concentration of P4 spanning from 0.9×10^{-9} to 27×10^{-6} mol/L, as well as the limit of detection of 0.28×10^{-9} mol/L, were determined. The construction and implementation of an AuNP/AMBI/rGO electrode for high-performance P4 sensing was reported, marking a significant advancement in the field. This novel electrode has promise for future development in other electrochemical sensing applications [26].

Huang et al. developed a novel electrochemical aptasensor they called the molecularly imprinted electrochemical aptasensor (MIEAS) for the specific detection of progesterone (P4). The MIEAS was made using SnO_2-graphene (SnO_2-Gr) nanomaterial and gold nanoparticles (AuNPs). The utilization of SnO_2-Gr, characterized by a significant specific surface area and superior conductivity, resulted in an enhanced adsorption capacity for P4. The biocompatible monomer, aptamer, was immobilized onto a modified electrode with the formation of an Au-S bond with gold nanoparticles (AuNPs). The electropolymerized molecularly imprinted polymer (MIP) film was composed of p-aminothiophenol as the chemical functional monomer and P4 as the template molecule. The combination of the molecularly imprinted polymer (MIP) and aptamer in their study resulted in a synergistic effect on the detection of P4. As a result, the developed molecularly imprinted electrochemical aptasensor (MIEAS) demonstrated enhanced selectivity compared with sensors that utilized either MIP or aptamer alone as the recognition element. The sensor that was constructed exhibited a limit of detection of 1.73×10^{-15} M across a broad linear range spanning from 10^{-14} M to 10^{-5} M. The successful recovery achieved in tap water and milk samples demonstrated the significant potential of this sensor in the examination of environmental and food samples [158].

A template-free method for the electrodeposition of tin nanorods onto glassy carbon electrodes (GCEs) was used to develop an electrochemical sensor for the detection of progesterone using voltammetric and amperometric techniques. The physicochemical parameters of the GCE modified with tin nanorods were assessed with various analytical techniques, including electron microscopy, UV-visible spectrum analysis, X-ray diffraction, and electrochemical impedance spectroscopy. The process by which progesterone is reduced was clarified with the application of quantum chemical calculations. Additionally, the impact of the cationic surfactant cetyltrimethylammonium bromide (CTAB) on the reduction of P4 was examined, and it was found that the addition of CTAB shifted the reduction peak to a more positive potential and enhanced the peak current. Progesterone estimation was conducted using DPV and amperometry techniques. The electrode demonstrated a linear range from 40 to 600 μM, with the lowest detectable concentration being 0.12 μM. DPV was used to examine the impact of additional interfering substances, including testosterone, 17β-estradiol, creatinine, uric acid, and ascorbic acid. The examination of the progesterone test in commercialized medicinal products was conducted and demonstrated a satisfactory level of concurrence [159].

The development of a straightforward, replicable, durable, and highly responsive electrochemical sensing system for progesterone (P4) was based on a nanocomposite of graphene quantum dots (GQDs). In the study, a modified glassy carbon electrode (Fe_3O_4@GQD/f–MWCNTs/GCE) was developed by including GQDs, Fe_3O_4 nanoparticles, and functionalized multi-walled carbon nanotubes (f–MWCNTs). The electrocatalytic characteristics of the modified electrode toward the oxidation of P4 were further evaluated. Graphene quantum dots (GQDs) with an approximate size of 15 nm were synthesized using a simple and cost-effective bottom-up approach. This involved carbonizing citric acid and subsequently scattering the resulting carbonized products into an alkaline solution. The characterization of the sensor was investigated using various analytical techniques, including transmission electron microscopy (TEM), X-ray diffraction (XRD), UV-visible spectroscopy, Fourier transform infrared spectroscopy (FT-IR), and voltammetry. The electrochemical investigations indicated that the sensor in its original state had several benefits, including a large effective surface area, a greater number of reactive sites, and exceptional electrochemical catalytic activity for the oxidation of P4. It was shown that the peak currents associated with P4 exhibited a linear relationship with its concentration within the broad ranges of 0.01–0.5 and 0.5–3 μM. The electrode's predicted detection limit and sensitivity were determined to be 2.18 nM and 16.84 μA μM^{-1}, respectively. The sensor demonstrated exceptional stability, selectivity, sensitivity, and repeatability. It proved to be effective in accurately determining the presence of P4 in human serum samples and pharmaceutical items, yielding great recoveries. Furthermore, it successfully avoided any interference from a range of species including K^+, Fe^{2+}, Na^+, NH_4^+, Mg^{2+}, SO_4^{2-}, Ca^{2+}, glucose, glycine, cysteine, alanine, ascorbic acid, and uric acid [160].

An electrochemical progesterone (P4) aptasensor without the need for labeling was achieved by chemically attaching a P4-specific aptamer, which was functionalized with NH_2 groups, to the surface of the electrode. The synthesis of NiO-Au hybrid nanofibers was achieved using the electrospinning process. The nanocomposite of GQDs-NiO-AuNFs was fabricated with the dispersion of electrospun NiO-AuNFs in a solution containing the produced graphene quantum dots (GQDs), followed by stirring for a duration of 24 h. The nano-architecture of GQDs-NiO-AuNFs, along with functionalized multiwalled carbon nanotubes (f-MWCNTs), was used to modify a screen-printed carbon electrode (SPCE). This modification aimed to create an immobilization matrix with a substantial number of carboxylic functional groups. The assembling process of the aptasensor was evaluated using cyclic voltammetry (CV) and electrochemical impedance spectroscopy (EIS). The creation of the aptamer–progesterone complex resulted in a slowed electron transfer process on the sensing interface, leading to a drop in the peak current of the redox probe. It was possible to detect progesterone in a quantitative manner by observing the decline in the DPV response of the $[Fe(CN)_6]^{3-/4-}$ peak current as the concentration of progesterone increased. When the experimental conditions were improved, the aptasensor demonstrated a dynamic concentration range spanning from 0.01 to 1000 nM, along with a detection limit of 1.86 pM. The aptasensor that was suggested in the study was effectively utilized to measure the concentration of progesterone in both human serum specimens as well as medicinal drugs [161].

An enhancement in electrode surfaces has been a focal point of investigation among numerous researchers with the aim of enhancing the analytical capabilities of electrochemical sensors. The novel application of an imidazole-functionalized graphene oxide (GO-IMZ) as a synthetic enzymatic active site for the electrochemical determination of progesterone (P4) represents the first instance of such utilization. The scanning electron microscopy technique was used to evaluate the morphology of the electrode that was modified with GOIMZ. Additionally, CV was used to assess the electrochemical performance of the modified electrode. Under circumstances of optimization, the sensor demonstrated a synergistic impact between the graphene oxide (GO) sheets and the imidazole groups attached to its backbone. This synergistic effect resulted in a substantial improvement in the electrochemical reduction of P4. The study yielded several performance indicators, including the linear dynamic response for P4 concentration spanning from 0.22 to 14.0 μM. Additionally, a limit of detection of 68 nM and a limit of quantification of 210 nM were determined. Furthermore, the modified electrode exhibited a significantly increased sensitivity of 426 nA μM^{-1} compared with the unmodified electrode. In general, the technology under consideration showed potential as a viable platform for the straightforward, expeditious, and immediate examination of progesterone [162].

The process of synthesizing a nanocomposite using a green approach, which involved the combination of poly(3,4-ethylenedioxythiophene) and zirconium oxide nanoparticles, exhibited favorable conductivity and a three-dimensional microporous network structure due to the inclusion of nanoparticles. The successful synthesis of the proposed nanocomposite was confirmed with the analysis of data obtained from scanning electron microscopy, field-emission scanning electron microscopy, energy dispersive spectrum, and Fourier transform infrared spectroscopy. The electrocatalytic activity of a nanocomposite-modified glassy carbon electrode was found to be highly effective in facilitating the oxidation of progesterone. The optimization of certain parameters influencing the response of the sensor was conducted, followed by the construction of a calibration curve. The suggested sensor achieved a detection limit of 0.32 nM and demonstrated two linear calibration ranges for progesterone determination: 1–100 nM and 100–6 $\times$ 10^3 nM. The method was effectively utilized to determine the concentration of progesterone in biological fluids and pharmaceutical items without the need for intricate sample preparation procedures [163]. Table 6 summarizes studied reported using nanostructure-modified electrodes to detect progesterone.

Table 6. Summary of electrochemical studies using nanostructure-modified electrodes to detect progesterone with the linear range and limit of detection.

Hormone	Electrochemical Method	Type of Nanostructures	Linear Range	LOD	Reference
Progesterone	EIS, CV, CA, DPV	Sn nanoparticles on GC electrodes	5–80 μM, 40–600 μM	0.12 μM	[159]
Progesterone	CV, LSV, DPV	Fe_3O_4@GQD/f–MWCNTs/GCE	0.01–0.5 μM and 0.5–3.0 μM	2.18 nM	[160]
Progesterone	SWV, CV	CS-gHEC/AuNCs	-	1 ng/mL	[157]
Progesterone	CV, EIS, DPV	GQDs–NiO-AuNFs/ f-MWCNTs/SPCE nanocomposite	0.01 to 1000 nM	1.86 pM	[161]
Progesterone	CV, SWV	GO-IMZ/GCE	0.22 and 14.0 μM	68 nM	[162]
Progesterone	CV	PEDOT/EDOTOH	1 fg/mL–0.1 ng/mL	0.1 fg/mL	[107]
Progesterone	EIS, CV	ssDNA	10 and 60 ng/mL	0.90 ng/mL	[152]
Progesterone	CV, SWV, SW adsorptive stripping voltammetry	BiFE	0.40–7.90 μM	0.18 μM	[153]
Progesterone	CV, DPV, EIS	Magnetic GO	0.01 pM–1000 nM	0.15 and 0.17 pM	[68]
Progesterone	CV, EIS, SWV	AuNP/AMBI/rGO/SPCE	0.9×10^{-9}–27×10^{-6} M	0.28 nM	[26]
Progesterone	CV, DPV	PEDOT/ZrO2-NPs/GCE	1–100 and 100–6×10^3 nM	0.32 nM	[163]
Progesterone	EIS, CV	AuNPs	0.2 to 125 nM	0.17 nM	[154]
Progesterone	CV, DPV	GQDs-PSSA/GO/GCE	0.001–6.0 μM	0.31 nM	[146]
Progesterone	CV, EIS, DPV	CNS/CFP	0.037 nM to 0.25 nM	0.012 nM	[84]

5.6. Applications in Insulin Detection

Insulin is a hormone that exerts significant influence on several metabolic processes and other physiological functions, such as vascular compliance [30]. It is classified as an endocrine peptide hormone, which exhibits the ability to bind to receptors located on the plasma membrane of target cells. This interaction plays a crucial role in coordinating a comprehensive anabolic response to the presence of nutrients. Insulin and insulin-like peptides (ILPs) have been detected in all animal species [164]. Insulin-like peptides (ILPs), namely, insulin, IGF1, and IGF2, have been extensively investigated in scientific research. These factors are evolutionarily conserved and are widely recognized as important regulators of energy metabolism and growth. They play critical roles in metabolic disorders related to insulin resistance, such as obesity as well as diseases like type 2 diabetes mellitus. Additionally, they are involved in immunological disorders associated with these conditions. There is an increasing body of evidence indicating that the effects of insulin and IGF1 receptors on the regulation of cell proliferation, differentiation, apoptosis, glucose transport, and energy metabolism are mediated by downstream signaling via insulin receptor substrate molecules. As a result, these receptors play a crucial role in determining the fate of cells. The significance of insulin-like growth factors (IGFs) in the development of cancer is associated with their ability to connect elevated energy intake, heightened cell proliferation, and inhibition of apoptosis to the risks of cancer. This has been suggested as the primary mechanism that links insulin resistance and cancer [165].

Insulin is synthesized by the β-cells of the pancreas; it plays a crucial role in the regulation of blood glucose levels and facilitates the cellular uptake of glucose for the synthesis of glycogen or triglycerides. Insufficient production of insulin has the potential to result in the medical conditions of hyperglycemia and diabetes. Hence, the detection of insulin holds significant importance in clinical diagnosis. One study involved the utilization of disposable gold (Au) electrodes that were subjected to modification using copper (II) benzene-1,3,5-tricarboxylate (Cu-BTC) and a leaf-like zeolitic imidazolate framework (ZIF-L) for the purpose of insulin detection. The aptamers were readily immobilized on the Cu-BTC/ZIF-L composite with physical adsorption, hence promoting the interaction between the aptamers and insulin as shown in Figure 14. The aptasensor developed using the Cu-BTC/ZIF-L composite had a broad linear range for detecting insulin concentrations, spanning from 0.1 pM to 5 μM. Additionally, it exhibited a low limit of detection of 0.027 pM. Furthermore, the aptasensor exhibited notable specificity, reliable repeatability and stability, and advantageous applicability when tested with human serum samples. To conduct in vivo experiments, electrodes modified with Cu-BTC/ZIF-L composites were implanted into both non-diabetic and diabetic mice. The quantification of insulin was performed using electrochemical and enzyme-linked immunosorbent assay techniques. Figure 8 shows the synthesis and manufacturing of a composite material consisting of Cu-BTC/ZIF-L, which was then used to modify a DGE (differential pulse voltammetry graphene electrode) for the purpose of insulin detection [166].

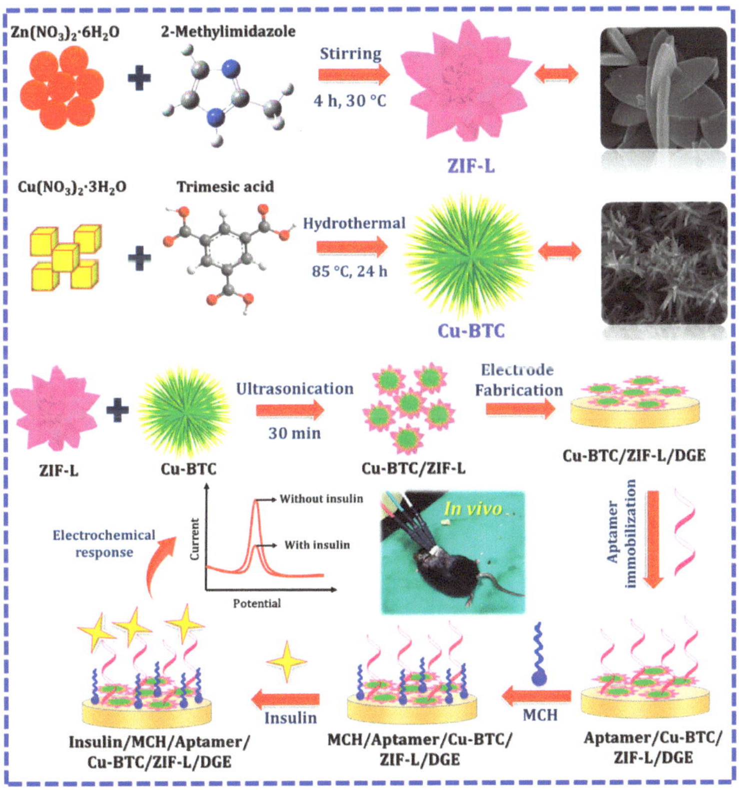

Figure 14. Cu-BTC/ZIF-L composite-modified DGE for insulin detection: synthesis and manufacturing. Reprinted with the permission from [166]. Copyright 2022 American Chemical Society.

A very sensitive electrochemical sensor was developed for the purpose of detecting insulin. This sensor was created by incorporating nickel nanoparticles (NiNPs) using ion implantation, a cost-effective and environmentally sustainable method. The appearance and structure of the nickel nanoparticles (NiNPs) were analyzed using SEM, which indicated sizes ranging from 4 to 8 nm. The performance of the insulin assay was assessed using three electrochemical techniques: CV, EIS, and chronoamperometry. The electrode enhanced with nickel nanoparticles on indium tin oxide (NiNPs/ITO) demonstrated remarkable analytical characteristics. These properties included an exceptionally high sensitivity of 2140 μA μM^{-1} for the detection of insulin at low concentrations, an extremely low detection limit of 10 pM, and a large dynamic range spanning from 100 pM to 2400 pM and 1 nM to 125 nM. Furthermore, the NiNPs/ITO electrode was used for the purpose of assessing the concentration of insulin in bovine insulin injections. The NiNPs/ITO electrode showed promise as a possible biosensor for insulin detection [167].

A novel electrochemical aptasensing platform without the need for labels was developed for the purpose of detecting insulin, which utilizes a functionalized mesoporous silica thin film (MSTF) coated onto a glassy carbon electrode. The coating process was achieved with a one-step electrochemically aided self-assembly (EASA) approach. The success of this approach relies on the formation of a covalent bond between a specific DNA oligonucleotide sequence, known as complementary DNA (cDNA), and the surface of mesoporous silica. This covalently attached cDNA sequence then undergoes hybridization with a labeled aptamer, which acts as a gating molecule. This hybridization event leads to the closure of mesochannels, thereby limiting the diffusion of the electroactive probe [$Fe(CN)_6{}^{3-/4-}$] toward the electrode surface. The introduction of insulin as the target molecule effectively disrupts the hybridization between the aptamer and cDNA. This disruption subsequently leads to the creation of nanochannels, which facilitates the diffusion of the redox probe toward the electrode. As a result, there is a significant increase in the signal observed during DPV. The aptasensor that was proposed demonstrated a broad detection range spanning from 10.0 to 350.0 nM, along with a detection limit of 3.0 nM, which was deemed appropriate. The presented technique provided a means for the efficient and expeditious identification of insulin, obviating the requirement for a cargo (such as a dye or fluorophore) to serve as an electrochemical indicator within the pore. This approach is characterized by its affordability and speedy modification process [168].

Pancreatic hormones, including insulin, are essential for the maintenance of glucose metabolism, which is vital for sustaining life. Considering the escalating prevalence of diabetic disorders and their consequential health complications on a global scale, as well as the detrimental impact of pathogenic viral infections on individuals with pre-existing conditions, there is a pressing need for the development of novel molecular diagnostic methods that are accessible, cost-effective, and capable of detecting minute quantities of biomarkers in human biofluids. These advancements are crucial in enhancing global health outcomes. The development of a cost-effective paper device that facilitates the convenient measurement of fasting blood levels of ultra-low picomolar insulin is a novel advancement, with potential for wider utilization in the analysis of other essential molecular targets. In the study, label-free electrochemical insulin aptasensing was integrated with a paper electrode device. This amalgamation offered a more straightforward, cost-effective, and dependable molecular diagnostic method for analyzing intricate serum samples. Additionally, the study included supplementary independent validation techniques to ensure the scientific integrity and suitability of the proposed approach. The aptasensor utilized in the reported study incorporated a surface design consisting of carboxylated graphene aptamers. This design enabled the accurate and measurable detection of picomolar levels of insulin in a 10-fold diluted neat serum. The detection was achieved by measuring the changes in interfacial capacitance, which were directly proportional to the concentration of insulin in the serum. The dynamic range of the aptasensor was from 5 to 500 pM, with a limit of detection of 1.5 pM. The application of concentrated serum samples was described. In addition, the study included an examination of a serum sample from a patient with diabetes and a comparison of the results obtained from the capacitance sensor with those obtained from the peroxidase antibody label-based insulin assay methods (namely, amperometric detection and a commercial enzyme-linked immunosorbent assay). These validation methods were used to ensure the accuracy and reliability of the capacitance sensor results [169].

Ebrahimiasl et al. aimed to enhance the properties of a pencil graphite electrode (PGE) by including a conductive polypyrrole (Ppy) and graphene (GF) nanocomposite.

The modified electrode was then utilized for the electrochemical analysis of insulin. The electrochemical characteristics of insulin on a platinum–graphite electrode (PGE) were assessed with the application of DPV, CV, and chronoamperometry techniques. The study examined many influential factors, such as pH, concentration, and scan rate, in the electrochemical alteration of electrodes. The research aimed to identify the most favorable conditions and afterward provided the optimal parameters. The study focused on investigating the kinetics of the oxidation reaction and determining the diffusion coefficient of the sensor. The steps that were executed enabled the quantification of insulin with a linear repeatability curve and satisfactory accuracy within the range of 0.225 to 1.235 μM. The limit of detection for insulin was determined to be 8.65 nM. The electron transfer coefficient between the modified electrode and insulin was determined to be 0.5, with an estimated range of 0.84 to 1 for the number of electrons exchanged during the oxidation of insulin [170].

A novel electrochemical aptasensor with high sensitivity and selectivity in detecting insulin utilizing laser-scribed graphene electrodes (LSGEs) as the sensing platform was reported in another study. Prior to utilizing disposable LSGEs, preliminary sensing tests were conducted on a research-grade glassy carbon electrode (GCE) to establish and validate the concept. The aptasensor utilized exonuclease I (Exo I) as a catalyst for the hydrolysis of single-stranded aptamers that were affixed to the electrode surface. It is important to note that the hydrolysis process was inhibited in the presence of insulin binding to the aptamer. Consequently, the aptamers that were not coupled to insulin were enzymatically cleaved by Exo I, but the aptamers that were bound to insulin remained attached to the electrode surface. In the subsequent stage, the gold nanoparticle–aptamer (AuNPs-Apt) probes were applied to the electrode surface, resulting in the formation of a 'sandwich' structure including the insulin and the aptamer connected to the surface. The utilization of the redox probe, methylene blue (MB), involves its intercalation into the guanine bases of aptamers. Additionally, the sandwich structure formed by the AuNPs-Apt/insulin/surface-bound aptamer served to enhance the electrochemical signal generated by MBs. The signal had a strong correlation with the amounts of insulin. The researchers discovered that the LSGE-based sensors exhibited a limit of detection of 22.7 fM, while the GCE-based sensors, which were utilized for comparison and initial sensor development, demonstrated a limit of detection of 9.8 fM. The findings illustrated the successful production of single-use and very sensitive sensors for insulin detection using LSGEs [171].

Another novel electrochemical immunoassay method for the quantification of insulin utilized a sandwich-type assay format. The experimental setup involved the utilization of a glassy carbon electrode that underwent modification with the addition of MoS2 nanosheets adorned with gold nanoparticles (AuNPs). This modification facilitated the immobilization of a substantial quantity of the primary antibody (Ab1). After being exposed to insulin, a secondary antibody (Ab2) that was cross-linked to a DNA initiator strand (T0) was introduced to form an Ab2@T0 conjugate. This conjugate was then used to undertake a sandwich immunoreaction. Following this, the formation of the long double-stranded DNA concatemer occurred with a hybridization chain reaction involving the interaction between Ab2@T0 and auxiliary probes (H1, H2). The electrochemical probe ruthenium (II) hexaammine was successfully incorporated into the double-stranded hairpin DNA amplification products with the electrostatic contact between the negatively charged DNA phosphate backbones and the positively charged probe. The electrochemical reaction, which is most accurately assessed at a potential of approximately −0.21 V (against Ag/AgCl), exhibited a wide range of variability spanning from 0.1 pM to 1 nM for insulin concentration. Furthermore, the lower limit of detection was exceptionally low, measuring at 50 fM. The assay demonstrated satisfactory levels of specificity, reproducibility, and stability. In our observation, it serves as a promising novel instrument for assessing this significant clinical parameter [172].

The preparation of composites consisting of carbon nanotubes (CNTs), nickel–cobalt oxide (CNT–nickel–cobalt oxide), and Nafion for electrode modification involved the direct combination of nickel–cobalt oxide powder and CNTs with a Nafion solution, which served as a binder. These composites were then deposited onto screen-printed electrodes. The electrochemical characteristics were examined using CV and amperometry methodologies in both NaOH and phosphate buffer solutions. The sensors were assessed for their electrocatalytic activity in relation to their sensitivity, detection limit, and stability in detecting insulin. The utilization of CNT–nickel–cobalt oxide/Nafion electrodes allowed for identifying the presence of insulin in aqueous solutions under physiological pH conditions. This detection method exhibited a favorable sensitivity of 22.57 A $\times$ mg^{-1} $\times$ mL^{-1}. It was demonstrated that an electrochemical activation process is required in an alkaline solution in order to obtain a threefold increase in sensitivity for the detection of insulin using a CNT–nickel–cobalt oxide composite electrode [173].

The utilization of CV facilitated the production of cobalt hydroxide nanoparticles onto a carbon ceramic electrode (CHN/CCE). The modified electrode underwent characterization with X-ray diffraction and scanning electron microscopy techniques. The experimental findings indicated that a single-layer structure of CHN was evenly coated on the surface of CCE. The electrocatalytic efficiency of the modified electrode in relation to the oxidation process of insulin was investigated using cyclic voltammetry (CV). The CHN/CCE technique was used in a self-made flow injection analysis device to quantify insulin levels. The study determined that the limit of detection, with a signal-to-noise ratio of 3, was 0.11 nM. Additionally, the sensitivity was measured to be 11.8 nA nM^{-1}. Furthermore, the sensor was used for the purpose of detecting insulin in samples of human serum. The sensor exhibited desirable characteristics including exceptional stability, repeatability, and notable selectivity [174].

Insulin assumes a pivotal function in the regulation of physiological glycolytic metabolism and is intricately linked to the pathogenesis of diabetes and its associated disorders. A novel dual-signaling electrochemical aptasensor that exhibits remarkable sensitivity and specificity for the detection of insulin utilized an insulin-binding aptamer (IBA) modified with methylene blue (MB) as a "signal-off" probe. Additionally, gold nanoparticles (GNPs) modified with a combination of DNA2, and ferrocene (Fc) were used as a "signal-on" probe. These probes were integrated with linker mDNA to create a modified Au electrode, which served as the sensing interface. The current reactions of MB and Fc were then measured as indicators of the signal. As anticipated, the incubation of insulin with the DNA2Fc@GNPs/mDNA/MB-IBA/Au electrode led to a decrease in the current responses of MB and an increase in the current responses of Fc. The successful detection of insulin was accomplished using this technique. The linear range for detection spanned from 10 pM to 10 nM, with a minimum detectable concentration of 0.1 pM. The aptasensor under consideration demonstrated a high level of specificity and accuracy in the measurement of insulin levels in serum samples. In addition, the utilization of dual signaling proved advantageous in enhancing the precision and consistency of detection by virtue of its inherent self-referencing capabilities. The aptasensor exhibited several advantageous characteristics, including simplicity, short response time, high sensitivity, specificity, and accuracy. These attributes are of great importance in enhancing the detection of disorders associated with insulin [175]. (See Table 7).

Table 7. Summary of electrochemical studies using nanostructure-modified electrodes to detect insulin with the linear range and limit of detection.

Hormone	Electrochemical Method	Type of Nanostructures	Linear Range	LOD	Ref.
Insulin	CV, DPV, CA	Si-CPE	90–1400 pM.	36 pM	[30]
Insulin	CV, EIS, CA	NiNPs/ITO	100 pM to 2400 pM, and 1 nM to 125 nM	10 pM	[167]
Insulin	DPV, EIS, DPV	MSTF/GCE	10.0 to 350.0 nM.	3.0 nM	[168]
Insulin	EIS, Amperometry	Carboxylated graphene/paper electrode	5–500 pM	1.5 pM	[169]
Insulin	CV, DPV, CA	Ppy-GF/PGE/GCE	0.225–1.235 μM	8.65 nM	[170]
Insulin	EIS, DPV	AuNPs-Apt/insulin/AuNPs-Apt/LSGE, AuNPs-Apt/insulin/AuNPs-Apt/GCE	0.1 pM to 1 μM	22.7 fM for LSGE-based sensor 9.8 fM for GCE-based sensor	[171]
Insulin	CV, amperometry	CHN/CCE	0.5–15-nM	0.11 nM	[174]
Insulin	CV, EIS, CA, Hydrodynamic amperometry	Carbon quantum dots (CQDs)	40–200 nM	2.24 nM	[90]
Insulin	DPV, EIS, CV	AuNP@MoS_2/GCE	0.1 pM–1 nM	50 fmol L^{-1}	[172]
Insulin	CV	CNT-$NiCoO_2$/Nafion electrodes	0.1–31.5 μg/mL	0.22 μg/mL	[173]
Insulin	CV, CA	NiONPs/chitosan-MWCNTs/SPCE	0.25–5 μM	94 nM	[176]
Insulin	CV, DPV	Cu-BTC/ZIF-L/DGE	0.1 pM to 5 μM	0.027 pM	[166]
Insulin	EIS, CV, SWV	DNA2Fc@GNPs/mDNA/MB-IBA/Au	10 pM to 10 nM	0.1 pM	[175]

5.7. Application in Oxytocin Detection

Oxytocin is a multifaceted peptide hormone that has wide-ranging ramifications for various aspects of general health, adaptability, development, reproduction, and social behavior. The presence of endogenous oxytocin and the activation of the oxytocin receptor contribute to the facilitation of development, resilience, and healing processes and can serve as a stress-regulating compound, an agent that reduces inflammation, and a substance that counteracts oxidative stress. These properties make it particularly beneficial in situations of adversity or trauma, where it exhibits protective effects and exerts an impact on both the autonomic nervous system and the immunological system. The characteristics of oxytocin have the potential to elucidate the advantages of favorable social interactions, thereby garnering interest in utilizing this compound as a potential remedy for many illnesses. Nevertheless, as outlined in this document, the distinctive chemical characteristics of oxytocin, such as the presence of active disulfide bonds and its ability to undergo chemical transformations and interact with other molecules, provide challenges in terms of manipulation and quantification of this compound [177]. Oxytocin, a nonapeptide, functions as a neuromodulator inside the central nervous system of humans. Oxytocin (OT) is a nonapeptide that is generated in the paraventricular nucleus and the supraoptic nucleus of the hypothalamus and then released by the pituitary gland. Research conducted on animals has demonstrated that oxytocin (OT) has a significant role in hormonal regulation of uterine contractions during childbirth and lactation in nursing females. Additionally, it has been found to have a crucial impact on social behavior and affiliation [178].

Asai et al. investigated the electrochemical detection of oxytocin utilizing boron-doped diamond (BDD) electrodes. The CV analysis of oxytocin in a phosphate buffer solution revealed the presence of an oxidation peak at +0.7 V (against Ag/AgCl). This peak was attributed to the oxidation of the phenolic group inside the tyrosyl moiety. Moreover, the current peaks observed in flow injection analysis (FIA) utilizing boron-doped diamond (BDD) microelectrodes exhibited a high degree of linearity over the concentration range of oxytocin, spanning from 0.1 to 10.0 μM. The detection limit for oxytocin was determined to be 50 nM, with a signal-to-noise ratio (S/N) of 3. The correlation coefficient (R2) for this linear relationship was found to be 0.995. The voltammograms of oxytocin and vasopressin were found to exhibit similarities when observed using both an as-deposited BDD electrode and a catholically reduced BDD electrode. However, a notable differentiation was observed when anodically oxidized BDD electrodes were used. This distinction can be attributed to the attractive interaction occurring between vasopressin and the oxidized BDD surface. With the utilization of this differentiation, the potential for selective assessments of oxytocin in situ or in vivo was proven by using chronoamperometry in conjunction with flow injection analysis at an optimal potential [29].

The utilization of bioelectrodes characterized by a reduced carbon footprint is a promising and inventive approach to address the escalating issue of electronic waste. Biodegradable polymers present environmentally friendly and renewable alternatives to artificial goods. A membrane composed of chitosan–carbon nanofibers (CNFs) was fabricated and modified for the purpose of electrochemical sensing. The membrane's surface was analyzed, revealing a crystalline structure with a consistent distribution of particles. The surface area was measured to be 25.52 m^2/g, while the pore volume was determined to be 0.0233 cm^3/g. The membrane was subjected to functionalization to create a bioelectrode that may be used for the purpose of detecting exogenous oxytocin in milk. The technique of electrochemical impedance spectroscopy was utilized to quantify the quantity of oxytocin within a linear range spanning from 10 to 10^5 ng/mL. The bioelectrode that was created demonstrated a limit of detection (LOD) of 24.98 $\pm$ 11.37 pg/mL and a sensitivity of 2.77×10^{-10} Ω/log ng mL^{-1}/mm^2 for the detection of oxytocin in milk samples. The recovery rate for oxytocin in these samples ranged from 90.85% to 113.34%. The utilization of the chitosan-CNF membrane demonstrates its ecological compatibility and presents novel opportunities for the development of environmentally sustainable disposable materials in the field of sensing applications [179].

The electrochemical properties of oxytocin were investigated utilizing DPV approach in conjunction with an electrochemical sensor. The sensor used a glassy carbon electrode (GCE) that was modified using multi-walled carbon nanotubes and polyfurfurylamine. The electrochemical properties of oxytocin were examined, focusing on the reduction peak of its cysteine component. Cysteine exhibited a distinct decrease peak at a potential of −0.4 V relative to Ag/AgCl saturated with the KCl reference electrode. The experimental conditions were optimized, and the impact of pH was investigated. Additionally, calibration curves were generated. Plotting current against concentrations resulted in linear relationships with R-squared values of 0.9938, 0.9835, 0.9924, and 0.9791 for the GCE-, GCE/PFA-, and GCE/MWCNTs/PFA-modified electrodes, respectively. The GCE/MWCNTs/PFA electrode exhibited approximately three times the current compared with the GCE electrode when exposed to a specific dose of oxytocin [180].

For several decades, the application of fast-scan cyclic voltammetry (FSCV) has involved the utilization of carbon fiber microelectrodes (CFMEs) to identify neurotransmitters and other biomolecules. This methodology quantifies neurotransmitters, such as dopamine and, more recently, physiologically significant neuropeptides. Oxytocin, a multifunctional peptide hormone, plays a significant role in various physiological processes such as adaptation, development, reproduction, and social behavior. The neuropeptide has multiple roles, including its role as a stress-coping agent, an anti-inflammatory substance, and an antioxidant with efficacy in mitigating the consequences of adversity or trauma. In one study, the measurement of tyrosine was conducted utilizing the modified sawhorse waveform (MSW) technique, which allowed for improved electrode sensitivity in detecting the presence of the amino acid and oxytocin peptide. The utilization of the MSW technique resulted in the reduction of surface fouling and facilitated the simultaneous detection of several monoamines. The detection of oxytocin was facilitated by the presence of tyrosine in its molecular structure, therefore prompting the utilization of mass spectrometry as a method for detection. The sensitivity of oxytocin detection was determined to be 3.99 ± 0.49 nA μM^{-1}, with a sample size (n) of 5. Furthermore, the utilization of the MSW technique on CFMEs enabled the monitoring of exogenously administered oxytocin in real time on the brain slices of rats. These investigations demonstrated the potential of innovative methods for detecting oxytocin in a rapid, sub-second timeframe. This could have important ramifications for measuring oxytocin levels in living organisms and advancing our knowledge of the physiological functions of oxytocin [181].

There is a significant need for an analytical technique that is dependable, economical, and prompt in its response for the analysis of oxytocin. In one study, oxytocin was electrochemically detected using a newly developed electrode modified with nitrogen, phosphorus co-doped coke-derived graphene (NPG). The electrooxidation behavior of oxytocin was examined using a square wave stripping voltammetry (SWSV) technique on an electrode modified with NPG. The investigation was conducted in a 0.1 M phosphate buffer solution with a pH of 7. The oxidation peak current exhibited a linear relationship within two distinct ranges, specifically ranging from 0.1 nM to 10 nM and 15 nM to 95 nM. The calculated limit of detection for the NPG electrode was determined to be 40 pM. The study investigated the practical implementation of the created sensor for the quantification of oxytocin in both consumable items and bodily fluids. This exploration highlights the potential for real-time monitoring and detection of oxytocin misuse [182]. (See Table 8).

Table 8. Summary of electrochemical studies using nanostructure-modified electrodes to detect oxytocin with linear range and limit of detection

Hormone	Electrochemical Method	Type of Nanostructures	Linear Range	LOD	Reference
Oxytocin	SWSV, CV	PG, NG, NPG-modified electrode	0.1 nM to 10 nM and 15 nM to 95 nM	40 pM	[182]
Oxytocin	FSCV, CV, MSW	CFMEs	0.5 μM to 10 μM	-	[181]
Oxytocin	DPV, CV	GCE/MWCNTs/PFA	$(15.9–79.6) \times 10^{-10}$ M and $(95.4–298.3) \times 10^{-10}$ M	-	[180]
Oxytocin	EIS	Chitosan-CNF	10 to 10^5 ng/mL	24.98 ± 11.37 pg/mL	[179]
Oxytocin	CV, CA	BDD microelectrode	0.1 to 10.0 M	50 nM	[29]

5.8. Applications in Thyroxine Detection

The thyroid gland produces thyroid hormones (THs), which regulate growth, energy generation, and homeostasis. Since most cellular activities are controlled by THs, thyroid gland dysfunction can cause or affect many medical disorders. Heart rate, arrhythmias, and cardiac output are raised in hyperthyroidism and lowered in hypothyroidism. Thyroid abnormalities alter immune system functions and are linked to cancer proliferation, apoptosis, and invasiveness [183]. Hyperthyroidism or hypothyroidism can occur when there are abnormally high or low levels of thyroid hormones [184].

Clinicians greatly benefit from measuring TSH and T4 together to identify hyper- and hypothyroidism. Karami et al. introduced a sandwich-type electrochemical immunoassay using Janus and magnetic nanoparticles for one-pot T4 and TSH detection. Janus cadmium

(CdO) and zinc oxide (ZnO) NPs coated with T4/TSH-specific molecularly imprinted polymers (MIPT4-CdO and MIPTSH-ZnO) formed the signaling probe. Coating magnetic Fe_3O_4 NPs with 1,3-Bis (3-carboxy propyl) tetramethyl disiloxane, activating with N-hydroxy succinimide (NHS) and 1-ethyl-3-(3-dimethyl aminopropyl) carbodiimide (EDC), and conjugating with T4/TSH-specific antibodies were used to produce the capture probe. MIPT4-CdO and MIPTSH-ZnO were added to the sample solutions, and the capture probes (Fe_3O_4-AbTSH and Fe_3O_4-AbT4) were added after incubation to assess T4 and TSH. An external magnetic field was used to separate the sandwiched nanosystem, then a dilute solution of nitric acid (HNO_3) dissolved CdO and ZnO NPs and freed Cd (II) and Zn (II) cations. On screen-printed electrodes (SPEs) were modified with multi-walled carbon nanotubes (MWCNTs), and a constant-current potentiometric stripping analysis (cc-PSA) was used assess the concentration of these cations. The findings for Cd (II) and Zn (II) correlated with T4 and TSH levels. The limits of detection (LOD) for the T4 and TSH analyses were 0.02 $ng \cdot dL^{-1}$ and 0.0002 $\mu U \cdot mL^{-1}$, respectively, with linear ranges of 0.05–50 $ng \cdot dL^{-1}$ and 0.001–100 $\mu U \cdot mL^{-1}$. The nanosystem's key benefits are great sensitiveness, specificity, and stability in detecting T4 and TSH in clinical samples [185].

In another study, an electrochemical aptasensor (MEA) based on a $MoS_2@Ti_3C_2Tx$ MXene hybrid was introduced for the sensitive and quick quantification of thyroxine (T4), which is an essential hormone that assumes a pivotal role in regulating multiple physiological processes within the human body. Consequently, there exists a significant need for a precise, responsive, and expeditious technique to identify T4. To fabricate the aptasensor, a nano-hybrid (NH) composed of Ti_3C_2Tx MXene and MoS_2 nanosheets (NSs) was produced and then deposited onto a carbon electrode surface. This was followed by the electroplating of gold nanostructures (GNs). The intelligent integration of Ti_3C_2Tx MXene and MoS2NS resulted in the augmentation of the physiochemical characteristics of the electrode surface while also serving as a fundamental component for the formation of three-dimensional graphene nanocomposites (3D GNs). The three-dimensional structure of the graphene nanosheets (GNs) provided a distinctive surface on which several T4 aptamer molecules could be immobilized, resulting in a significant enhancement in the signal by approximately six times. The quantification of thyroxine was performed using the MEA method, which demonstrated a limit of detection (LOD) of 0.39 pg/mL. This method exhibited a dynamic range spanning from 7.8×10^1 pg/mL to 7.8×10^6 pg/mL, and the quantification process was completed within a time frame of 10 min. Additionally, the MEA demonstrated the effective detection of T4 in samples of human serum. Finally, the outcomes derived from the aptasensor were juxtaposed with those obtained using the conventional ELISA standard technique. The comparative analysis revealed a high level of compatibility between the two methodologies [186].

Thyroxine (T4) is a prominent thyroid hormone that is generated within the thyroid gland and consists of four iodine atoms. Dysregulation of T4 levels inside the body is implicated in the pathogenesis of many endocrine disorders. In one study, the development and purpose of an electrochemical biosensor consisting of a multi-functional DNA structure/rhodium nanoplate heterolayer was to accurately detect the quantity of T4. The study involved the construction of a DNA three-way junction (3WJ) structure, which served as a versatile bioprobe capable of performing multiple activities concurrently. These functions encompassed target recognition, electrochemical signal reporting, and immobilization. The confirmation of the binding between T4 and the T4 DNA aptamer was achieved with the utilization of enzyme-linked aptamer assays (ELAAs) and filtration procedures. The immobilization of multi-functional DNA was achieved by attaching it to porous rhodium nanoplates (pRhNPs) that were modified with a heterolayer on an electrode made of gold micro-gaps. The utilization of pRhNPs resulted in an increase in the surface area and an enhancement in the electrochemical signal. The techniques used for the detection of T4 involved the utilization of CV and EIS. Under ideal circumstances, the minimum amount of T4 that could be detected was determined to be 10.33 pM. In addition, clinical samples exhibited the detection of T4 levels up to 11.41 pM. The cited study presents evidence for the potential of label-free detection of T4 using a multi-functional DNA/pRhNPs heterolayer. This detection method shows promise for future applications in small molecule detection platforms [187].

Another study developed a novel electrochemical immunosensor capable of detecting thyroxine at extremely low concentrations. The sensor utilized a cascade catalysis mechanism involving cytochrome c (Cyt c) and glucose oxidase (GO_X) to boost the signal amplification. Additionally, the sensor was designed to be reusable, allowing for multiple measurements. It is noteworthy to mention that a considerable number of Cyt c and GO_X enzymes were initially immobilized onto the double-stranded DNA polymers using a hybridization chain reaction (HCR). Subsequently, the amplified responses were attained with the sequential catalytic actions of Cyt c and GO_X, facilitated by glucose recycling. In addition, a multi-functionalized magnetic graphene sphere was synthesized and used as a signal tag. This material had favorable mechanical properties, a substantial surface area, and an exceptional electron transfer rate characteristic of graphene. Furthermore, it exhibited remarkable redox activity and acceptable magnetic properties. The suggested cascade catalysis amplification technique has the potential to significantly boost the sensitivity for detecting thyroxine with a sandwich-type immunoreaction. The immunosensor exhibited a broad linear range spanning from 0.05 picograms per milliliter (pg mL^{-1}) to 5 nanograms per milliliter (ng mL^{-1}) and a low detection limit as low as 15 femtograms per milliliter (fg mL^{-1}) under ideal circumstances. Significantly, the suggested technology holds potential for conducting repeatable and inexpensive analysis of biological material [188].

A sensor based on a molecularly imprinted polymer (MIP) was created for the purpose of detecting thyroxine (T4), which is a crucial biomarker. The achievement described involves the incorporation of T4 into a polyaniline matrix, which is then supported on indium tin oxide-coated glass electrodes. The process of synthesizing polyaniline involved oxidative polymerization, which was then refined using response surface technology. The optimal parameters for the extraction of T4 from a polyaniline matrix using ultrasonication were found to be a duration of 15 min at a temperature of 30 °C, using a 75 mM NaOH solution. The observed imprinting factor was determined to be 1.98. The MIP-based sensor was subjected to characterization using various analytical techniques, including chromatography, spectroscopy, and electrochemistry. The calibration of the sensor was achieved by measuring T4 concentrations ranging from 5 to 50 pg/mL. The predicted limit of detection for the sensor was determined to be 6.16 pg/mL. The repeatability studies demonstrated a relative standard deviation of 2.45%. The study observed a range of 96 to 115.2% for the recovery rate of saliva. The electroanalytical sensor that was built using MIP-based technology demonstrated a notable level of selectivity against a range of interfering substances. This characteristic makes it highly promising for potential applications in point-of-care settings in the future [189].

An electrochemical sensor that utilizes a nanocomposite for the purpose of accurately detecting thyroxine (T4) was developed. In the study, hydrodynamic amperometry was conducted using a nanocomposite electrode that consisted of a dispersion of a hybrid-nanomaterial filler, based on graphene, within an insulating epoxy resin. This composition ratio is ideal, as it is close to the percolation composition. The hybrid-nanomaterial under consideration was composed of reduced graphene oxide that was modified with gold nanoparticles, along with the inclusion of a biorecognition agent known as thiolated β-cyclodextrin. The identification of T4 was achieved with the utilization of supramolecular chemistry because of the establishment of an integration complex between β-cyclodextrin and T4. The amperometric device functioned at a potential of +0.85 V compared with the Ag/AgCl reference electrode, facilitating the oxidation of T4 on the surface of the electrode. The sensor could detect T4 concentrations ranging from 1.00 nM to 14 nM in a 0.1 M HCl solution. It had a detection limit of 1.00 ± 0.02 nM. The sensor's settings can be readily reset with the process of polishing. This particular electrochemical electrode demonstrated the most minimal detection limit compared with previously documented methods for determining T4 [190]. (See Table 9).

Table 9. Summary of electrochemical studies using nanostructure-modified electrodes to detect thyroxin with the linear range and limit of detection.

Hormone	Electrochemical Method	Type of Nanostructures	Linear Range	LOD	Reference
Thyroxin	DPV	MoS_2@Ti_3C_2Tx/SPCE	7.8×10^{-1}– 7.8×10^{6} pg/mL	0.39 pg/mL	[186]
Thyroxin	EIS, CV	DNA3WJ/pRhNPs/Au	100 nM–1 pM	10.33 pM	[187]
Thyroxin	EIS, CV, DPV	S1-SA-Ab2-MFMGRS/GCE	0.05 pg/mL–5 ng/mL	15 fg/msL	[188]
Thyroxin	CV	β–CD-SH/Au–NPs@rGO hybrid nanomaterial	1.00 nM to 14 nM	1.00 ± 0.02 nM	[190]
Thyroxin	CV, DPV	Cu-MOF@PANI composite	10–10^5 pM	0.33 pM–0.17 pM	[44]
Thyroxin	CV	MIP/ITO	5–50 pg/mL	7.96 pM (6.16 pg/mL)	[189]

6. Conclusions and Future Perspectives

The development of electrochemical biosensors for hormone detection using nanostructure-modified electrodes continues at a rapid pace. These new sensor designs should prove useful for creating improved medical and clinical sensors in the future. Hormones like progesterone, estradiol, testosterone, cortisol, VD, and prostaglandin play an important role in maintaining human life, and electrochemical sensors are easy to apply, so research into developing electrochemical sensors for hormone detection continues to gain interest. The field of steroid hormone electrochemical sensors is further along in terms of development and maturity than that of electrochemical sensors for fatty acid derivative hormones. The only recognition elements used in existing electrochemical biosensors for fatty acid derivative hormones are antibodies. Electrochemical sensors have great future promise for research and practical applications in lipid hormone detection, and this is an area in need of future research. New functional nanomaterials and analytical technologies present a promising window of interest for advancing electrochemical sensor and biosensor platform development. Researchers should keep looking into areas like new electrode materials with increased selectivity and sensitivity, smaller, more wearable sensors, and sensors with immediate usage [191].

This review article provides a brief overview of the electrochemical sensors that have been shown to be effective in detecting both biomolecules (such as DNA, enzymes, and hormones) as well as cellular complexes. According to Suhito et al., electrochemical techniques can be used to assess a wide variety of analytes quickly, accurately, and non-destructively [192]. Nanomaterials (such as metal nanoparticles, graphene, and graphene derivatives) and functional peptides (also known as aptamers) have been utilized to boost sensitivity. Results reported using carbon nanomaterials such as graphene show some of the lowest detection limits. Improved combinations of carbon nanomaterials and aptamers in composite electrodes seems to be a promising avenue for future study. During electrochemical measurement, a measurable read-out signal is produced because of interactions among targets and designated probes or composites. Considering this, strategies are being examined to create electrochemical tools that can quickly and accurately sense hormones in live cells for use in assessment and drug development. Electrochemical devices have been shown to have the potential for monitoring highly proliferative cells, such as cancer cell viability, stem cell potential, and differentiation states based on redox behaviors. The development of a cell-friendly method for the analysis of high-precision healthcare diagnosis and point-of-care testing may be facilitated by future advances in biosensing technologies. The use of electrochemical sensors is a recent breakthrough in the field of biology. Small biomolecules (such as DNA and proteins, enzymes, and hormones) all the way up to the cellular level (equating to cell viability) can be detected. Future industries will benefit from its sensitivity, selectivity, and processing speed. Therefore, advanced large-scale systems for disease detection and quality assurance of stem cell-based goods may include quick, non-destructive, and adaptable electrochemical sensors [192].

The examples presented in this review are merely a preview of what is to come in this exciting field. The growth and success of this field will depend on the abilities of researchers to keep exploring and working together to tackle the current hurdles and possibilities in electrochemical biosensing in areas such as salivary evaluation. Electrochemical biosensing platforms for saliva-based diagnosis, even in the form of an authentic simplified device for on-site medical diagnostics, are expected to emerge in the very near future as a result of the rapidly developing fields of proteomics, bioengineering, and electroanalytical chemistry [193].

According to Joshi et al. [194], hormones can be detected quickly, accurately, and sensitively using electrochemical sensing techniques developed with the use of nanomaterial technology. The sensitivity of the sensor can be greatly increased with different approaches (such as chemical functionalization, hybrid composite fabrication, and modulation of surface architecture of these nanomaterials), as confirmed with a thorough comparison of various nanostructured materials. However, more work needs to be completed to overcome the major obstacles in this area of study, such as regulated synthesis (in terms of size, shape, and surface area); developing new functional nanometric interfaces, analyzing the links between structure, composition, and reactivity of nanomaterials; and biocompatibility of nanomaterials with the biorecognition molecules and traverse influence with existing technologies. It is therefore anticipated that nanomaterial-based electrochemical sensing technologies will provide incredibly accurate on-site evaluations of hormone in both complicated biological settings and in the real world. Technical difficulties with these cutting-edge sensing technologies can be overcome by combining multifunctional nanomaterials, recognition elements, and electrochemical approaches. In addition, the efficient networking and incorporation of electrochemical sensors with communication tools (such smartphones and tablets) may pave the way to the development of the desired lightweight gadgets. The assumption is that the current state of hormone diagnostics might be vastly improved with the successful merging of two independent areas of research [194].

According to research conducted by Mathew et al., the future development of wearable sensors should prioritize the following areas: 1. Diversifying sensing devices beyond electrolytes and metabolites to encompass a broad range of biomarkers. This can be achieved with the implementation of noninvasive immunoassays that enable simultaneous tracking of the wearers' health at the molecular level. 2. Advancing microfluidic platforms to effectively monitor biomolecule concentrations that are extremely low. This will enhance the sensitivity and accuracy of the sensors in detecting and measuring these biomolecules. 3. Ensuring the safety and reliability of biosensing diagnostic platforms by incorporating valuable information obtained from population studies and establishing correlations with blood-based clinical assays. This will enhance the diagnostic capabilities of the sensors and enable more accurate health monitoring. 4. Developing appropriate surface coatings to prevent biofouling, which refers to the accumulation of biological substances on a sensor's surface. These coatings will help maintain the functionality and accuracy of sensors over extended periods of use [61].

The utilization of carbon nanotubes (CNTs) is facilitated by their distinct electronic characteristics, as well as the exceptional electrochemical reactivity exhibited by their edge-plane defects. These properties, in conjunction with the inherent benefits of affinity and enzymatic recognition, direct charge transfer, and catalysis has enabled the creation of numerous (bio)sensing methodologies. These approaches have proven to be highly effective in accurately and selectively quantifying biomarkers. Despite the evident efficacy of electrochemical (bio)sensors using carbon nanotubes (CNTs) for detecting biomarkers, there are certain demands within the fields of medical research and clinical chemistry in the 21st century that warrant attention. One of the most pressing difficulties in the field is the creation of compact, non-intrusive, rapid, straightforward, and effective devices that may be utilized for point-of-care (POC) assessments [195].

Two-dimensional-based electrochemical sensors provide convenience, speed, simplicity, cheap cost, and quick time consumption in substance detection, biological science, ecological surveillance, pharmaceutical businesses, pesticide residues, heavy metal ions, and small biological molecule sensing. Electrochemical sensor applications using 2D nanomaterials still have issues. Some nanomaterials are difficult to prepare, have limited repeatability, and have a short service life, which affects test reliability. They have weak binding strength, and this impacts sensor test stability since many nano-sensing materials and electrodes are physically bound [196]. Nanoparticles improve several benefits of immunochemical biosensors, including small dimensions, flexibility, appropriate testing costs, simplified usage, easy to comprehend results, excellent selectivity and sensitivity, and increased surface area by improving specific steps in sensing and immunorecognition. The development of repeatable devices is hindered by several factors, including variation at the nanoscale, stability of nanomaterials, significant variation in immunosensors due to undefined properties, and immobilizing issues [197].

Hormone detection is crucial for early diagnosis and treatment of various health issues, including fertility, chronic illnesses, mental health, and weight management. It allows for customized treatment strategies, improved therapeutic efficacy, and reduced negative effects. Hormone monitoring aids fertility and reproductive health assessment, enabling couples to conceive and manage prescription doses and hormone levels. Early identification can reduce depression and anxiety symptoms, improve well-being, and guide hormone replacement treatment. Hormone monitoring during menopause can guide weight-management strategies and reduce symptoms. Early identification can also aid in birth control and assisted reproductive technology choices. Having quick hormone detection procedures can alleviate stress and worry about hormonal health. In conclusion, accessible hormone detection helps people make health choices early, leading to improved health, lower healthcare costs, and a higher quality of life.

Electrochemical hormone detection utilizing nanostructured electrodes promises improved sensitivity, selectivity, and practicality. One approach is to build new nanostructures with optimized characteristics to increase electrochemical performance. To improve hormone sensing, researchers are investigating MOFs, two-dimensional structures, and hybrid nanomaterials. Multimodal nanomaterial incorporation is another promising area. Constructing nanostructures with catalytic, electrical, and optical functions might improve sensor performance and synergy. Novel sensing technologies and increased analytical capabilities for hormone detection at lower doses may result from this approach. Researchers are also adding smart materials to nanostructured electrodes. These materials react dynamically to physiological alterations, thus improving sensor performance. In vivo sensors require this versatility to work in complicated biological contexts and monitor live creatures in real-time. Nanostructured electrodes depend on nanofabrication advances. Nanostructure dimensions, form, and makeup must be precisely controlled for manufacturing repeatability and adaptability. As nanofabrication processes improve, nanostructured electrodes become easier to make, making electrochemical hormone sensors increasingly popular.

Miniaturizing sensors and creating wearable devices for real-time hormone monitoring are crucial. Nanostructured electrodes in wearable platforms provide discrete and personalized health monitoring. These technologies may help people understand their hormone patterns and identify health risks early, thus revolutionizing healthcare. The intersection between electrochemical detection and machine learning is growing. The coupling of machine learning techniques with electrochemical data should improve hormone detection accuracy and reliability. Machine learning can evaluate complicated information, find subtle trends, and improve electrochemical sensor intelligence and adaptability. Adapting nanostructured electrodes for portable, user-friendly sensors is important in environmental and point-of-care applications. Researchers are developing sensors for speedy and consistent hormone detection in different contexts outside of labs. This affects environmental monitoring and decentralized healthcare. Exploring novel recognition aspects is another focus. Peptide-based compounds and molecularly imprinted polymers are being studied to improve nanostructured electrode selectivity. These novel identification components strengthen hormone sensors for precise and specific detection in complicated biological or environmental samples. Electrochemical hormone detection utilizing nanostructured electrodes will require advances in sensor integration, nanofabrication, materials science, and machine learning. Hormone detection technologies that are more efficient, versatile, and accessible might revolutionize medical care, environmental surveillance, and the field of analytical chemistry.

Author Contributions: Conceptualization, N.H. and K.J.S.; writing—original draft preparation, N.H.; writing—review and editing, N.H. and K.J.S. All authors have read and agreed to the published version of the manuscript.

Funding: This research received no funding.

Conflicts of Interest: The authors declare no conflict of interest.

References

1. Sarlis, N.; Gourgiotis, L. Hormonal effects on drug metabolism through the CYP system: Perspectives on their potential significance in the era of pharmacogenomics. *Curr. Drug Targets-Immune Endocr. Metab. Disord.* **2005**, *5*, 439–448. [CrossRef] [PubMed]
2. Chen, W.-C.; Zouboulis, C.C. Hormones and the pilosebaceous unit. *Derm.-Endocrinol.* **2009**, *1*, 81–86. [CrossRef] [PubMed]
3. Kelkar, N.; Prabhu, A.; Prabhu, A.; Nandagopal, M.G.; Mani, N.K. Sensing of body fluid hormones using paper-based analytical devices. *Microchem. J.* **2022**, *174*, 107069. [CrossRef]
4. Cifrić, S.; Nuhić, J.; Osmanović, D.; Kišija, E. Review of Electrochemical Biosensors for Hormone Detection. In *Cmbebih 2019*; IFMBE Proceedings: Banja Luka, Bosnia and Herzegovina, 2020; pp. 173–177.
5. Chen, W.Y. Exogenous and endogenous hormones and breast cancer. *Best Pract. Res. Clin. Endocrinol. Metab.* **2008**, *22*, 573–585. [CrossRef] [PubMed]
6. Khelifa, L.; Hu, Y.; Jiang, N.; Yetisen, A.K. Lateral flow assays for hormone detection. *Lab A Chip* **2022**, *22*, 2451–2475. [CrossRef] [PubMed]
7. Hamid, A.A.; Issa, M.B.; Nizar, N.N.A. Hormones. In *Preparation and Processing of Religious and Cultural Foods*; Elsevier: Amsterdam, The Netherlands, 2018; pp. 253–277.
8. Sanchez-Almirola, J.; Gage, A.; Lopez, R.; Yapell, D.; Mujawar, M.; Kamat, V.; Kaushik, A. Label and bio-active free electrochemical detection of testosterone hormone using MIP-based sensing platform. *Mater. Sci. Eng. B* **2023**, *296*, 116670. [CrossRef]
9. Bahadir, E.B.; Sezginturk, M.K. Electrochemical biosensors for hormone analyses. *Biosens. Bioelectron.* **2015**, *68*, 62–71. [CrossRef]
10. Raval, J.B.; Mehta, V.N.; Jha, S.; Singhal, R.K.; Basu, H.; Kailasa, S.K. Functional nanostructures in analytical chemistry: New insights into the optical and electrochemical sensing of animal hormones in food, environmental and biological samples. *Sens. Diagn.* **2023**, *2*, 815–836. [CrossRef]
11. Høj, P.H.; Møller-Sørensen, J.; Wissing, A.L.; Alatraktchi, F.A.A. Electrochemical biosensors for monitoring of selected pregnancy hormones during the first trimester: A systematic review. *Talanta* **2023**, *258*, 124396. [CrossRef]
12. Naqvi, S.M.Z.A.; Zhang, Y.; Tahir, M.N.; Ullah, Z.; Ahmed, S.; Wu, J.; Raghavan, V.; Abdulraheem, M.I.; Ping, J.; Hu, X. Advanced strategies of the in-vivo plant hormone detection. *TrAC Trends Anal. Chem.* **2023**, *166*, 117186. [CrossRef]
13. Duarte, L.C.; Baldo, T.A.; Silva-Neto, H.A.; Figueredo, F.; Janegitz, B.C.; Coltro, W.K. 3D printing of compact electrochemical cell for sequential analysis of steroid hormones. *Sens. Actuators B Chem.* **2022**, *364*, 131850. [CrossRef]
14. Ghoshal, K. Recent advances in biosensing technologies for detecting hormones. In *Advanced Sensor Technology*; Elsevier: Amsterdam, The Netherlands, 2023; pp. 261–295, ISBN 9780323902229. [CrossRef]
15. Qin, Q.; Feng, D.; Hu, C.; Wang, B.; Chang, M.; Liu, X.; Yin, P.; Shi, X.; Xu, G. Parallel derivatization strategy coupled with liquid chromatography-mass spectrometry for broad coverage of steroid hormones. *J. Chromatogr. A* **2020**, *1614*, 460709. [CrossRef]
16. El-Ansary, A.; Faddah, L.M. Nanoparticles as biochemical sensors. *Nanotechnol. Sci. Appl.* **2010**, *3*, 65–76. [CrossRef]
17. Hutchinson, J.; Burholt, S.; Hamley, I. Peptide hormones and lipopeptides: From self-assembly to therapeutic applications. *J. Pept. Sci.* **2017**, *23*, 82–94. [CrossRef]
18. Reiher, W.; Shirras, C.; Kahnt, J.; Baumeister, S.; Isaac, R.E.; Wegener, C. Peptidomics and peptide hormone processing in the Drosophila midgut. *J. Proteome Res.* **2011**, *10*, 1881–1892. [CrossRef]
19. Collins, J.J., III; Hou, X.; Romanova, E.V.; Lambrus, B.G.; Miller, C.M.; Saberi, A.; Sweedler, J.V.; Newmark, P.A. Genome-wide analyses reveal a role for peptide hormones in planarian germline development. *PLoS Biol.* **2010**, *8*, e1000509. [CrossRef]
20. Zavala, E. Misaligned hormonal rhythmicity: Mechanisms of origin and their clinical significance. *J. Neuroendocrinol.* **2022**, *34*, e13144. [CrossRef]
21. Sanghavi, B.J.; Moore, J.A.; Chávez, J.L.; Hagen, J.A.; Kelley-Loughnane, N.; Chou, C.-F.; Swami, N.S. Aptamer-functionalized nanoparticles for surface immobilization-free electrochemical detection of cortisol in a microfluidic device. *Biosens. Bioelectron.* **2016**, *78*, 244–252. [CrossRef]
22. Moraes, F.C.; Rossi, B.; Donatoni, M.C.; de Oliveira, K.T.; Pereira, E.C. Sensitive determination of 17β-estradiol in river water using a graphene based electrochemical sensor. *Anal. Chim. Acta* **2015**, *881*, 37–43. [CrossRef]

23. Vanschoenbeek, K.; Vanbrabant, J.; Hosseinkhani, B.; Vermeeren, V.; Michiels, L. Aptamers targeting different functional groups of 17β-estradiol. *J. Steroid Biochem. Mol. Biol.* **2015**, *147*, 10–16. [CrossRef]
24. Jodar, L.V.; Santos, F.A.; Zucolotto, V.; Janegitz, B.C. Electrochemical sensor for estriol hormone detection in biological and environmental samples. *J. Solid State Electrochem.* **2018**, *22*, 1431–1438. [CrossRef]
25. Bozdoğan, B. Electrochemical testosterone biosensor based on pencil graphite electrode electrodeposited with copper oxide nanoparticles. *Meas. Sci. Technol.* **2023**, *34*, 105106. [CrossRef]
26. Zhao, X.; Zheng, L.; Yan, Y.; Cao, R.; Zhang, J. An electrocatalytic active AuNPs/5-Amino-2-mercaptobenzimidazole/rGO/SPCE composite electrode for ultrasensitive detection of progesterone. *J. Electroanal. Chem.* **2021**, *882*, 115023. [CrossRef]
27. Duan, D.; Ding, Y.; Li, L.; Ma, G. Rapid quantitative detection of melatonin by electrochemical sensor based on carbon nanofibers embedded with FeCo alloy nanoparticles. *J. Electroanal. Chem.* **2020**, *873*, 114422. [CrossRef]
28. Mardani, L.; Vardini, M.T.; Es'haghi, M.; Kalhor, E.G. Design and construction of a carbon paste electrode modified with molecularly imprinted polymer-grafted nanocomposites for the determination of thyroxin in biological samples. *Anal. Methods* **2020**, *12*, 333–344. [CrossRef]
29. Asai, K.; Ivandini, T.A.; Einaga, Y. Continuous and selective measurement of oxytocin and vasopressin using boron-doped diamond electrodes. *Sci. Rep.* **2016**, *6*, 32429. [CrossRef]
30. Jaafariasl, M.; Shams, E.; Amini, M.K. Silica gel modified carbon paste electrode for electrochemical detection of insulin. *Electrochim. Acta* **2011**, *56*, 4390–4395. [CrossRef]
31. Nasri, I.F.M.A.; Johnson, N.; Sharp, G.; Richard, M.; Sorel, M.; Gauchotte-Lindsay, C. Detection of estrogenic hormones using plasmonic nanostructures. *ChemRxiv* **2020**, *preprint*. [CrossRef]
32. Cherian, A.R.; Keerthana, P.; Bhat, V.S.; Sirimahachai, U.; Varghese, A.; Hegde, G. Label free electrochemical detection of stress hormone cortisol using sulphur doped graphitic carbon nitride on carbon fiber paper electrode. *New J. Chem.* **2022**, *46*, 19975–19983. [CrossRef]
33. Li, Z.; Zhou, J.; Dong, T.; Xu, Y.; Shang, Y. Application of electrochemical methods for the detection of abiotic stress biomarkers in plants. *Biosens. Bioelectron.* **2021**, *182*, 113105. [CrossRef]
34. Yagati, A.K.; Go, A.; Chavan, S.G.; Baek, C.; Lee, M.-H.; Min, J. Nanostructured Au-Pt hybrid disk electrodes for enhanced parathyroid hormone detection in human serum. *Bioelectrochemistry* **2019**, *128*, 165–174. [CrossRef]
35. Nur Topkaya, S.; Cetin, A.E. Electrochemical aptasensors for biological and chemical analyte detection. *Electroanalysis* **2021**, *33*, 277–291. [CrossRef]
36. Spychalska, K.; Zając, D.; Cabaj, J. Electrochemical biosensor for detection of 17β-estradiol using semi-conducting polymer and horseradish peroxidase. *RSC Adv.* **2020**, *10*, 9079–9087. [CrossRef] [PubMed]
37. Yuksel, M.; Luo, W.; McCloy, B.; Mills, J.; Kayaharman, M.; Yeow, J.T. A precise and rapid early pregnancy test: Development of a novel and fully automated electrochemical point-of-care biosensor for human urine samples. *Talanta* **2023**, *254*, 124156. [CrossRef] [PubMed]
38. Streeter, I.; Compton, R.G. Numerical simulation of potential step chronoamperometry at low concentrations of supporting electrolyte. *J. Phys. Chem. C* **2008**, *112*, 13716–13728. [CrossRef]
39. Chen, G.-C.; Liu, C.-H.; Wu, W.-C. Electrochemical immunosensor for serum parathyroid hormone using voltammetric techniques and a portable simulator. *Anal. Chim. Acta* **2021**, *1143*, 84–92. [CrossRef]
40. Kim, H.-U.; Kim, H.Y.; Seok, H.; Kanade, V.; Yoo, H.; Park, K.-Y.; Lee, J.-H.; Lee, M.-H.; Kim, T. Flexible MoS_2–polyimide electrode for electrochemical biosensors and their applications for the highly sensitive quantification of endocrine hormones: PTH, T3, and T4. *Anal. Chem.* **2020**, *92*, 6327–6333. [CrossRef]
41. Cherian, A.R.; Benny, L.; George, A.; Sirimahachai, U.; Varghese, A.; Hegde, G. Electro fabrication of molecularly imprinted sensor based on Pd nanoparticles decorated poly-(3 thiophene acetic acid) for progesterone detection. *Electrochim. Acta* **2022**, *408*, 139963. [CrossRef]
42. Dai, Y.; Liu, C.C. Detection of 17 β-estradiol in environmental samples and for health care using a single-use, cost-effective biosensor based on differential pulse voltammetry (DPV). *Biosensors* **2017**, *7*, 15. [CrossRef]
43. Yang, B.; Liu, D.; Zhu, L.; Liu, Y.; Wang, X.; Qiao, L.; Zhang, W.; Liu, B. Sensitive detection of thyroid stimulating hormone by inkjet printed microchip with a double signal amplification strategy. *Chin. Chem. Lett.* **2018**, *29*, 1879–1882. [CrossRef]
44. Mradula; Raj, R.; Mishra, S. Voltammetric immunosensor for selective thyroxine detection using Cu-MOF@ PANI composite. *Electrochem. Sci. Adv.* **2022**, *2*, e2100051. [CrossRef]
45. Paimard, G.; Shamsipur, M.; Gholivand, M.B.; Shahlaei, M. Simultaneous electrochemical investigation and detection of two glucocorticoids; interactions with human growth hormone, somatropin. *Results Chem.* **2022**, *4*, 100324. [CrossRef]
46. Pali, M.; Garvey, J.E.; Small, B.; Suni, I.I. Detection of fish hormones by electrochemical impedance spectroscopy and quartz crystal microbalance. *Sens. Bio-Sens. Res.* **2017**, *13*, 1–8. [CrossRef]
47. Fan, Y.; Guo, Y.; Shi, S.; Ma, J. An electrochemical immunosensor based on reduced graphene oxide/multiwalled carbon nanotubes/thionine/gold nanoparticle nanocomposites for the sensitive testing of follicle-stimulating hormone. *Anal. Methods* **2021**, *13*, 3821–3828. [CrossRef]
48. Levent, A. Electrochemical determination of melatonin hormone using a boron-doped diamond electrode. *Diam. Relat. Mater.* **2012**, *21*, 114–119. [CrossRef]

49. Malla, P.; Liao, H.-P.; Liu, C.-H.; Wu, W.-C. Electrochemical immunoassay for serum parathyroid hormone using screen-printed carbon electrode and magnetic beads. *J. Electroanal. Chem.* **2021**, *895*, 115463. [CrossRef]
50. da Silveira, J.P.; Piovesan, J.V.; Spinelli, A. Carbon paste electrode modified with ferrimagnetic nanoparticles for voltammetric detection of the hormone estriol. *Microchem. J.* **2017**, *133*, 22–30. [CrossRef]
51. Kim, S.; Yu, G.; Kim, T.; Shin, K.; Yoon, J. Rapid bacterial detection with an interdigitated array electrode by electrochemical impedance spectroscopy. *Electrochim. Acta* **2012**, *82*, 126–131. [CrossRef]
52. Ino, K.; Kitagawa, Y.; Watanabe, T.; Shiku, H.; Koide, M.; Itayama, T.; Yasukawa, T.; Matsue, T. Detection of hormone active chemicals using genetically engineered yeast cells and microfluidic devices with interdigitated array electrodes. *Electrophoresis* **2009**, *30*, 3406–3412. [CrossRef]
53. Ochiai, L.M.; Agustini, D.; Figueiredo-Filho, L.C.; Banks, C.E.; Marcolino-Junior, L.H.; Bergamini, M.F. Electroanalytical thread-device for estriol determination using screen-printed carbon electrodes modified with carbon nanotubes. *Sens. Actuators B Chem.* **2017**, *241*, 978–984. [CrossRef]
54. Adeniji, T.M.; Stine, K.J. Nanostructure Modified Electrodes for Electrochemical Detection of Contaminants of Emerging Concern. *Coatings* **2023**, *13*, 381. [CrossRef]
55. Yi, Y.; Weinberg, G.; Prenzel, M.; Greiner, M.; Heumann, S.; Becker, S.; Schlögl, R. Electrochemical corrosion of a glassy carbon electrode. *Catal. Today* **2017**, *295*, 32–40. [CrossRef]
56. Dushna, O.; Dubenska, L.; Marton, M.; Hatala, M.; Vojs, M. Sensitive and selective voltammetric method for determination of quinoline alkaloid, quinine in soft drinks and urine by applying a boron-doped diamond electrode. *Microchem. J.* **2023**, *191*, 108839. [CrossRef]
57. Macpherson, J.V. A practical guide to using boron doped diamond in electrochemical research. *Phys. Chem. Chem. Phys.* **2015**, *17*, 2935–2949. [CrossRef]
58. Uçar, M.; Levent, A. Novel voltammetric strategy for determination and electrochemical evaluation of progesterone by CPT-BDD electrode. *Diam. Relat. Mater.* **2021**, *117*, 108459. [CrossRef]
59. Yang, Z.; Li, M.; Li, H.; Li, H.; Li, C.; Yang, B. Polycrystalline boron-doped diamond-based electrochemical biosensor for simultaneous detection of dopamine and melatonin. *Anal. Chim. Acta* **2020**, *1135*, 73–82. [CrossRef]
60. Liu, X.; Sakthivel, R.; Chen, Y.-C.; Chang, N.; Dhawan, U.; Li, Y.; Zhao, G.; Lin, C.; Chung, R.-J. Tin disulfide–graphene oxide-β-cyclodextrin mediated electro-oxidation of melatonin hormone: An efficient platform for electrochemical sensing. *J. Mater. Chem. B* **2020**, *8*, 7539–7547. [CrossRef]
61. Mathew, M.; Rout, C.S. Electrochemical biosensors based on $Ti_3C_2T_x$ MXene: Future perspectives for on-site analysis. *Curr. Opin. Electrochem.* **2021**, *30*, 100782. [CrossRef]
62. Nayak, M.K.; Kumari, P.; Patel, M.K.; Kumar, P. Functional nanomaterials based opto-electrochemical sensors for the detection of gonadal steroid hormones. *TrAC Trends Anal. Chem.* **2022**, *150*, 116571.
63. Su, S.; Sun, Q.; Gu, X.; Xu, Y.; Shen, J.; Zhu, D.; Chao, J.; Fan, C.; Wang, L. Two-dimensional nanomaterials for biosensing applications. *TrAC Trends Anal. Chem.* **2019**, *119*, 115610. [CrossRef]
64. Balasubramaniam, B.; Singh, N.; Kar, P.; Tyagi, A.; Prakash, J.; Gupta, R.K. Engineering of transition metal dichalcogenide-based 2D nanomaterials through doping for environmental applications. *Mol. Syst. Des. Eng.* **2019**, *4*, 804–827. [CrossRef]
65. Sulleiro, M.V.; Dominguez-Alfaro, A.; Alegret, N.; Silvestri, A.; Gómez, I.J. 2D Materials towards sensing technology: From fundamentals to applications. *Sens. Bio-Sens. Res.* **2022**, *38*, 100540. [CrossRef]
66. Cesarino, I.; Cincotto, F.H.; Machado, S.A. A synergistic combination of reduced graphene oxide and antimony nanoparticles for estriol hormone detection. *Sens. Actuators B Chem.* **2015**, *210*, 453–459. [CrossRef]
67. Kim, M.-Y.; Park, H.; Lee, J.-Y.; Lee, J.-Y.; Myung, N.V.; Lee, K.H. Hierarchically palladium nanoparticles embedded polyethyleneimine–reduced graphene oxide aerogel (RGA–PEI–Pd) porous electrodes for electrochemical detection of bisphenol a and H_2O_2. *Chem. Eng. J.* **2022**, *431*, 134250. [CrossRef]
68. Kumari, P.; Nayak, M.K.; Kumar, P. An electrochemical biosensing platform for progesterone hormone detection using magnetic graphene oxide. *J. Mater. Chem. B* **2021**, *9*, 5264–5271.
69. Mia, A.K.; Meyyappan, M.; Giri, P. Two-dimensional transition metal dichalcogenide based biosensors: From fundamentals to healthcare applications. *Biosensors* **2023**, *13*, 169. [CrossRef]
70. Tajik, S.; Dourandish, Z.; Nejad, F.G.; Beitollahi, H.; Jahani, P.M.; Di Bartolomeo, A. Transition metal dichalcogenides: Synthesis and use in the development of electrochemical sensors and biosensors. *Biosens. Bioelectron.* **2022**, *216*, 114674. [CrossRef] [PubMed]
71. Jiang, F.; Zhao, W.-S.; Zhang, J. Mini-review: Recent progress in the development of $MoSe_2$ based chemical sensors and biosensors. *Microelectron. Eng.* **2020**, *225*, 111279. [CrossRef]
72. Tian, L.; Jiang, M.; Su, M.; Cao, X.; Jiang, Q.; Liu, Q.; Yu, C. Sweat cortisol determination utilizing MXene and multi-walled carbon nanotube nanocomposite functionalized immunosensor. *Microchem. J.* **2023**, *185*, 108172. [CrossRef]
73. Sakthivel, R.; Keerthi, M.; Chung, R.-J.; He, J.-H. Heterostructures of 2D materials and their applications in biosensing. *Prog. Mater. Sci.* **2023**, *132*, 101024. [CrossRef]
74. Lu, D.; Zhao, H.; Zhang, X.; Chen, Y.; Feng, L. New horizons for MXenes in biosensing applications. *Biosensors* **2022**, *12*, 820. [CrossRef]

75. San Nah, J.; Barman, S.C.; Zahed, M.A.; Sharifuzzaman, M.; Yoon, H.; Park, C.; Yoon, S.; Zhang, S.; Park, J.Y. A wearable microfluidics-integrated impedimetric immunosensor based on $Ti_3C_2T_x$ MXene incorporated laser-burned graphene for noninvasive sweat cortisol detection. *Sens. Actuators B Chem.* **2021**, *329*, 129206.
76. Lu, L.; Wang, M.; Zhang, D.; Zhang, H. Establishment of an immunofiltration strip for the detection of 17β-estradiol based on the photothermal effect of black phosphorescence. *Analyst* **2019**, *144*, 6647–6652. [CrossRef] [PubMed]
77. Muzaffar, N.; Afzal, A.M.; Hegazy, H.; Iqbal, M.W. Recent advances in two-dimensional metal-organic frameworks as an exotic candidate for the evaluation of redox-active sites in energy storage devices. *J. Energy Storage* **2023**, *64*, 107142. [CrossRef]
78. Chia, H.L.; Mayorga-Martinez, C.C.; Pumera, M. Doping and decorating 2D materials for biosensing: Benefits and drawbacks. *Adv. Funct. Mater.* **2021**, *31*, 2102555. [CrossRef]
79. Ren, S.; Li, Y.; Guo, Q.; Peng, Y.; Bai, J.; Ning, B.; Gao, Z. Turn-on fluorometric immunosensor for diethylstilbestrol based on the use of air-stable polydopamine-functionalized black phosphorus and upconversion nanoparticles. *Microchim. Acta* **2018**, *185*, 429. [CrossRef] [PubMed]
80. Vessally, E.; Farajzadeh, P.; Najafi, E. Possible sensing ability of boron nitride nanosheet and its Al–and Si–doped derivatives for methimazole drug by computational study. *Iran. J. Chem. Chem. Eng.* **2021**, *40*, 1001–1011.
81. Xia, Y.; Liu, Y.; Hu, X.; Zhao, F.; Zeng, B. Dual-Mode Electrochemical Competitive Immunosensor Based on Cd^{2+}/Au/Polydopami ne/Ti_3C_2 Composite and Copper-Based Metal–Organic Framework for 17β-Estradiol Detection. *ACS Sens.* **2022**, *7*, 3077–3084. [CrossRef]
82. Raymundo-Pereira, P.A.; Campos, A.M.; Vicentini, F.C.; Janegitz, B.C.; Mendonça, C.D.; Furini, L.N.; Boas, N.V.; Calegaro, M.L.; Constantino, C.J.; Machado, S.A. Sensitive detection of estriol hormone in creek water using a sensor platform based on carbon black and silver nanoparticles. *Talanta* **2017**, *174*, 652–659. [CrossRef]
83. Panahi, Z.; Custer, L.; Halpern, J.M. Recent advances in non-enzymatic electrochemical detection of hydrophobic metabolites in biofluids. *Sens. Actuators Rep.* **2021**, *3*, 100051. [CrossRef]
84. Akshaya, K.; Bhat, V.S.; Varghese, A.; George, L.; Hegde, G. Non-enzymatic electrochemical determination of progesterone using carbon nanospheres from onion peels coated on carbon fiber paper. *J. Electrochem. Soc.* **2019**, *166*, B1097.
85. Hareesha, N.; Manjunatha, J.G. Surfactant and polymer layered carbon composite electrochemical sensor for the analysis of estriol with ciprofloxacin. *Mater. Res. Innov.* **2020**, *24*, 349–362. [CrossRef]
86. Peng, C.; Ji, H.; Wang, Z. An Electrochemical Biosensor Based on Gold Nanoparticles/Carbon Nanotubes Hybrid for Determination of recombinant human erythropoietin in human blood plasma. *Int. J. Electrochem. Sci.* **2022**, *17*, 221127. [CrossRef]
87. Bolat, G.; Yaman, Y.T.; Abaci, S. Highly sensitive electrochemical assay for Bisphenol A detection based on poly (CTAB)/MWCNTs modified pencil graphite electrodes. *Sens. Actuators B Chem.* **2018**, *255*, 140–148. [CrossRef]
88. Erkmen, C.; Demir, Y.; Kurbanoglu, S.; Uslu, B. Multi-Purpose electrochemical tyrosinase nanobiosensor based on poly (3, 4 ethylenedioxythiophene) nanoparticles decorated graphene quantum dots: Applications to hormone drugs analyses and inhibition studies. *Sens. Actuators B Chem.* **2021**, *343*, 130164. [CrossRef]
89. Xu, G.; Lin, G.; Lin, S.; Wu, N.; Deng, Y.; Feng, G.; Chen, Q.; Qu, J.; Chen, D.; Chen, S. The reproductive toxicity of CdSe/ZnS quantum dots on the in vivo ovarian function and in vitro fertilization. *Sci. Rep.* **2016**, *6*, 37677. [CrossRef]
90. Abazar, F.; Sharifi, E.; Noorbakhsh, A. Antifouling properties of carbon quantum dots-based electrochemical sensor as a promising platform for highly sensitive detection of insulin. *Microchem. J.* **2022**, *180*, 107560. [CrossRef]
91. Kumari, P.; Nayak, M.K.; Kumar, P. A bio-sensing platform based on graphene quantum dots for label free electrochemical detection of progesterone. *Mater. Today Proc.* **2022**, *48*, 583–586.
92. Cardoso, R.M.; Pereira, T.S.; dos Santos, D.M.; Migliorini, F.L.; Mattoso, L.H.; Correa, D.S. Laser-induced graphitized electrodes enabled by a 3D printer/diode laser setup for voltammetric detection of hormones. *Electrochim. Acta* **2023**, *442*, 141874. [CrossRef]
93. Velasco, A.; Ryu, Y.K.; Hamada, A.; de Andrés, A.; Calle, F.; Martinez, J. Laser-Induced Graphene Microsupercapacitors: Structure, Quality, and Performance. *Nanomaterials* **2023**, *13*, 788. [CrossRef]
94. Zhang, Z.; Zhu, H.; Zhang, W.; Zhang, Z.; Lu, J.; Xu, K.; Liu, Y.; Saetang, V. A review of laser-induced graphene: From experimental and theoretical fabrication processes to emerging applications. *Carbon* **2023**, *214*, 118356. [CrossRef]
95. Yoon, H.; Nah, J.; Kim, H.; Ko, S.; Sharifuzzaman, M.; Barman, S.C.; Xuan, X.; Kim, J.; Park, J.Y. A chemically modified laser-induced porous graphene based flexible and ultrasensitive electrochemical biosensor for sweat glucose detection. *Sens. Actuators B Chem.* **2020**, *311*, 127866. [CrossRef]
96. He, P.; Wang, W.; Du, L.; Dong, F.; Deng, Y.; Zhang, T. Zeolite A functionalized with copper nanoparticles and graphene oxide for simultaneous electrochemical determination of dopamine and ascorbic acid. *Anal. Chim. Acta* **2012**, *739*, 25–30. [CrossRef] [PubMed]
97. Anusha, T.; Bhavani, K.S.; Kumar, J.S.; Brahman, P.K. Synthesis and characterization of novel lanthanum nanoparticles-graphene quantum dots coupled with zeolitic imidazolate framework and its electrochemical sensing application towards vitamin D3 deficiency. *Colloids Surf. A Physicochem. Eng. Asp.* **2021**, *611*, 125854. [CrossRef]
98. Oladipo, A.A.; Oskouei, S.D.; Gazi, M. Metal-organic framework-based nanomaterials as opto-electrochemical sensors for the detection of antibiotics and hormones: A review. *Beilstein J. Nanotechnol.* **2023**, *14*, 631–673. [CrossRef] [PubMed]
99. Leite, J.P.; Figueira, F.; Mendes, R.F.; Almeida Paz, F.A.; Gales, L. Metal–Organic Frameworks as Sensors for Human Amyloid Diseases. *ACS Sens.* **2023**, *8*, 1033–1053. [CrossRef] [PubMed]

100. Fathi, F.; Rashidi, M.-R.; Omidi, Y. Ultra-sensitive detection by metal nanoparticles-mediated enhanced SPR biosensors. *Talanta* **2019**, *192*, 118–127. [CrossRef] [PubMed]
101. Donini, C.A.; da Silva, M.K.L.; Simões, R.P.; Cesarino, I. Reduced graphene oxide modified with silver nanoparticles for the electrochemical detection of estriol. *J. Electroanal. Chem.* **2018**, *809*, 67–73. [CrossRef]
102. Zare, I.; Yaraki, M.T.; Speranza, G.; Najafabadi, A.H.; Shourangiz-Haghighi, A.; Nik, A.B.; Manshian, B.B.; Saraiva, C.; Soenen, S.J.; Kogan, M.J. Gold nanostructures: Synthesis, properties, and neurological applications. *Chem. Soc. Rev.* **2022**, *51*, 2601–2680. [CrossRef]
103. Jazayeri, M.H.; Aghaie, T.; Avan, A.; Vatankhah, A.; Ghaffari, M.R.S. Colorimetric detection based on gold nano particles (GNPs): An easy, fast, inexpensive, low-cost and short time method in detection of analytes (protein, DNA, and ion). *Sens. Bio-Sens. Res.* **2018**, *20*, 1–8. [CrossRef]
104. Sharma, N.; Reddy, A.S.; Yun, K. Electrochemical detection of hydrocortisone using green-synthesized cobalt oxide nanoparticles with nafion-modified glassy carbon electrode. *Chemosphere* **2021**, *282*, 131029. [CrossRef]
105. Naqvi, S.M.Z.A.; Zhang, Y.; Ahmed, S.; Abdulraheem, M.I.; Hu, J.; Tahir, M.N.; Raghavan, V. Applied surface enhanced Raman Spectroscopy in plant hormones detection, annexation of advanced technologies: A review. *Talanta* **2022**, *236*, 122823. [CrossRef] [PubMed]
106. Sheng, Y.; Qian, W.; Huang, J.; Wu, B.; Yang, J.; Xue, T.; Ge, Y.; Wen, Y. Electrochemical detection combined with machine learning for intelligent sensing of maleic hydrazide by using carboxylated PEDOT modified with copper nanoparticles. *Microchim. Acta* **2019**, *186*, 543. [CrossRef] [PubMed]
107. Luo, S.C.; Thomas, J.L.; Guo, H.Z.; Liao, W.T.; Lee, M.H.; Lin, H.Y. Electrosynthesis of Nanostructured, Imprinted Poly (hydroxymethyl 3, 4-ethylenedioxythiophene) for the Ultrasensitive Electrochemical Detection of Urinary Progesterone. *ChemistrySelect* **2017**, *2*, 7935–7939. [CrossRef]
108. Sun, Y.; Jiang, J.; Liu, Y.; Wu, S.; Zou, J. A facile one-pot preparation of Co_3O_4/g C_3N_4 heterojunctions with excellent electrocatalytic activity for the detection of environmental phenolic hormones. *Appl. Surf. Sci.* **2018**, *430*, 362–370. [CrossRef]
109. Siontorou, C.G.; Nikoleli, G.-P.; Nikolelis, M.-T.; Nikolelis, D.P. Challenges and future prospects of Nanoadvanced sensing technology. In *Advanced Biosensors for Health Care Applications*; Elsevier: Amsterdam, The Netherlands, 2019; pp. 375–396, ISBN 9780128157435. [CrossRef]
110. Shirtcliff, E.A.; Peres, J.C.; Dismukes, A.R.; Lee, Y.; Phan, J.M. Hormones: Commentary: Riding the physiological roller coaster: Adaptive significance of cortisol stress reactivity to social contexts. *J. Personal. Disord.* **2014**, *28*, 40–51. [CrossRef] [PubMed]
111. Karuppaiah, G.; Velayutham, J.; Sethy, N.K.; Manickam, P. DNA aptamer and gold-nanofiller integrated hybrid hydrogel network for electrochemical detection of salivary cortisol. *Mater. Lett.* **2023**, *342*, 134310. [CrossRef]
112. Singh, N.K.; Chung, S.; Sveiven, M.; Hall, D.A. Cortisol Detection in Undiluted Human Serum Using a Sensitive Electrochemical Structure-Switching Aptamer over an Antifouling Nanocomposite Layer. *ACS Omega* **2021**, *6*, 27888–27897. [CrossRef]
113. Whitworth, J.A.; Williamson, P.M.; Mangos, G.; Kelly, J.J. Cardiovascular consequences of cortisol excess. *Vasc. Health Risk Manag.* **2005**, *1*, 291–299. [CrossRef]
114. Sharma, V.; Sharma, T.K.; Kaur, I. Electrochemical detection of cortisol on graphene quantum dots modified electrodes using a rationally truncated high affinity aptamer. *Appl. Nanosci.* **2021**, *11*, 2577–2588. [CrossRef]
115. Singh, A.; Kaushik, A.; Kumar, R.; Nair, M.; Bhansali, S. Electrochemical sensing of cortisol: A recent update. *Appl. Biochem. Biotechnol.* **2014**, *174*, 1115–1126. [CrossRef]
116. Stalder, T.; Kirschbaum, C. Analysis of cortisol in hair–state of the art and future directions. *Brain Behav. Immun.* **2012**, *26*, 1019–1029. [CrossRef]
117. Sharma, V.; Sharma, T.K.; Kaur, I. Electrochemical detection of cortisol using a structure-switching aptamer immobilized on gold nanoparticles-modified screen-printed electrodes. *J. Appl. Electrochem.* **2023**, *53*, 1765–1776. [CrossRef]
118. Karuppaiah, G.; Velayutham, J.; Hansda, S.; Narayana, N.; Bhansali, S.; Manickam, P. Towards the development of reagent-free and reusable electrochemical aptamer-based cortisol sensor. *Bioelectrochemistry* **2022**, *145*, 108098. [CrossRef] [PubMed]
119. Yeasmin, S.; Wu, B.; Liu, Y.; Ullah, A.; Cheng, L.-J. Nano gold-doped molecularly imprinted electrochemical sensor for rapid and ultrasensitive cortisol detection. *Biosens. Bioelectron.* **2022**, *206*, 114142. [CrossRef] [PubMed]
120. Shama, N.A.; Aşır, S.; Göktürk, I.; Yılmaz, F.; Türkmen, D.; Denizli, A. Electrochemical Detection of Cortisol by Silver Nanoparticle-Modified Molecularly Imprinted Polymer-Coated Pencil Graphite Electrodes. *ACS Omega* **2023**, *8*, 29202–29212. [CrossRef] [PubMed]
121. Madhu, S.; Ramasamy, S.; Magudeeswaran, V.; Manickam, P.; Nagamony, P.; Chinnuswamy, V. SnO_2 nanoflakes deposited carbon yarn-based electrochemical immunosensor towards cortisol measurement. *J. Nanostructure Chem.* **2023**, *13*, 115–127. [CrossRef]
122. Laochai, T.; Yukird, J.; Promphet, N.; Qin, J.; Chailapakul, O.; Rodthongkum, N. Non-invasive electrochemical immunosensor for sweat cortisol based on L-cys/AuNPs/MXene modified thread electrode. *Biosens. Bioelectron.* **2022**, *203*, 114039. [CrossRef]
123. Zubarev, A.; Cuzminschi, M.; Iordache, A.-M.; Iordache, S.-M.; Rizea, C.; Grigorescu, C.E.; Giuglea, C. Graphene-Based Sensor for the Detection of Cortisol for Stress Level Monitoring and Diagnostics. *Diagnostics* **2022**, *12*, 2593. [CrossRef]
124. Duan, D.; Lu, H.; Li, L.; Ding, Y.; Ma, G. A molecularly imprinted electrochemical sensors based on bamboo-like carbon nanotubes loaded with nickel nanoclusters for highly selective detection of cortisol. *Microchem. J.* **2022**, *175*, 107231. [CrossRef]
125. Castagnola, E.; Woeppel, K.; Golabchi, A.; McGuier, M.; Chodapaneedi, N.; Metro, J.; Taylor, I.M.; Cui, X.T. Electrochemical detection of exogenously administered melatonin in the brain. *Analyst* **2020**, *145*, 2612–2620. [CrossRef]

126. Santhan, A.; Hwa, K.-Y. Rational design of nanostructured copper phosphate nanoflakes supported niobium carbide for the selective electrochemical detection of melatonin. *ACS Appl. Nano Mater.* **2022**, *5*, 18256–18269. [CrossRef]
127. Gomez, F.J.V.; Martín, A.; Silva, M.F.; Escarpa, A. Microchip electrophoresis-single wall carbon nanotube press-transferred electrodes for fast and reliable electrochemical sensing of melatonin and its precursors. *Electrophoresis* **2015**, *36*, 1880–1885. [CrossRef] [PubMed]
128. Thenrajan, T.; Girija, S.; Sangeetha, S.; Alwarappan, S.; Wilson, J. Electrochemical Detection of Melatonin at Tungsten Oxide Nanospheres Decorated Chitosan Electrode. *J. Electrochem. Soc.* **2023**, *170*, 077510. [CrossRef]
129. Lete, C.; López-Iglesias, D.; García-Guzmán, J.J.; Leau, S.-A.; Stanciu, A.E.; Marin, M.; Palacios-Santander, J.M.; Lupu, S.; Cubillana-Aguilera, L. A sensitive electrochemical sensor based on sonogel-carbon material enriched with gold nanoparticles for melatonin determination. *Sensors* **2021**, *22*, 120. [CrossRef] [PubMed]
130. Jin, X.; Nodehi, M.; Baghayeri, M.; Xu, Y.; Hua, Z.; Lei, Y.; Shao, M.; Makvandi, P. Development of an impedimetric sensor for susceptible detection of melatonin at picomolar concentrations in diverse pharmaceutical and human specimens. *Environ. Res.* **2023**, *238*, 117080. [CrossRef]
131. Amjadi, S.; Akhoundian, M.; Alizadeh, T. A Simple Method for Melatonin Determination in the Presence of High Levels of Tryptophan using an Unmodified Carbon Paste Electrode and Square Wave Anodic Stripping Voltammetry. *Electroanalysis* **2023**, *35*, e202200210. [CrossRef]
132. Sebastian, N.; Yu, W.-C.; Balram, D.; Chen, Q.; Shiue, A.; Noman, M.; Amor, N. Porous hematite embedded C and Fe codoped graphitic carbon nitride for electrochemical detection of pineal gland hormone melatonin. *Mater. Today Chem.* **2023**, *29*, 101406. [CrossRef]
133. Selvam, S.P.; Hansa, M.; Yun, K. Simultaneous differential pulse voltammetric detection of uric acid and melatonin based on a self-assembled Au nanoparticle–MoS_2 nanoflake sensing platform. *Sens. Actuators B Chem.* **2020**, *307*, 127683. [CrossRef]
134. Zhou, H.; Ma, X.; Sailjoi, A.; Zou, Y.; Lin, X.; Yan, F.; Su, B.; Liu, J. Vertical silica nanochannels supported by nanocarbon composite for simultaneous detection of serotonin and melatonin in biological fluids. *Sens. Actuators B Chem.* **2022**, *353*, 131101. [CrossRef]
135. Rahmati, R.; Hemmati, A.; Mohammadi, R.; Hatamie, A.; Tamjid, E.; Simchi, A. Sensitive voltammetric detection of melatonin in pharmaceutical products by highly conductive porous graphene-gold composites. *ACS Sustain. Chem. Eng.* **2020**, *8*, 18224–18236. [CrossRef]
136. Deplewski, D.; Rosenfield, R.L. Role of hormones in pilosebaceous unit development. *Endocr. Rev.* **2000**, *21*, 363–392. [CrossRef] [PubMed]
137. Crulhas, B.P.; Basso, C.R.; Parra, J.P.; Castro, G.R.; Pedrosa, V.A. Reduced graphene oxide decorated with AuNPs as a new aptamer-based biosensor for the detection of androgen receptor from prostate cells. *J. Sens.* **2019**, *2019*, 5805609. [CrossRef]
138. Moura, S.L.; De Moraes, R.R.; Dos Santos, M.A.P.; Pividori, M.I.; Lopes, J.A.D.; de Lima Moreira, D.; Zucolotto, V.; dos Santos Júnior, J.R. Electrochemical detection in vitro and electron transfer mechanism of testosterone using a modified electrode with a cobalt oxide film. *Sens. Actuators B Chem.* **2014**, *202*, 469–474. [CrossRef]
139. Mundaca, R.; Moreno-Guzmán, M.; Eguílaz, M.; Yáñez-Sedeño, P.; Pingarrón, J. Enzyme biosensor for androsterone based on 3α-hydroxysteroid dehydrogenase immobilized onto a carbon nanotubes/ionic liquid/NAD+ composite electrode. *Talanta* **2012**, *99*, 697–702. [CrossRef] [PubMed]
140. Yao, B.; Zhu, S.; Xu, X.; Feng, N.; Tian, Y.; Zhou, N. Ultrasensitive detection of the androgen receptor through the recognition of an androgen receptor response element and hybridization chain amplification. *Analyst* **2019**, *144*, 2179–2185. [CrossRef]
141. Knowlton, A.; Lee, A. Estrogen and the cardiovascular system. *Pharmacol. Ther.* **2012**, *135*, 54–70. [CrossRef]
142. Liang, J.; Shang, Y. Estrogen and cancer. *Annu. Rev. Physiol.* **2013**, *75*, 225–240. [CrossRef]
143. Yaşar, P.; Ayaz, G.; User, S.D.; Güpür, G.; Muyan, M. Molecular mechanism of estrogen–estrogen receptor signaling. *Reprod. Med. Biol.* **2017**, *16*, 4–20. [CrossRef]
144. Klikarová, J.; Chromá, M.; Sýs, M. Simultaneous voltammetric determination of female hormones using different carbonaceous electrodes in a non-aqueous environment. *Microchem. J.* **2023**, *193*, 109219. [CrossRef]
145. Han, Q.; Shen, X.; Zhu, W.; Zhu, C.; Zhou, X.; Jiang, H. Magnetic sensing film based on Fe_3O_4@ Au-GSH molecularly imprinted polymers for the electrochemical detection of estradiol. *Biosens. Bioelectron.* **2016**, *79*, 180–186. [CrossRef]
146. Arvand, M.; Hemmati, S. Analytical methodology for the electro-catalytic determination of estradiol and progesterone based on graphene quantum dots and poly (sulfosalicylic acid) co-modified electrode. *Talanta* **2017**, *174*, 243–255. [CrossRef] [PubMed]
147. Li, J.; Jiang, J.; Zhao, D.; Xu, Z.; Liu, M.; Deng, P.; Liu, X.; Yang, C.; Qian, D.; Xie, H. Facile synthesis of Pd/N-doped reduced graphene oxide via a moderate wet-chemical route for non-enzymatic electrochemical detection of estradiol. *J. Alloys Compd.* **2018**, *769*, 566–575. [CrossRef]
148. Tanrıkut, E.; Özcan, İ.; Sel, E.; Köytepe, S.; Savan, E.K. Simultaneous electrochemical detection of estradiol and testosterone using nickel ferrite oxide doped mesoporous carbon nanocomposite modified sensor. *J. Electrochem. Soc.* **2020**, *167*, 087509. [CrossRef]
149. Uliana, C.V.; Peverari, C.R.; Afonso, A.S.; Cominetti, M.R.; Faria, R.C. Fully disposable microfluidic electrochemical device for detection of estrogen receptor alpha breast cancer biomarker. *Biosens. Bioelectron.* **2018**, *99*, 156–162. [CrossRef] [PubMed]
150. Zhu, Y.; Liu, X.; Jia, J. Electrochemical detection of natural estrogens using a graphene/ordered mesoporous carbon modified carbon paste electrode. *Anal. Methods* **2015**, *7*, 8626–8631. [CrossRef]
151. Jesu Amalraj, A.J.; Narasimha Murthy, U.; Wang, S.-F. Pt nanoparticle-decorated se rods for electrochemical detection of 17β-estradiol and methanol oxidation. *ACS Appl. Nano Mater.* **2022**, *5*, 1944–1957. [CrossRef]

152. Contreras Jiménez, G.N.; Eissa, S.; Ng, A.; Alhadrami, H.; Zourob, M.; Siaj, M. Aptamer-based label-free impedimetric biosensor for detection of progesterone. *Anal. Chem.* **2015**, *87*, 1075–1082. [CrossRef]
153. de Lima, C.A.; Spinelli, A. Electrochemical behavior of progesterone at an ex situ bismuth film electrode. *Electrochim. Acta* **2013**, *107*, 542–548. [CrossRef]
154. Laza, A.; Godoy, A.; Pereira, S.; Aranda, P.R.; Messina, G.A.; Garcia, C.D.; Raba, J.; Bertolino, F.A. Electrochemical determination of progesterone in calf serum samples using a molecularly imprinted polymer sensor. *Microchem. J.* **2022**, *183*, 108113. [CrossRef]
155. Taraborrelli, S. Physiology, production and action of progesterone. *Acta Obstet. Gynecol. Scand.* **2015**, *94*, 8–16. [CrossRef]
156. Trabert, B.; Sherman, M.E.; Kannan, N.; Stanczyk, F.Z. Progesterone and breast cancer. *Endocr. Rev.* **2020**, *41*, 320–344. [CrossRef] [PubMed]
157. Velayudham, J.; Magudeeswaran, V.; Paramasivam, S.S.; Karruppaya, G.; Manickam, P. Hydrogel-aptamer nanocomposite based electrochemical sensor for the detection of progesterone. *Mater. Lett.* **2021**, *305*, 130801. [CrossRef]
158. Huang, Y.; Ye, D.; Yang, J.; Zhu, W.; Li, L.; Ding, Y. Dual recognition elements for selective determination of progesterone based on molecularly imprinted electrochemical aptasensor. *Anal. Chim. Acta* **2023**, *1264*, 341288. [CrossRef] [PubMed]
159. Das, A.; Sangaranarayanan, M. A sensitive electrochemical detection of progesterone using tin-nanorods modified glassy carbon electrodes: Voltammetric and computational studies. *Sens. Actuators B Chem.* **2018**, *256*, 775–789. [CrossRef]
160. Arvand, M.; Hemmati, S. Magnetic nanoparticles embedded with graphene quantum dots and multiwalled carbon nanotubes as a sensing platform for electrochemical detection of progesterone. *Sens. Actuators B Chem.* **2017**, *238*, 346–356. [CrossRef]
161. Samie, H.A.; Arvand, M. Label-free electrochemical aptasensor for progesterone detection in biological fluids. *Bioelectrochemistry* **2020**, *133*, 107489. [CrossRef] [PubMed]
162. Gevaerd, A.; Blaskievicz, S.F.; Zarbin, A.J.; Orth, E.S.; Bergamini, M.F.; Marcolino-Junior, L.H. Nonenzymatic electrochemical sensor based on imidazole-functionalized graphene oxide for progesterone detection. *Biosens. Bioelectron.* **2018**, *112*, 108–113. [CrossRef]
163. Arvand, M.; Elyan, S.; Ardaki, M.S. Facile one-pot electrochemical synthesis of zirconium oxide decorated poly (3, 4-ethylenedioxythiophene) nanocomposite for the electrocatalytic oxidation and detection of progesterone. *Sens. Actuators B Chem.* **2019**, *281*, 157–167. [CrossRef]
164. Petersen, M.C.; Shulman, G.I. Mechanisms of insulin action and insulin resistance. *Physiol. Rev.* **2018**, *98*, 2133–2223. [CrossRef]
165. Djiogue, S.; Nwabo Kamdje, A.H.; Vecchio, L.; Kipanyula, M.J.; Farahna, M.; Aldebasi, Y.; Seke Etet, P. Insulin resistance and cancer: The role of insulin and IGFs. *Endocr. Relat. Cancer* **2013**, *20*, R1–R17. [CrossRef]
166. Sakthivel, R.; Lin, L.-Y.; Duann, Y.-F.; Chen, H.-H.; Su, C.; Liu, X.; He, J.-H.; Chung, R.-J. MOF-derived Cu-BTC nanowire-embedded 2D leaf-like structured ZIF composite-based aptamer sensors for real-time in vivo insulin monitoring. *ACS Appl. Mater. Interfaces* **2022**, *14*, 28639–28650. [CrossRef] [PubMed]
167. Yu, Y.; Guo, M.; Yuan, M.; Liu, W.; Hu, J. Nickel nanoparticle-modified electrode for ultra sensitive electrochemical detection of insulin. *Biosens. Bioelectron.* **2016**, *77*, 215–219. [CrossRef] [PubMed]
168. Asadpour, F.; Mazloum-Ardakani, M.; Hoseynidokht, F.; Moshtaghioun, S.M. In situ monitoring of gating approach on mesoporous silica nanoparticles thin-film generated by the EASA method for electrochemical detection of insulin. *Biosens. Bioelectron.* **2021**, *180*, 113124. [CrossRef] [PubMed]
169. Niroula, J.; Premaratne, G.; Krishnan, S. Lab-on-paper aptasensor for label-free picomolar detection of a pancreatic hormone in serum. *Biosens. Bioelectron. X* **2022**, *10*, 100114. [CrossRef]
170. Ebrahimiasl, S.; Fathi, E.; Ahmad, M. Electrochemical detection of insulin in blood serum using Ppy/GF nanocomposite modified pencil graphite electrode. *Nanomed. Res. J.* **2018**, *3*, 219–228.
171. Liu, J.; Zhu, B.; Dong, H.; Zhang, Y.; Xu, M.; Travas-Sejdic, J.; Chang, Z. A novel electrochemical insulin aptasensor: From glassy carbon electrodes to disposable, single-use laser-scribed graphene electrodes. *Bioelectrochemistry* **2022**, *143*, 107995. [CrossRef] [PubMed]
172. Sun, H.; Wu, S.; Zhou, X.; Zhao, M.; Wu, H.; Luo, R.; Ding, S. Electrochemical sandwich immunoassay for insulin detection based on the use of gold nanoparticle-modified MoS_2 nanosheets and the hybridization chain reaction. *Microchim. Acta* **2019**, *186*, 6. [CrossRef] [PubMed]
173. Arvinte, A.; Westermann, A.C.; Sesay, A.M.; Virtanen, V. Electrocatalytic oxidation and determination of insulin at CNT-nickel–cobalt oxide modified electrode. *Sens. Actuators B Chem.* **2010**, *150*, 756–763. [CrossRef]
174. Habibi, E.; Omidinia, E.; Heidari, H.; Fazli, M. Flow injection amperometric detection of insulin at cobalt hydroxide nanoparticles modified carbon ceramic electrode. *Anal. Biochem.* **2016**, *495*, 37–41. [CrossRef]
175. Zhao, Y.; Xu, Y.; Zhang, M.; Xiang, J.; Deng, C.; Wu, H. An electrochemical dual-signaling aptasensor for the ultrasensitive detection of insulin. *Anal. Biochem.* **2019**, *573*, 30–36. [CrossRef]
176. Šišoláková, I.; Hovancova, J.; Oriňaková, R.; Oriňak, A.; Trnková, L.; Garcia, D.R.; Radoňak, J. Influence of a polymer membrane on the electrochemical determination of insulin in nanomodified screen printed carbon electrodes. *Bioelectrochemistry* **2019**, *130*, 107326. [CrossRef] [PubMed]
177. Carter, C.S.; Kenkel, W.M.; MacLean, E.L.; Wilson, S.R.; Perkeybile, A.M.; Yee, J.R.; Ferris, C.F.; Nazarloo, H.P.; Porges, S.W.; Davis, J.M. Is oxytocin "nature's medicine"? *Pharmacol. Rev.* **2020**, *72*, 829–861. [CrossRef] [PubMed]
178. Shamay-Tsoory, S.G.; Abu-Akel, A. The social salience hypothesis of oxytocin. *Biol. Psychiatry* **2016**, *79*, 194–202. [CrossRef] [PubMed]

179. Mehrotra, S.; Rai, P.; Gautam, K.; Saxena, A.; Verma, R.; Lahane, V.; Singh, S.; Yadav, A.K.; Patnaik, S.; Anbumani, S. Chitosan-carbon nanofiber based disposable bioelectrode for electrochemical detection of oxytocin. *Food Chem.* **2023**, *418*, 135965. [CrossRef] [PubMed]
180. Al-Taee, A.T.; Al-Hyali, R.H. Electrochemical Behavior of Oxytocin Hormone Through Its Cysteine Reduction Peak Using Glassy Carbon Electrode Modified with Poly Furfurylamine and Multi-Walled Carbon Nanotubes. *Egypt. J. Chem.* **2021**, *64*, 5831–5837. [CrossRef]
181. Liu, F.A.; Ardabili, N.; Brown, I.; Rafi, H.; Cook, C.; Nikopoulou, R.; Lopez, A.; Zou, S.; Hartings, M.R.; Zestos, A.G. Modified Sawhorse Waveform for the Voltammetric Detection of Oxytocin. *J. Electrochem. Soc.* **2022**, *169*, 017512. [CrossRef] [PubMed]
182. Thomas, R.; Balachandran, M. Heteroatom engineered graphene-based electrochemical assay for the quantification of high-risk abused drug oxytocin in edibles and biological samples. *Food Chem.* **2023**, *400*, 134106. [CrossRef]
183. David, M.; Şerban, A.; Enache, T.A.; Florescu, M. Electrochemical quantification of levothyroxine at disposable screen-printed electrodes. *J. Electroanal. Chem.* **2022**, *911*, 116240. [CrossRef]
184. Bhatia, A.; Nandhakumar, P.; Kim, G.; Lee, N.-S.; Yoon, Y.H.; Yang, H. Simple and fast Ag deposition method using a redox enzyme label and quinone substrate for the sensitive electrochemical detection of thyroid-stimulating hormone. *Biosens. Bioelectron.* **2022**, *197*, 113773. [CrossRef]
185. Karami, P.; Gholamin, D.; Johari-Ahar, M. Electrochemical immunoassay for one-pot detection of thyroxin (T4) and thyroid-stimulating hormone (TSH) using magnetic and Janus nanoparticles. *Anal. Bioanal. Chem.* **2023**, *415*, 4741–4751. [CrossRef]
186. Kashefi-Kheyrabadi, L.; Koyappayil, A.; Kim, T.; Cheon, Y.-P.; Lee, M.-H. A MoS_2@ Ti_3C_2Tx MXene hybrid-based electrochemical aptasensor (MEA) for sensitive and rapid detection of Thyroxine. *Bioelectrochemistry* **2021**, *137*, 107674. [CrossRef] [PubMed]
187. Park, S.Y.; Kim, J.; Yim, G.; Jang, H.; Lee, Y.; Kim, S.M.; Park, C.; Lee, M.-H.; Lee, T. Fabrication of electrochemical biosensor composed of multi-functional DNA/rhodium nanoplate heterolayer for thyroxine detection in clinical sample. *Colloids Surf. B Biointerfaces* **2020**, *195*, 111240. [CrossRef] [PubMed]
188. Han, J.; Zhuo, Y.; Chai, Y.; Yu, Y.; Liao, N.; Yuan, R. Electrochemical immunoassay for thyroxine detection using cascade catalysis as signal amplified enhancer and multi-functionalized magnetic graphene sphere as signal tag. *Anal. Chim. Acta* **2013**, *790*, 24–30. [CrossRef] [PubMed]
189. Singh, D.; Roy, S.; Mahindroo, N.; Mathur, A. Design and development of an electroanalytical sensor based on molecularly imprinted polyaniline for the detection of thyroxine. *J. Appl. Electrochem.* **2023**. [CrossRef]
190. Muñoz, J.; Riba-Moliner, M.; Brennan, L.J.; Gun'ko, Y.K.; Céspedes, F.; González-Campo, A.; Baeza, M. Amperometric thyroxine sensor using a nanocomposite based on graphene modified with gold nanoparticles carrying a thiolated β-cyclodextrin. *Microchim. Acta* **2016**, *183*, 1579–1589. [CrossRef]
191. Zhang, T.; Du, X.; Zhang, Z. Advances in electrochemical sensors based on nanomaterials for the detection of lipid hormone. *Front. Bioeng. Biotechnol.* **2022**, *10*, 993015. [CrossRef] [PubMed]
192. Suhito, I.R.; Koo, K.-M.; Kim, T.-H. Recent advances in electrochemical sensors for the detection of biomolecules and whole cells. *Biomedicines* **2020**, *9*, 15. [CrossRef]
193. Campuzano, S.; Yánez-Sedeño, P.; Pingarrón, J.M. Electrochemical bioaffinity sensors for salivary biomarkers detection. *TrAC Trends Anal. Chem.* **2017**, *86*, 14–24. [CrossRef]
194. Joshi, A.; Kim, K.-H. Recent advances in nanomaterial-based electrochemical detection of antibiotics: Challenges and future perspectives. *Biosens. Bioelectron.* **2020**, *153*, 112046. [CrossRef]
195. Rivas, G.A.; Rodriguez, M.C.; Rubianes, M.D.; Gutierrez, F.A.; Eguílaz, M.; Dalmasso, P.R.; Primo, E.N.; Tettamanti, C.; Ramírez, M.L.; Montemerlo, A. Carbon nanotubes-based electrochemical (bio) sensors for biomarkers. *Appl. Mater. Today* **2017**, *9*, 566–588. [CrossRef]
196. Li, T.; Shang, D.; Gao, S.; Wang, B.; Kong, H.; Yang, G.; Shu, W.; Xu, P.; Wei, G. Two-dimensional material-based electrochemical sensors/biosensors for food safety and biomolecular detection. *Biosensors* **2022**, *12*, 314. [CrossRef] [PubMed]
197. Farka, Z.; Jurik, T.; Kovar, D.; Trnkova, L.; Skládal, P. Nanoparticle-based immunochemical biosensors and assays: Recent advances and challenges. *Chem. Rev.* **2017**, *117*, 9973–10042. [CrossRef] [PubMed]

coatings

MDPI

Article

Exploring Preliminary Biocompatibility Testing in Coating Development

Rick M. Postema [1], Cory B. Sims [1], Michael J. Fyfe [1], Xiaohong Tan [1,*], Hans Wildschutte [2,*] and Joseph C. Furgal [1,*]

[1] Department of Chemistry and Center for Photochemical Sciences, Bowling Green State University, Bowling Green, OH 43403, USA; prick@bgsu.edu (R.M.P.); cbsims@bgsu.edu (C.B.S.); mfyfe@bgsu.edu (M.J.F.)
[2] Department of Biological Sciences, Bowling Green State University, Bowling Green, OH 43403, USA
* Correspondence: tanx@bgsu.edu (X.T.); hansw@bgsu.edu (H.W.); furgalj@bgsu.edu (J.C.F.)

Abstract: Material testing, particularly in biological applications, can be an extensive endeavor leading to a significant investment of resources. This article details a simple material and coating testing assay series that provides insights into leaching, antibacterial, antifouling, and foul-release characteristics. The results of these methods can guide future research, applications, and development efforts by providing data from which to make informed decisions. A material or coating can be quickly assessed in industrial and academic settings with minimal resources by employing a set of benign, single-species direct-contact toxicity assays and simple spectroscopic and microscopic analysis methods. Herein, we demonstrate how this series of biological assays may be utilized and the potential interpretations of the results by using two-hybrid organo-silicon-based coatings.

Keywords: hydrophobicity; roughness; leaching; toxicity; inhibition; fouling; cell analysis

1. Introduction

Testing the toxicity of materials is becoming increasingly important, particularly in cases where the material comes into direct contact with organisms, waterways, and the environment [1,2]. Traditional methods of toxicity examination do not apply well to advanced coatings and materials, especially when the material characteristics are complex and originate from volatile precursors that culminate in a potentially inert final product. In such instances, scrutinizing individual components through common methods such as agar diffusion, agar overlay, or broth dilution methods does not realistically represent final product characteristics and effects [3,4]. In this context, one of the most productive, economically viable, and readily implementable methodologies for the preliminary evaluation of the toxicity of a coating or nanomaterial is through a set of direct contact assays [5–8]. The direct contact method is a well-established approach utilized for the expeditious assessment of contact toxicity for materials and polymers [4,6,9–12]. These methodologies typically entail using microorganisms, including *Escherichia coli* and other bacterial strains, and then moving on to more advanced cultures such as mammalian cells or small organisms. Such approaches find widespread application in assessing the biocompatibility of materials, coatings, and medical devices [4,8,13–16]. Examining nanomaterials or coatings via direct contact methods yields results reflective of what is expected to be observed from a fully formed and protected surface interacting with natural biological systems. This is particularly pertinent in testing medical devices for implantation, medical grade coatings, antifoul marine coatings, and foul-release coatings. It is important to understand any antibacterial, biofilm formation/inhibition, and environmentally compatible features that are highly desirable within these application niches [4,17–21]. Concerning protective coatings, many exhibit notable hydrophobic characteristics including substantial contact angles, designed surface topography, or smooth surface finishes, with silicon-based coatings becoming evermore popular to reduce dependencies on fluoro-polymers [21–23].

The direct contact methods below are tailored to assess coatings and materials, facilitate insights into surface-to-surface interactions, and uncover surface-related effects. To demonstrate how these assays are performed and how they can provide insights about the material's biocompatibility, we have utilized two coatings that were previously developed and characterized by Sims et al. and will be referenced as NF-1 (non-fluorocarbon containing) and FS-1 (fluorocarbon containing, tridecafluoro-1,1,2,2tetrahydrooctyl) [24]. These coatings exhibited characteristics including non-wetting, photo and thermal stability, oxidative and flame resistance, and chemical inertness when applied to various materials. Due to these features, they have shown usefulness as protective coatings in preservation applications, but they are also likely suitable as candidates for medical implant, antifoul, or foul-releasing coatings [22]. In marine or aqueous environments, it is imperative to establish that the coating or material does not elicit toxic reactions and verify that the components are not leaching from the polymer matrix.

Employing a straightforward array of direct contact assays complimented by photochemical analysis enables the generation of initial insights that can guide subsequent investigations toward more intricate explorations of application-specific functionalities. Coatings NF-1 and FS-1 incorporate photo-activatable polymerization mechanisms and subsequent curing; therefore, it is crucial to investigate any toxic effects from surface interactions or leaching before possible applications are defined. This arises from two known hazardous photochemical molecules used to initiate polymerization. A radical initiator, Omnirad 819, as well as a photoacid generator, diphenyliodonium hexafluorophosphate; both are UV-activated by sunlight, as shown in Figure 1. These substituents and their byproducts may cause adverse environmental and toxicological effects on living organisms [24–26].

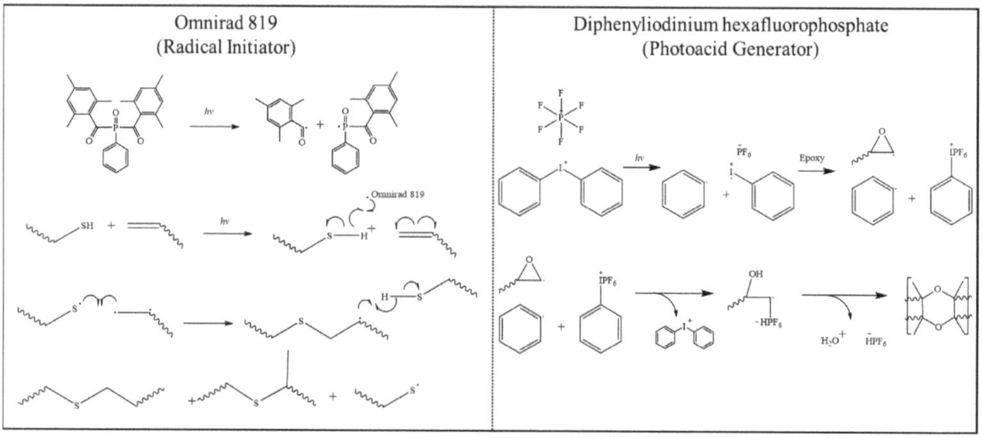

Figure 1. The photoinitiated radical thiol-ene reaction occurs in the hybrid organosilane coatings when Omnirad 819 is irradiated with UV light [27,28]. When diphenlyiodonium hexafluorophosphate is exposed to UV light, it triggers a photoinitiated reaction that generates a Lewis acid. This particular example illustrates the homogenous self-polymerization of an epoxy [29,30]; however, additional mechanisms exist that are ionic and generate H^+ and biphenyl byproducts as well to aid in amine–epoxy reactions [31,32].

Direct contact experiments with *E. coli* can be used to determine toxicological effects on substrates or molecules (i.e., unreacted photoinitiators), but many of these methods are challenging. For example, antibacterial characteristics can be determined through the well-established and optimized agar overlay method. This involves standard media (with 1.5% agar) overlaid with top agar (0.5% agar). Any antibacterial features or leaching will result in a ring of inhibited growth around and above the material [33–35]. The agar overlay method was adapted for coated surfaces to facilitate direct contact with the coating interface to determine if *E. coli* can form colonies while grown in contact with the cured coating to observe any antibacterial effects due to surface features [34]. While the visual assessment of colony counting provides straightforward results, a more analytical method should be used to give a quantitative comparison for a full understanding. The examination of a culture's growth rate by observing the optical density at 600 nanometers, OD_{600}, utilizing absorbance, can be directly correlated to the number of cells growing in a solution and provides insights into the rate of growth when monitored over time. This offers a more systematic evaluation of inhibitory effects. Furthermore, this testing method involves a 100% liquid interface, enabling a greater opportunity for contact with the coated surface while providing a medium for any unpolymerized material components to become mobile, which gives insights into leaching effects in aqueous environments. This method offers the ability to examine materials for acute and minor toxic effects, growth rate inhibition, and component leaching in contrast to the direct contact method, which may not provide a complete understanding of these interactions.

Biofilm production is a crucial characteristic for bacteria colonizing a surface and can lead to enhanced antibacterial resistance and persistent infections. Inhibiting biofilm formation holds significant implications for material use, facilitating surface sterilization and preventing large groups of organisms from building up on surfaces [36]. Biofilm production assays represent a standard approach to determining antibacterial properties but can provide novel information on biofouling characteristics. Coatings believed to be antibacterial typically inhibit biofilm formation through two means—directly inhibiting bacterial proliferation on the coating surface or by diminishing adherence of the fouling cultures, thereby resulting in reduced biofilm formation. This is a trait commonly observed in antimicrobial, antifouling, and foul-release coatings [37,38]. Given that adhesion for many organisms is protein-based, biofilm formation assays cannot only shed light on antibacterial properties but also provide insights into protein binding to surfaces.

A comparable approach can be applied in mammalian models, akin to the analysis using *E. coli*. However, observing OD_{600} does not reflect mammalian cells' colonization strength or growth rate. Instead, a supplementary method must be employed to examine and quantify any change to cell growth by assessing cell viability over a standard growth period, with findings ideally aligning with observations made in bacterial studies. Percent viability is used to determine whether a culture is healthy enough to continue to be passaged, where $\geq$95% cell viability is considered the standard for health and is directly related to the number of live cells compared to the total cell count of a culture [39]. Traditionally, percent viability is examined via cell counting using a hemocytometer and trypan blue staining, where trypan blue is excluded from living cells. Studies have shown this to be an effective means for tracking viability compared to other staining methods, lending validation to its reliability [40–43]. Adapting techniques from Sjollema et al., adherent HEK cells can be used in direct contact methodologies and to colonize coated surfaces [34].

Coatings that impede or significantly diminish the attachment strength of biological organisms have considerable significance as they allow for surfaces to be easily cleaned or prevent initial adhesion [37,44,45]. Bacteria, algae, and some small organisms, such as mussels and barnacles, utilize protein-to-surface interactions to adhere to various surfaces. When considering the coating traits employed to prevent this adhesion, hydrophobicity provides a way to repel water, reducing the ease of adhesion. At the same time, surface roughness can indicate a preference to prevent adhesion of different-sized organisms due to the available surface area for the organism's attachment sites. Generally, smoother surfaces, with closer peak-to-peak distances, are more susceptible to binding by microorganisms, while rougher surfaces are preferred by larger species that need more surface area. Antifouling coatings achieve this with biocidal effects (i.e., metals) and have been regulated due to their unintentional release of biocidal agents (i.e., copper salts) in aqueous systems. An alternative solution to biofouling is foul-releasing coatings, which have unfavorable surfaces for adherence, enabling easier release, cleaning, or prevention of biological buildup without directly harming the organism [37,38,46,47]. This method is therefore gaining preference over antifouling coatings [48,49]. Understanding the combined surface properties of roughness, hydrophobicity, and their interactions with microbes may suggest if a coating should be analyzed for interactions with larger organisms, potentially reducing unnecessary economic costs.

In this study, we delineate methods for differentiating coating or material characteristics and potential uses. This is achieved through assays of physical properties, photochemical analysis, and direct contact methods. These methods allow for the identification of appropriate applications and product targeting while also defining environmental compatibility.

2. Experimental Methods

2.1. Materials

Bacterial culture materials were all sourced from reliable *suppliers* and unmodified unless specified. Neb-5-alpha *E. coli* cells (C2988J) and super optimal broth with catabolite repression (B9035S), SOC, were sourced from New England Biolabs (Ipswich, MA, USA). Premixed Miller Broth (NCM0088A) was sourced from Neogen (Lansing, MI, USA). Terrific Broth premix (IB49141) was acquired from IBI Scientific (Dubuque, IA, USA). Mammalian cell culture supplies are also sourced from reliable vendors. Dulbecco's Modified Eagle Medium (DMEM) with glutamine XL (112-300-101) was sourced from Quality Biological (Gaithersburg, MD, USA). Fetal Bovine Serum (FBS) (00-106) was acquired from GeminiBio (West Sacramento, California). Tissue culture plates of 60 mm (10062-890) and 100 mm (10062-880), the 0.4% trypan blue solution (K940), and crystal violet were obtained from VWR (Radnor, PA, USA). Acetic acid and the microscope cover glass (12-542B) were purchased from Fisher Scientific Co. (Waltham, MA, USA). Casamino acid (Costar REF 2797) and Falcon polystyrene (Falcon REF 351172) multiwell plates were obtained from Becton, Dickinson, and Company (Franklin Lakes, NJ, USA). Mixed cellulose ester and MCE syringe filters, 0.22 µm, were purchased from Lab Instruments Co. A microplate reader was obtained from BioTek Synergy HT.

All chemicals used in the study were sourced from reputable commercial suppliers and, unless otherwise noted in the methods section, were used without further purification. Specifically, Gelest Inc. (Morrisville, PA, USA) provided 3-aminopropyltriethoxysilane, (3-glycidoxypropyl)trimethoxysilane, 3-mercaptopropyltrimethoxysilane, (tridecafluoro-1,1,2,2tetrahydrooctyl)triethoxysilane, and vinyltriethoxysilane. Isopropyl alcohol, cyclohexanes, and methanol were purchased from EMD Millipore Corporation. Benzyl alcohol ($\geq$99%) was procured from Alfa Aesar (Ward Hill, MA, USA). Diiodomethane was obtained from Eastman Chemical Company (Kingsport, TN, USA). Diphenyliodonium hexafluorophosphate was sourced from Acros Organics and Omnirad 819 from IGM Resins USA, Inc. (Charlotte, NC, USA). Anhydrous acetonitrile (99.8%), toluene ($\geq$95%), n-hexanes (98.5%), monopotassium phosphate (KH_2PO_4), dipotassium phosphate (K_2HPO_4), ammonium sulfate $(NH_4)_2SO_4$), magnesium sulfate ($MgSO_4$), glucose, and agar (A1296) were procured from Sigma-Aldrich. The 10 µm syringe used in the experiments was obtained from Hamilton Company (Franklin, MA, USA). The *Pseudomonas aeruginosa* strain AU10241 was kindly provided by Dr. John LiPuma at the University of Michigan. HEK 293 cells were obtained from ATCC. NF-1 and FS-1 were provided by Dr. Furgal, Bowling Green State University.

2.2. Analytical Methods

2.2.1. Coating Application

In controlled laboratory conditions, we directly dipped glass slide cover slips into a coating solution and exposed them to UV C light (18.7 mW/cm^2, 200-Watt Mercury Arc Lamp with a 320–390 nm filter, OmniCure Series 1500, Excelitas Technologies, Lumen Dynamics Group Inc., Mississauga, ON, Canada) for two minutes. After the coating process, we allowed the samples to sit for thirty minutes. For obtaining lamp output measurements, we used a PM100D Energy Meter Console (Thorlabs, Newton, NJ, USA) equipped with an

S121C standard photodiode power sensor (Si, 400–1100 nm, 500 mW, Thorlabs, Newton, NJ, USA) with utmost precision and accuracy.

For the biofilm assay, plastic well dishes (Falcon multiwell, polystyrene) were filled with coating solutions, the excess was poured off, and the dishes were exposed to UV C light for 2 min and then inverted while curing.

2.2.2. Fragmentation and Coating Preparation

Two samples for each coating, NF-1 and FS-1, were applied in excess on glass and cured, causing the intentional delamination of the coating. The coating was then scraped and collected via a razor blade. The freed coating samples were then split in half, retaining moderately large samples between 0.1 and 1 cm as "flakes", and the other half was ground to a fine powder via a mortar and pestle and examined by light microscopy and ATR-FTIR. These samples were stirred in 15 mL of water for a week and shaken daily before final analysis. Leaching samples were then filtered to remove flakes or powdered coating and reduced in a 47 °C vacuum oven from 15 mL to 3 mL. This process was repeated with samples placed in water and heated to 45 °C for a week, allowing any potential leached material to concentrate as the water volume decreased from 15 mL to 3 mL at 47 °C in a vacuum oven.

2.2.3. Attenuated Total Reflection Fourier Transform Infrared Spectroscopy (ATR-FTIR)

ATR-FTIR spectra were recorded on a Thermo Scientific ATR-FTIR (Nicolet iS5 Fourier Transform Infrared Spectrometer iD7 Attenuated Total Reflection, SN: ASB1817610) utilizing a ZnSe crystal platform. Spectral analysis was facilitated by OMNIC software (version 9.8.372, Thermo Scientific, Waltham, MA, USA). The sample platform was first cleaned using isopropyl alcohol and zeroed for background, followed by small solid samples of cured NF-1 and FS-1 coating flakes and powder, and all samples were scanned from 4000 to 400 cm^{-1} for 16 scans and transformed in the OMNIC program.

2.2.4. Contact Angle, Light Microscopy, and Surface Energy

To apply water to the surfaces, a syringe was used to add drops of approximately 10 μL in volume. For capturing images, we utilized a Zeiss Stemi 2000-C light microscope equipped with an AxioCam ERc5s camera manufactured by Carl Zeiss Meditec Inc. located in Dublin, CA, USA. We conducted contact angle analysis using ImageJ software (version 1.53e), developed by the National Institutes of Health (in Bethesda, MD, USA), using the manual points procedure in the contact angles plug-in. We calculated the contact angle values by taking the mean of the right and left angles; we utilized three samples for each coating and took four sets of measurements for each. From these twelve readings, we calculated the average and standard deviation. We also obtained standard static contact angles from four samples and derived the average and standard deviation for uncoated glass slide coverslips. Additionally, previously prepared coating flakes and powder of both NF-1 and FS-1 were analyzed and recorded by light microscope to show relative size and consistency between both coating types. This analysis verifies similar consistency and size between the samples.

Employing Zisman's method, the surface energy of NF-1 and FS-1 was determined through the observation of contact angles of solvents with varying surface tension forces. NF-1 and FS-1 contact angles were recorded with n-hexanes, toluene, cyclohexane, benzyl alcohol, diiodomethane, glycerol, and water. Each measurement was conducted in triplicate, and the data were treated the same as the static contact angles described above. Then, for each solvent, the cosine (contact angle) was plotted against the known surface tension forces, and the material's surface energy was found as the x value where the cosine (CA) is equal to 1 [50].

2.2.5. Surface Roughness Analysis

To determine the arithmetic mean roughness (Ra) and root-mean-square roughness (Rq), we utilized the Alpha-Step IQ Surface Profiler from KLA Tencor in Milpitas, CA, USA. Glass slides were prepared using the dip-coat method, and twenty data sets were collected for each of the two coatings by taking five measurements in opposing directions over the interface of each coverslip. Our step height analysis was conducted with a stylus force between 24.2 and 25.7 mg, a 2000 μm scan length, a 50 μm/s scan speed, a 50 Hz sampling rate, a 40 s scan time, a 400 μm/23.8 pm sensor range, a center bias adjustment, a 1 μm resolution, a contact speed of 5, and a required radius of 5.0 μm. Averages were generated from two samples of each coating, and the standard deviations were calculated.

2.2.6. Preparation of Accelerated Leaching Test

Sample leaching may occur through multiple avenues; components can slowly leak from the coating surface or leach as the coating degrades over time. Of the available components, two curing agents are of interest, Omnirad 819 and diphenyliodonium hexafluorophosphate, as possible hazardous leaching components. Coating flakes and powder were generated according to Section 2.2.2 and were then added to nanopure water and left in a dark place for a week. After one week, samples are filtered using a 0.22 μm syringe filter and examined, initially diluted by UV-vis spectroscopy in a quartz cuvette. Leaching samples are then reduced in a vacuum at 47 °C from 15 mL to 5 mL and re-examined by UV-Vis.

2.2.7. UV-Vis Spectroscopy Analysis

An Agilent Cary 60 UV/Vis instrument (Agilent, Santa Clara, CA, USA) was used to collect a range of absorbance spectra. The coating and reagent samples were dissolved in anhydrous acetonitrile (99.8%) and diluted to a concentration of less than 100 μM in a quartz cuvette. Scans were taken after two, five, and ten minutes of UV C light [18.7 mW/cm^2, 200-Watt Mercury Arc Lamp with a 320–390 nm filter, OmniCure Series 1500 (Excelitas Technologies, Mississauga, ON, Canada)], and a reading from 200 to 800 nm was taken to verify any changes in absorption by photo-active molecules. Excel was used to plot absorbance versus wavelength for the conformational change. Further analysis was conducted using water as a solvent to test for any solvent-specific interactions with the materials.

2.2.8. Direct Contact Toxicity Assay

A simple but effective method of testing contact toxicity is the direct contact method, adapted from Sjollema et al. [34]. This method is cost-effective, time-efficient, and can provide reliable preliminary toxicity data. Furthermore, direct contact best replicates the manner in which a surface coating will be interacted with. The direct contact method uses uncoated and FS-1- and NF-1-coated microscope slides to ensure coating-to-culture interfacing and a negative control uncoated surface for comparison.

E. coli strain DH5α, a Gram-negative bacterium, is preserved in its original state without any genetic alterations conferring antibiotic resistance. Initially, 25 μL of stock Neb DH5α cells were incubated with 100 μL of super optimal broth for one hour. Subsequently, agar plates were prepared from 1.5% agarose in Miller Liquid Broth by Neogen culture media, followed by the addition and spreading of 25 μL of DH5α cells. The seeded agar plate was grown overnight for 24 h at 37 °C, and the following day, one isolated colony was picked and used to inoculate 5 mL of a liquid culture and incubated overnight at 37 °C.

Once fully colonized, two liquid cultures underwent serial dilution; one sample was serial diluted six times, and the second was diluted seven times. The resulting culture dilutions were employed to seed an agar plate to promote single colony growth. Agar plates were seeded with 50 μL and 75 μL of *E. coli* solution to ensure even growth, resulting in two sets of cultured samples each, one diluted to 10^6 and another diluted to 10^7, on which UV sterilized FS-1, NF-1, and uncoated microscope slides were placed. Growth was observed after 24 h, and the colonies were counted and recorded.

2.2.9. Direct Contact Growth Rate Assay

Liquid broth is cultured in contact with a coated surface, and any effects it may have on growth rate can be observed by monitoring OD_{600}. Glass test tubes are internally coated and cured with FS-1 and NF-1 in triplicate. Terrific broth medium was inoculated with a pure isolated colony of *E. coli* DH5α and grown overnight with shaking at 37 °C. The culture was used to seed 10 mL of LB broth at 1% in FS-1, NF-1, and uncoated test tubes at 37 °C. Cultures were monitored versus a negative control at regular intervals and recorded utilizing a CO8000 cell density meter by Biowave (Norwalk, CT, USA) and Disposable UV cuvettes from BRAND (Wertheim, Germany). Analyzing 1 mL of the culture allowed for the tracking of the growth rate, and any adverse effects of the interfaced coatings could be recorded and quantified compared to those of the control.

2.2.10. Biofilm Production Assay

The *Pseudomonas aeruginosa* strain AU10241 was kindly provided by Dr. John LiPuma at the University of Michigan. The strain was inoculated into 3 mL of LB medium and incubated overnight with shaking at 37 °C. The culture was diluted at 1:100 into an M63 minimal medium prepared according to the methods of O'Toole [51]. A 5 × M63 medium was prepared by dissolving 15 g KH_2PO_4, 35 g K_2HPO_4, and 10 g $(NH_4)_2SO_4$ in 1 L of water. The stock was diluted at 1:5, and supplements were added for final concentrations of 1 mM magnesium sulfate, 0.2% glucose, and 0.5% casamino acids. A total of 100 μL of the diluted cells were added per well in a 96-well U-bottom microtiter plate (Costar REF 2797). The plate was incubated at 37 °C for 24 h. Eight single-well replicates were performed.

After incubation, the cells and media were dumped out by inverting the plate. The plate was then submerged two times in water to remove unbound cells. In total, 125 μL of a 0.1% solution of crystal violet in water was added to each well containing cells and incubated at room temperature for 10–15 min—crystal violet stains biofilm production. The wells were rinsed 3–4 times with water and then air-dried for four hours at room temperature. A total of 125 μL of 30% acetic acid in water was added to each well and incubated for 15 min. The solubilized solution was transferred to a new flat-bottomed 96-well microtiter plate (Falcon REF 351172), and absorbance was measured in a plate reader at 550 nm using 30% acetic acid in water as the blank. Absorbance at 550 nm indicates biofilm formation.

2.2.11. Direct Contact HEK Cell Viability

A second direct contact method adapted from Sjollema et al. examines possible effects on mammalian cells [34]. Due to the photocuring method of FS-1 and NF-1 coatings, polystyrene plates modified by poly-lysine lead to a chemical reaction and subsequent decomposition of plate material. To remedy this, co-incubation with FS-1- and NF-1-coated slides will allow for stable cell growth and an analysis of percent viability after a standard passage period of 3 days. Initial cells are thawed and cultured on 60 mm tissue culture plates in DMEM containing 10% FBS at 37 °C in 5% CO_2. The initial culture is passaged 4 times after 90% confluency is reached to ensure cell health and growth rate after thawing, with the final passage used to seed a 100 mm tissue culture plate to provide a large enough sample to seed nine 60 mm tissue culture plates to perform direct contact analysis in triplicate. Unmodified microscope slides coated in FS-1 and NF-1 and the uncoated negative control are UV sterilized, placed in seeded DMEM media, and cultured for three days. Utilizing trypan blue and hemocytometer, cells are counted employing the Olympus SZX10 microscope from Olympus Lifesciences (Hachioji, Tokyo, Japan), and any cell death effects can be observed with viability compared to the negative control.

3. Results and Discussion

3.1. Hydrophobicity and Surface Roughness

Many organisms can foul surfaces including bacteria, lichens, algae, mollusks, and barnacles, where many of the adhesion mechanisms occur from protein binding or mechanical forces like friction. When considering the adhesion of organisms or the formation of biofilms on surfaces, the materials' hydrophobicity and surface roughness are often used as initial indicators for fouling resistance. Many silicon-based coatings are actively used as foul-releasing finishes, where the material is non-toxic but enables the detachment of organisms via a low applied force through reduced adhesion. One commonly used technique for quantifying hydrophobicity is a static contact angle (CA), which investigates a material's ability to repel water by measuring the angle of water droplets at the point where they contact the surface of interest. A water droplet on a coated surface with a CA 90°–180° is considered non-wetting and suggests the surface may be classified as hydrophobic or superhydrophobic [52–55]. Several coatings reported to have foul-releasing characteristics also exhibit water contact angles ranging from 90 to 160° [37,46]. The coatings used to simulate the biocompatibility analyses herein have demonstrated average contact angles within this range, doubling the observed CA for glass, to make the material non-wetting, as demonstrated in Figure 2. As shown in Table 1, NF-1 and FS-1 exhibit average static contact angles around 100°, suggesting that they may reduce fouling through hydrophobic means and that the effects on adhesion and biofilm formation should be investigated. If the material does not have a water contact angle of >90°, it may not exhibit foul-releasing or foul-resistance characteristics through hydrophobic means. In this analysis, the model coatings had contact angles exceeding 90°, suggesting they have potential in foul-releasing or foul-resistance applications, and further testing was warranted.

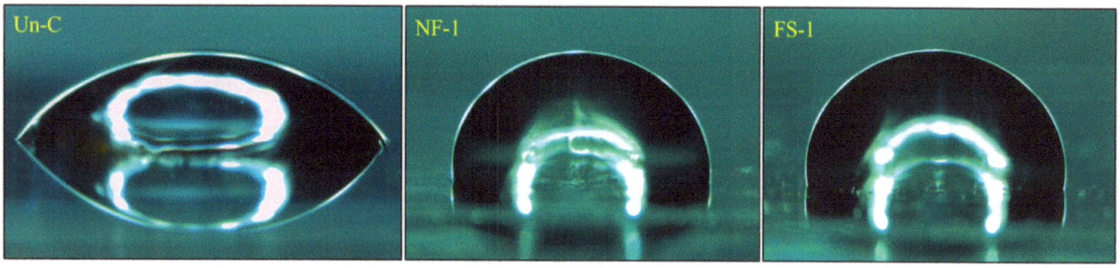

Figure 2. Examples of the average static contact angles observed in the uncoated microscope slides (58.9°, **left**), NF-1-coated slide (99.4°, **middle**), and FS-1-coated slide (101.4°, **right**).

Table 1. The average contact angles of the glass slide cover slip uncoated and coated with NF-1 and FS-1. The arithmetic mean roughness (Ra) and root-mean-square roughness (Rq) values for the coated samples are included.

Sample	Average Contact Angle	Arithmetic Mean Roughness (Ra)	Root-Mean-Square Roughness (Rq)
Uncoated	49.9° (±6.5)	-	-
NF-1	99.6° (±7.4)	7.26 nm (±2.8)	9.51 nm (±3.5)
FS-1	101.5° (±1.82)	3.65 nm (±1.8)	4.91 nm (±2.9)

Another manner in which surfaces may inhibit biofouling or exhibit hydrophobic characteristics is through surface roughness, typically a result of surface engineering. Designed surfaces may utilize specified peak-to-peak roughness to interfere with the adhesion mechanisms of the fouling organism, where microorganisms can adhere to smoother surfaces with shorter (<25 μm) peak-to-peak distances and larger organisms adhering to rougher materials [37]. To measure the roughness profile of a surface, the Ra (average roughness) value is commonly used. It refers to the deviation of the surface from its arithmetic mean height and provides an idea of how much the material surface varies across a given area. This can be supplemented by the root-mean-square deviation (Rq), which indicates the root mean square along the sampling length [38,47,56–58]. These foul-resistant materials may be rough and exhibit an Ra of 0.5–1.0 μm or utilize a super smooth surface, making the material challenging to adhere to [38,56]. The coatings being analyzed herein are considered smooth, with measurements showing values <8 nm for Ra and <10 nm for Rq (Table 1), indicating that the cured surfaces vary topographically, less than 10 nm across the test length. This suggests that the material may have a foul-resistant character through a smooth finish, making it preferable to prevent adhesion from larger organisms, as commonly reported with siloxane-based foul-releasing coatings [37,38,46,47,56–60]. Due to the smooth coating finish, we would expect less foul resistance to microorganisms such as bacteria, but a likely resistance to fouling by larger organisms. The studies herein will seek to justify or disprove microorganism claims through a series of simple experiments.

3.2. Verification of Coating Samples and Surface Energy

While coating characterization is not the focus of testing herein, it is vital to verify that the tested materials, NF-1 and FS-1, match the previously reported coating [24]. Images were taken of flaked and powdered coating to show the tested material's consistency, as well as relative size, with flakes appearing at sizes $\geq$ 50 μm and powder appearing at sizes $\leq$ 10 μm (Figure S1). All samples show a small alcohol peak near 3300 cm^{-1} and Si-O-Si bonding between 1100 and 990 cm^{-1}, which are expected in an alkoxysilane sol–gel coating system. Peaks at 2970–2860 cm^{-1} show the aliphatic carbons of this coating's carbon side chains (epoxy, amine, thiol, and vinyl), and a lack of peaks in the C=C and thiol regions suggests the material's thiol-ene polymerization was completed. The absence of primary amines also suggests the completion of the epoxy–amine curing. This spectrum reflects the original analysis detailed in a previous publication (Figure S2) [24].

Surface energy was determined using Zisman's method, indicating that NF-1 exhibits a surface energy of 33.58 mJ/m^2 while FS-1 exhibited 15.01 mJ/m^2 found from a linear regression of the trendline in Figure 3. These findings suggest that FS-1's inclusion of fluorinated side chains significantly affects surface energy as lower surface tension solvents could still exhibit recordable contact angles. A CA of zero was observed for n-hexanes, toluene, and cyclohexanes with NF-1 samples, indicating extensive solvent dispersion on coated surfaces beyond measurability. Conversely, FS-1 samples provided CAs for toluene and cyclohexanes, but no CA was observed with n-hexanes as this solvent also dispersed widely on the FS-1-coated surface. This method indicates that both NF-1 and FS-1 may be considered low-energy surfaces on par with what one would expect to observe in polymers such as polydimethylsiloxane (PDMS), polytetrafluoroethylene (PTFE), polyvinylidene fluoride (PVDF), and polystyrene (PS) [61]. This is another typical characteristic that may be observed with foul release coatings as low surface energy correlates to low surface binding ability similar to the surface energies exhibited by NF-1 and FS-1 [22,62]. These coatings may be considered low surface energy materials and suggest the potential for foul-releasing applications.

3.3. Accelerated Leaching Test

The accelerated leaching tests are designed to mimic the abrasive breakdown of coatings, starting with delamination or chipping, and ultimately forming fine powders from the material in an aqueous environment (i.e., microplastics), simulated via submersion of mechanically generated samples in nanopure water. This will allow any soluble free components to leach outward into the water, which can then be subjected to analytical methods such as UV-Vis spectroscopy to compare with the components of interest. Examining chips and powdered NF-1 and FS-1 will cover the full range of eventual abrasive breakdown for the coatings from realistic degradation seen with delamination or the hyper-realistic degradation of complete pulverization, providing insights into material behavior under these extremes.

NF-1	Surface Tension Energy (mN/m)	Cos (Contact Angle)	Contact Angle (°)
Water	72.80	-0.09	95.37 (± 3.46)
Glycerol	63.40	-0.14	98.33 (± 1.03)
Diiodomethane	50.80	0.56	55.68 (± 2.87)
Benzyl Alcohol	39.00	0.83	34.12 (± 2.67)
Toluene	28.40	1.00	0.00
Cyclohexanes	25.50	1.00	0.00
n-Hexane	18.40	1.00	0.00

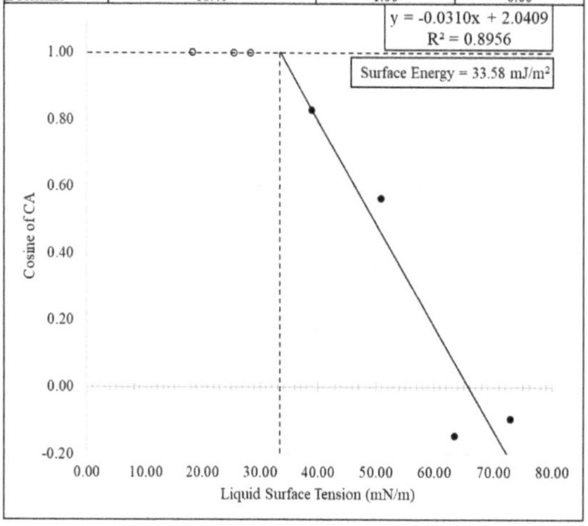

FS-1	Surface Tension Energy (mN/m)	Cos (Contact Angle)	Contact Angle (°)
Water	72.80	-0.18	100.17 (± 1.76)
Glycerol	63.40	-0.19	100.82 (± 2.05)
Diiodomethane	50.80	0.47	61.87 (± 3.71)
Benzyl Alcohol	39.00	0.49	60.83 (± 4.24)
Toluene	28.40	0.73	43.15 (± 4.48)
Cyclohexanes	25.50	0.72	44.25 (± 2.34)
n-Hexane	18.40	1.00	0

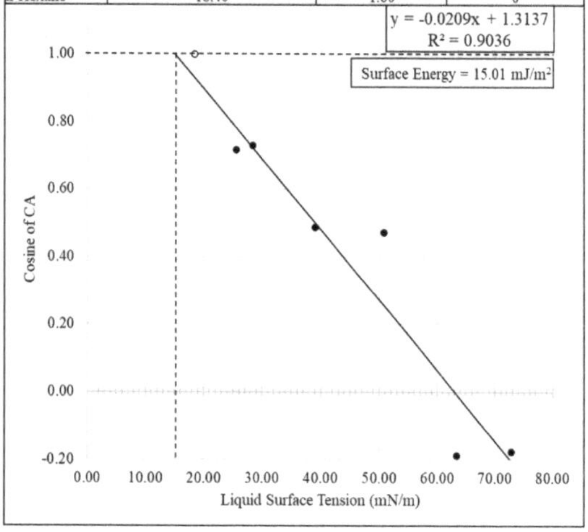

Figure 3. Zisman plots of surface energy for NF-1 (**left**) exhibiting 33.58 mJ/m^2 and FS-1 (**right**) exhibiting 15.01 mJ/m^2. The trend lines are shown in solid black, the solvents with non-zero contact angles are depicted as solid black dots, and solvents exhibiting no contact angle are shown as empty circles. Dashed lines show the maximum value for Cos θ at 1.00, and the corresponding X-intercept which represents the estimated surface energy.

The primary components of interest for the NF-1 and FS-1 coatings are the photo-activated curing agents, the radical initiator (RI—Omnirad 819), and the photoacid generator (PAG—diphenyliodonium hexafluorophosphate). The photoacid generator and radical initiator will ultimately undergo termination once reactivity has ceased; however, verifying that the coating remains inert throughout its consumer lifetime is essential. To verify this, the pure UV-Vis spectra of both the RI and PAG were analyzed in water and acetonitrile from 200 to 800 nm. Spectra were derived for the photo-active region (200−500 nm) to compare to the coating samples for leaching, as presented in Figure 4. Exposure to additional UV light (0, 2, 5, and 10 min) was utilized to look for a photo response (change in absorbance) in the photoinitiators compared to any leached particulate. Omnirad 819 is insoluble in water and therefore was not expected to contribute to significant leaching. Despite this, there is a notable shift in the absorption seen in Omnirad 819 over 10 min of UV exposure in ACN, showing a loss of absorption at 300 nm and weakened absorption at 245 nm. The UV-Vis spectra of Omnirad in ACN appears as expected per the manufacturer [27]. In the photoacid generator samples, the solvent did not affect the presence of the absorption seen from 220 to 230 nm and the shoulder at 275 nm. Upon irradiation, the PAG indicates a minimal shift in absorption intensity around these peaks, but there is no activity loss. An investigation into the effects of additional UV exposure provides insights into the degree of degradation of these starting materials, whether unconstrained or when utilized by the coating. For photoinitiators, we hypothesize that if peaks are observed and decrease over time with additional exposure, the initiators degrade or terminate their reactive processes. If peaks are observed but no change occurs over time with exposure, then the initiators are present or already in a degraded state.

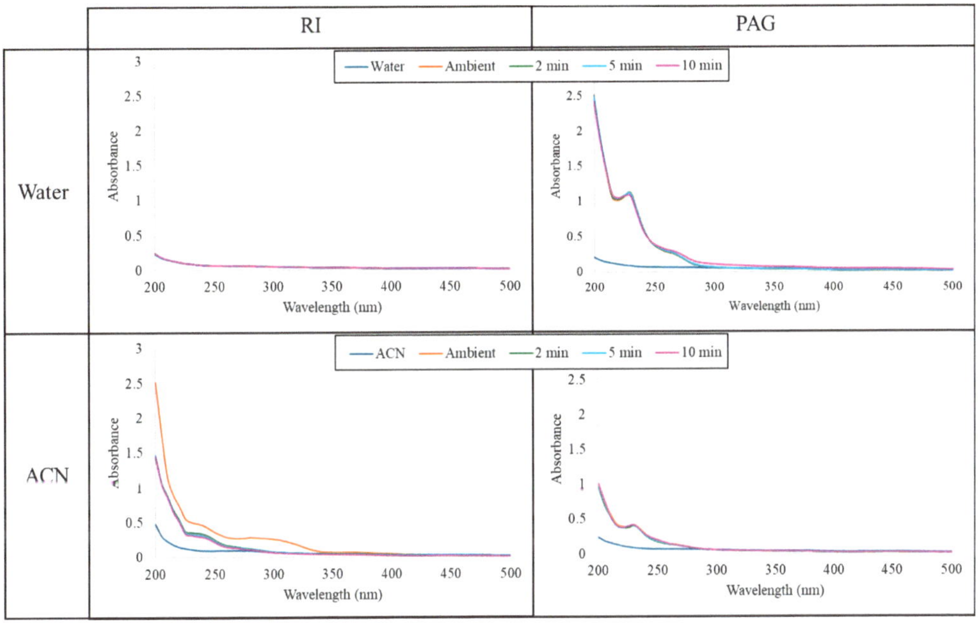

Figure 4. The UV-Vis spectra of the Omnirad 819 radical initiator (RI) and diphenyliodonium hexafluorophosphate photoacid generator (PAG) in water and acetonitrile (ACN) before and after exposure to a UV light source for 2, 5, and 10 min. All absorbances are shown from 0 to 3.0 for direct comparison.

NF-1 and FS-1 coatings were investigated in two degraded forms that offer insight into the effect of differing exposed surface areas (see Experimental Methods), including delaminated flakes (lower surface area) and powdered coatings (higher surface area). These samples were incubated in water and shaken daily to redistribute coating particulates and then filtered after one week. This was performed to ensure that any unpolymerized or freely moving reagents had the opportunity to diffuse into the solution while removing any large, polymerized materials before analysis. The filtered samples were then left to concentrate in a vacuum oven evaporatively before UV-Vis analysis. The coating samples utilized herein were cured by an initial 2 min of UV light exposure to begin the polymerization process; however, these coatings may be used outdoors where continuous sunlight ensures a complete cure where all reactants are expected to be polymerized. For this test, freshly lamp-cured coatings are used to provide insight into any leaching that would occur from artificial light sources with a limited exposure time. If present, these coatings are expected to show any traces of the photoinitiators. An investigation into changes occurring in response to additional UV exposure (2, 5, and 10 min) was used to assess any reduction in absorption or similarity to the initiators, which may suggest the degradation of any leached material. The samples generated by flakes of NF-1 and FS-1 show a slight increase (Δ0.2) in absorbance from 200 to 220 nm compared to the water blank and cannot definitively say that low surface area samples (flakes) of NF-1 and FS-1 exhibit component leaching due to the lack of absorption characteristics seen in Figure 5 for flaked samples.

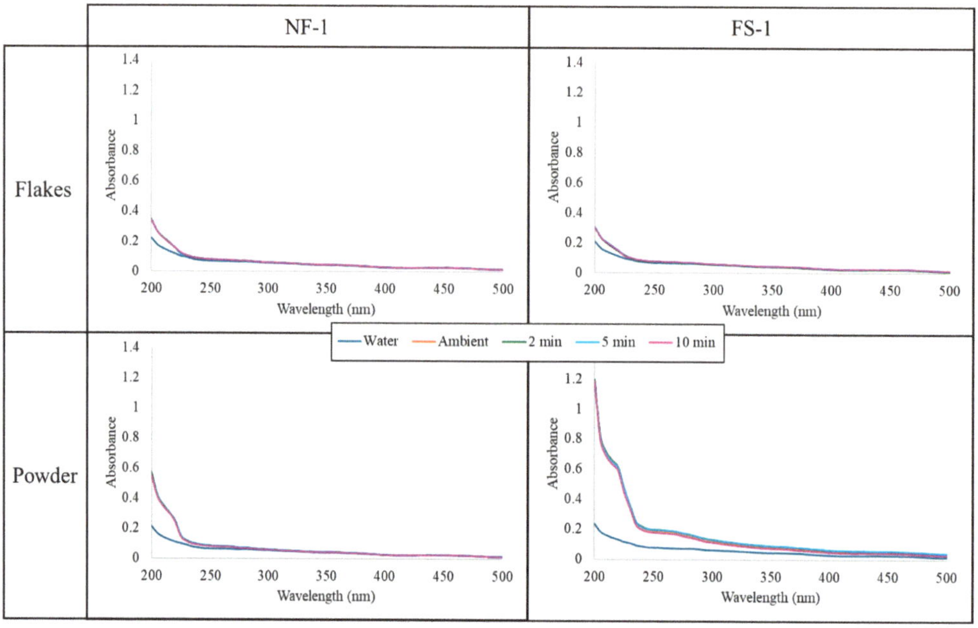

Figure 5. The UV-Vis spectra of the aqueous solutions that had been exposed to NF-1 and FS-1 coating samples after removal from a substrate as flakes and after being finely ground into a powder while heating. The spectra were taken under ambient conditions, and after exposing the solution to a UV light source, the solution was exposed for an additional 2, 5, and 10 min. Absorption is given in a range of intensity from 0 to 1.4 for comparative representation.

In contrast, the powdered NF-1 and FS-1 samples show some peak characteristics like the PAG in water, with a significant increase in the absorbance intensity compared to the flakes observed in Figure 4, showing absorption at 220 nm in both samples, with a shoulder at 275 nm in the FS-1 sample. A slight change in absorption intensity occurs over the final 5 min of UV exposure, and this, in combination with Omnirad's inability to dissolve in water, leads us to believe that powdered samples with a larger exposed surface area of material are potentially releasing PAG or its byproducts. While the observed peaks show a blue shift from ~230 nm to ~220 nm, this may indicate that leaching components are not the starting material but one of the byproducts generated through the photoinitiation process depicted in Figure 1. If peaks are observed but decrease over the time of exposure, then the initiators are theorized to become inactive with extra exposure (more than a 2 min lamp cure). If peaks are observed but do not change, this could result from degraded products or unreacted thiol-ene products remaining present. In these analyses, lower surface area samples do not exhibit pronounced leaching, but the high surface area samples show a shoulder broadening at 220 nm, which is likely from the PAG and/or its byproducts (i.e., iodo-biphenyls).

In addition to room temperature water leaching experiments, coating samples were also submerged in nanopure water and heated at 47 °C for 7 days to mimic extreme use conditions. Our main focus was on the powdered coating as these displayed the largest potential for leaching based on the surface area. Powdered samples are baked at 47 °C for 7 days in nanopure water and then analyzed at ambient and after two, five, and ten minutes of UV light to verify the effects of heat on the degradation and leaching of unreacted materials seen in Figure 6. Heated powdered samples showed no significant difference from those tested at an ambient temperature, suggesting minimal effects in regard to extreme temperature, which could improve solubility. If increasing the temperature of the sample resulted in more notable absorbance, then additional heating would result in more leaching in this material. Both coatings showed no change in absorbance over 10 min, suggesting no soluble photo-active components remained.

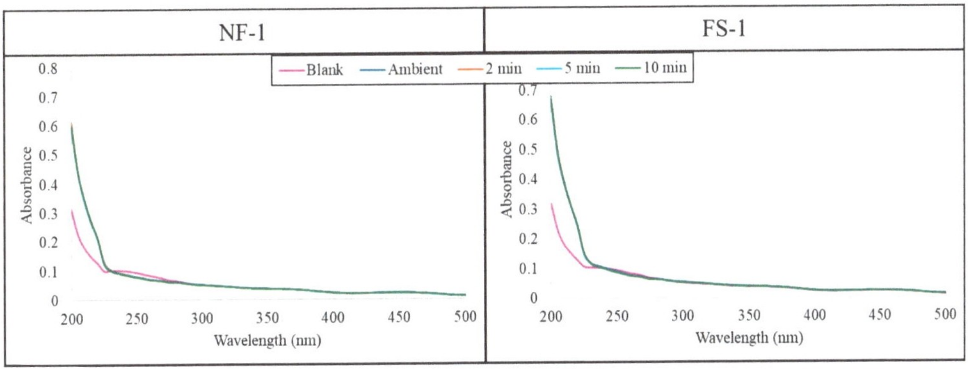

Figure 6. The UV-Vis spectra of the powdered NF-1 and FS-1 coatings after being held at 47 °C for 7 days. The samples were analyzed in water throughout ten minutes of UV exposure.

3.4. Direct Contact Colony Formation Toxicity Assay

Direct surface interactions are unavoidable with coating materials, and it is essential to understand if there are any surface-based reactions, leaching, or absorption that can occur when organisms encounter the material in a semi-dry environment (agar and air). Analyzing these materials for toxicity, biocidal, or inhibitory effects on prokaryotic cells provides a way to understand the nature of a material and its interactions with simple organisms such as *E. coli*, which are self-sufficient and rapidly reproduce to form colonies. Researchers can observe acute cell toxicity and growth inhibition by utilizing direct contact culturing to examine the materials for possible antimicrobial or biocidal activity.

In this case, two cell cultures of *E. coli* were produced at different concentrations of cells (10^7 and 10^6 dilute). These were plated, and coverslips were added. Each plate contained an unaltered agar segment, a control glass slide, and NF-1-coated and FS-1-coated samples. If a material has antimicrobial character, we expect to see inhibited or reduced colony formation in the area contacting the coating or material being tested. This would result in an observed and measurable zone of inhibition around the materials' outer edges, suggesting that the material or any potentially leaching components create an antimicrobial local environment. Regions with no culture growth around a test site are called the zone of inhibition and are measurable regions used to compare antimicrobial properties and cell toxicity adapted for this analysis. A reduction in colony count around the material indicates an inhibited growth rate, and a lack of colonies indicates antimicrobial/biocidal characteristics. If the cells grow adjacent to or touch the coverslip, there is no zone of inhibition or reduction in growth rate—indicating no microbial toxic agents or antimicrobial character leaching to the surface.

NF-1- and FS-1-coated slides displayed colony growth under each slide while in contact with the coatings in both traditional media and rich media environments in a similar frequency to an uncoated negative control. No zone of inhibition was observed around any of the coated slides, and the colony count was similar in frequency to that of the control, as seen in Figure 7. Colony growth under the slides is observable in less frequency and size than the surrounding environments, attributed to a lack of air exchange as colonies of similar morphology grew in contact with the uncoated negative control. To be sure this was the case, a direct contact method employing growth rate monitoring in liquid media was used to demonstrate that coated slides were not toxic but did not allow for the adequate exchange of air which will be discussed in Section 3.5. Cells grew directly in contact with the edges of slides in all examples–with higher concentrations leading to an outline of the slides– and since samples were dip-coated, all edges are coated. This indicates that the material does not leach in a harmful manner in dry/semi-dry conditions, suggesting it is not biocidal or toxic in a semi-dry environment.

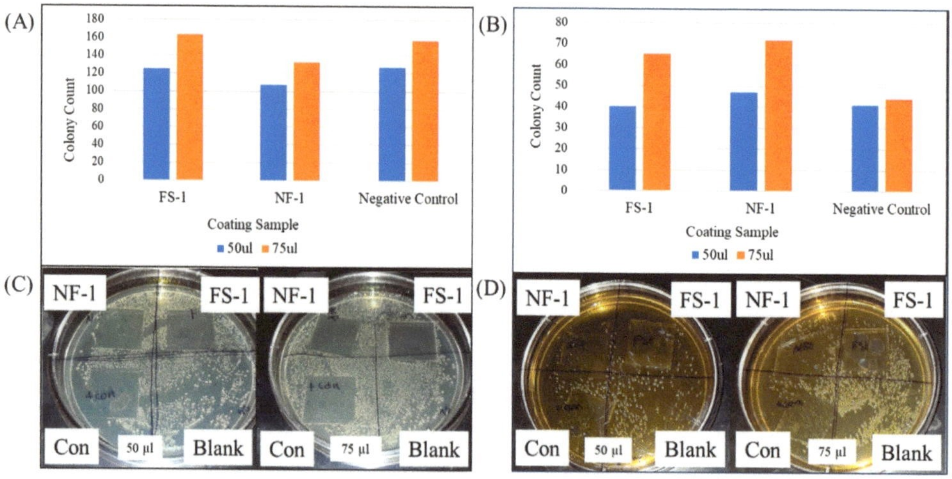

Figure 7. (**A**) Direct contact methods on Miller LB diluted 10^6 seeded with 50 μL (blue) and 75 μL (orange) colony count under coated surface. (**B**) Terrific LB diluted 10^7 seeded with 50 μL (blue) and 75 μL (orange) colony count under coated surface. (**C**) Miller LB plates seeded with 50 μL and 75 μL. (**D**) Terrific broth plates seeded with 50 μL and 75 μL.

3.5. Liquid Culture Direct Contact Growth Rate

While colonies grew in contact with cured FS-1 and NF-1 coatings, liquid culture testing is vital to inspect for the occurrence of leaching or toxicological effects, which may occur when the material is in contact with a solution that may allow unpolymerized or unreacted components to mobilize. The two primary efforts are determining if minute levels of inhibitory or toxic agents are present that reduce growth rates without inhibiting them completely, and the second is to test the material for leaching in an aqueous environment and to understand if there were effects or leaching, which occurs only in non-dry environments when a medium is present, which allows molecules to disperse. The liquid culture test described herein used *E. coli*-cultured LB (liquid broth) and six cultures for each testing group (control, NF-1, and FS-1), taking one measurement at each time interval to generate an average. The interior of the test tubes was coated, and the cured material was inspected for leaching or toxicological effects when cells were cultured in an aqueous solution to observe effects on growth throughout the exponential growth phase. Any minute toxicity effects are expected to show a deviation in OD_{600} corresponding to a change in growth rate compared to the control. Readings are taken until an optical density of 2.0 at 600 nm is achieved (7–8 h) as the cultures exit exponential growth into the stationary phase (proliferation–death equilibrium point).

The results showed that no significant deviation in growth rate occurred as compared to an uncoated negative control when all cells were grown from the same mother culture to reduce genetic effects. Figure 8 shows a strong overlap of the control and coated substrates occurring for the duration of the analysis. NF-1 and FS-1 show little deviation from the control, with all graph lines retaining the same shape and little to no deviation in culture absorbance throughout the experiment. If there was an inhibition of growth due to the leaching of biocidal or toxic components, there would be a difference in the time it took to reach an OD_{600} of 2.0, which was not observed, and the extrapolated graphs would be notably different, with reduced growth rates (slope).

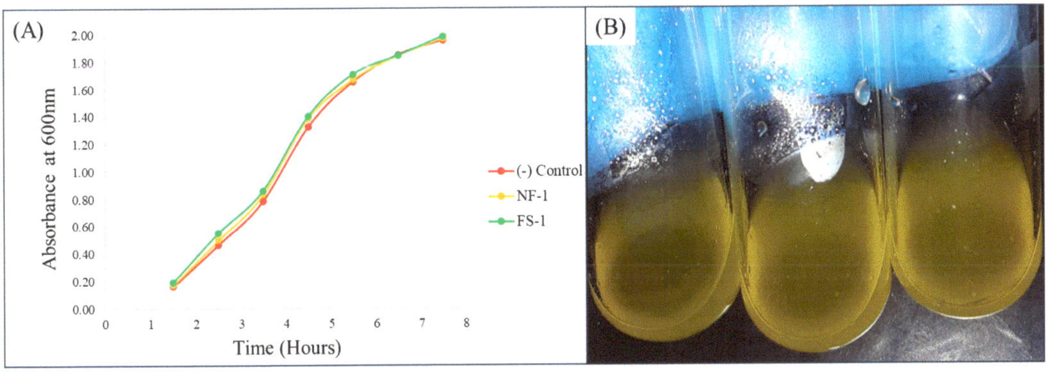

Figure 8. Optical density measured at 600 nm over 8 h: FS-1 (green), NF-1 (orange), and control (red). (**A**) Growth rate over 8 hrs. (**B**) *E. coli* culture in coated test tubes.

These results support the findings of the direct contact growth analysis in Section 3.3, which indicates that NF-1 and FS-1 have minimal interface effects on cell health or proliferation. The uninhibited growth rate observed in the liquid culture test suggests that the morphological colony changes observed in the direct contact method were due to a lack of air exchange and not a toxic effect due to contact with the cured coating. The combined results propose that neither coating has a direct biocidal, toxic, or inhibitory effect. This demonstrates that there is no leaching of material components that inhibit cell growth or cause cell death in prokaryotic organisms.

3.6. Biofilm Production Assay

Biofilm production is a characteristic of some bacteria when colonizing a surface, which may lead to enhanced antibiotic resistance and a persistent infection due to the protective extracellular matrix. In hospital settings, inhibiting the formation of biofilms on medical devices promotes sterile surfaces and helps retain a clean and uncontaminated environment. Testing to understand if a surface prevents or inhibits biofilm formation can be conducted by looking for a reduction in or prevention of protein adherence to the surface, which is quantifiable by utilizing a microplate reader and absorbance measured at 550 nm for the detection of crystal violet, which binds to proteins and DNA, staining the live cells.

To investigate the effect of coatings on cell adherence and biofilm formation, 96-well plates were dip-coated and then utilized in a culture and staining analysis with a cystic fibrosis-derived *P. aeruginosa* strain, and the cultures were inspected for differences. If there was a reduction in the absorbance in the coated sample sets, it would suggest a decrease in cell adherence to the surface, indicating that the coatings exhibited biofilm inhibitory effects. Both of the coatings analyzed exhibit hydrophobic character and a smooth surface, traits seen in other anti-biofilm coatings as previously mentioned in Section 3.1. As a result, NF-1 and FS-1 were both theorized to be unfavorable for attachment compared to the control. When analyzed, the NF-1 sample performed similarly to the bare microplate samples, indicating no biofilm inhibition. The microplate coated with FS-1 showed a slight increase in biofilm formation in Figure 9 over the control, indicating that it may provide a preferential surface for bacterial adhesion.

While these results did not show an anti-biofilm character, the successful cell growth suggests that the coating exhibits no biocidal effects on these prokaryotic cells. This implies that any material that may leach from the surface did not cause direct cell growth inhibition or toxicity, supporting the results of the direct contact toxicity assay and liquid culture growth rate analysis, which both utilized a different bacterium, *E. coli.*, suggesting these results were non-species specific. The low variations in absorbance between the control, NF-1, and FS-1 systems insinuate no effect on biofilm formation. Forming a biofilm implies that the material is not antibacterial or does not inhibit growth—suggesting that it is not toxic to cells and permits adhesion. If biofilms were inhibited, then the mechanism of inhibition needs to be validated through other methods, including leaching, toxicity, and surface adherence, as described herein. Neither coating analyzed prevents the formation of biofilms from P. aeruginosa despite having considerably low surface energy, indicating that these materials are not antifouling against bacteria; however, the reason for this manuscript was not necessarily to verify the fouling resistance of these coatings, but to provide a model of simple methods to prove or disprove this without expensive testing methods.

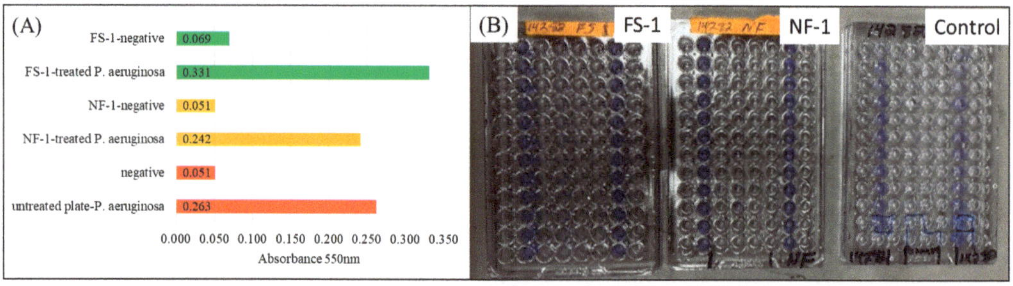

Figure 9. (**A**) Biofilm formation assay with *P. aeruginosa*. Absorbance at 550 nm for FS-1 (green), NF-1 (yellow), and control (red). (**B**) Here, 96-well coated plates of FS1, NF1, and control from left to right.

3.7. Assessment of Direct Contact HEK 293 Cell Viability

Although no acute toxic effects were detected with *E. coli*, it is noteworthy to consider that interactions with the treated NF-1 and FS-1 could yield different outcomes when exposed to more complex eukaryotic cells. Such observations could offer valuable insights into their potential biomedical applications. Mammalian adherent HEK 293 cells are highly sensitive to environmental changes such as pH like the PAG initiator and hazardous materials like Omnirad 819, making them valuable for investigating non-toxic properties. These cells also utilize proteins to adhere to surfaces, similar to *P. aeruginosa*, forming a monolayer of cells on the substrate—a process necessary for their proliferation [39]. This analysis offers a means to validate potential discrepancies between prokaryotic (simple) and eukaryotic (complex) organisms concerning contact and environmental toxicity, as well as the inhibition of cell adhesion by coated substrates. These validations are essential for confirming the reliability of assays outlined in Sections 3.3–3.5. We hypothesized that in the event of acidic material or leaching components, a change in media color would be anticipated, accompanied by the death of pH-sensitive mammalian cells. Additionally, if the materials decreased adhesion, it would result in the inhibition of culture proliferation.

Adherent HEK 293 cells grown in contact with NF-1- or FS-1-coated slides exhibited no observable change in media color, which typically shifts to yellow when pH reaches 6.8 or lower. Consequently, the study as shown in Figure 10 gives no discernible pH alteration observed over the course of 3 days of culturing. The cells' sensitivity to pH underscores the absence of corrosive effects from the material, as evidenced by a sustained cell viability of over 97%, surpassing the threshold of 95% indicative of a healthy interaction with sufficient adhesion and pH stability. Moreover, consistent cell morphology further supports the notion that adhered cells remain unaffected by the coated substrates, affirming their benign nature and lack of antifouling properties. The substrates neither impede adhesion nor induce pH alterations detrimental to sensitive mammalian (eukaryotic) cells and therefore exhibit non-toxic characteristics.

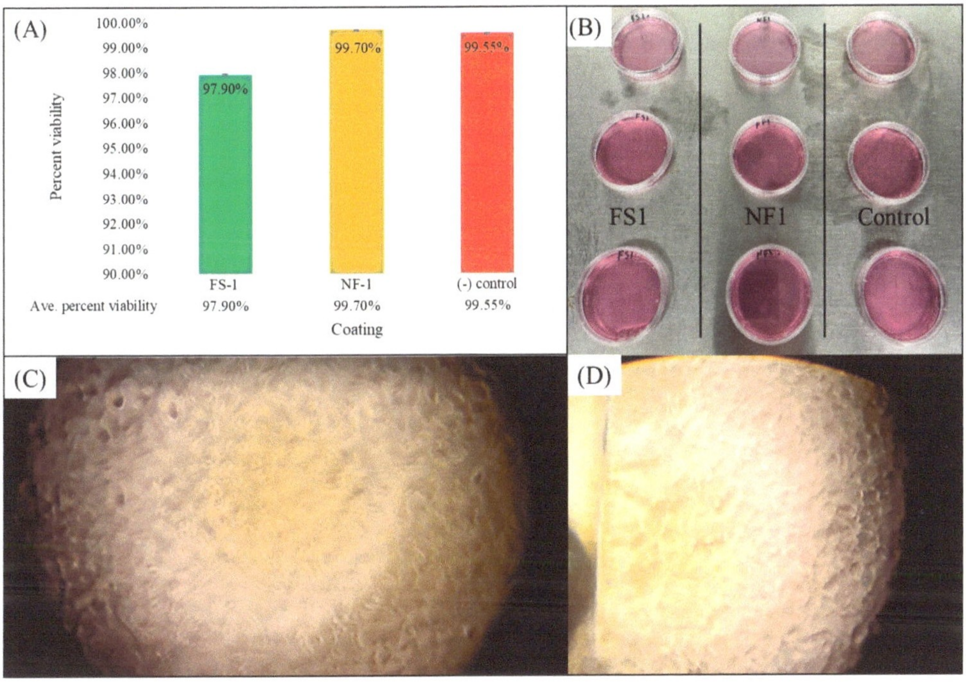

Figure 10. (**A**) Average percent viability. (**B**) Culture plates in triplicate. (**C**) NF-1-coated HEK cell-colonized slide. (**D**) FS-1-coated HEK cell-colonized slide.

4. Conclusions

Protective coatings are widely used in the modern world, from industrial applications to everyday consumer products. As the coatings industry strives to be more eco-friendly, it becomes increasingly important to assess both the toxicity and environmental effects of materials carefully. By utilizing the range of biological assays and accelerated leaching tests described herein, we can efficiently and cost-effectively determine the compatibility of materials with the environment and various organisms. Combining these analytical methods as a single, modifiable set of assays can provide insights into materials' hydrophobicity, leaching, toxicity, adhesion, and fouling interactions of surface coatings. While some of the example analyses involved multiple organisms, these methods can be tailored to focus on a single organism for biocompatibility characterization within a BSL-1 lab. These assays provide a simple and effective means of material characterization in a manner that is readily accessible to researchers, small laboratories, and businesses who wish to understand the potential applications of coatings or thin films either commercially available or designed by themselves.

We chose the two investigated organo-silane-based protective coatings (NF-1 and FS-1) for their surface characteristics, makeup, and hypothesized applications for antimicrobial, antifouling, or foul-releasing uses. Prior research on these coatings focused on material characterization and their use as protective coatings for a variety of substrates, with no known advantage or disadvantage for biocompatibility, making them an unbiased material for these analyses. Testing bacterial interactions shows they were not biocidal or antifouling but display benign bacterial cellular interactions. We also used an accelerated leaching analysis to simulate natural degradation over time to investigate photo-activatable agents and their ability to leach. Diphenlyiodonium hexafluorophosphate was quite soluble in water and demonstrated similar peak patterns to the acetonitrile control. When inspecting samples with a lower surface area, representative of delaminated or chipped material, no significant leaching of photo-active components was observed. However, in the higher surface area, the finely powdered samples, some absorptive characteristics of diphenlyiodonium hexafluorophosphate were notable at 220 nm, which may be due to the nature of the photoacid generator and its byproducts, as no significant pH change was observed in the HEK cell analysis. This gave confidence that no significant component leaching occurs when the coating delaminates, as analysis of highly degraded material suggests the minimal presence of the initiators seen in absorption measurements in Figure 5. No antibacterial traits were detected, as evidenced by the high cell viability observed in all cell cultures seen in Figure 7, Figure 8, and Figure 10 as compared to controls. While the biofilm assay was able to show the buildup and formation of a biofilm on each surface, this is not representative of foul release testing and may be due to surface interactions since bacteria typically have minimal adherence to substrates with surface energies around 25 mJ/m^2 and may occur due to the coating surface makeup or the gentle culture workup [63]. The benign biocompatibility interactions and surface characterization, such as CA, surface roughness, and surface energy, will steer future investigations regarding NF-1 and FS-1 toward foul-releasing studies as opposed to antibacterial and antifouling implementations as these characteristics were not observed. By thoroughly evaluating biocompatibility, leaching, and surface properties, these assays can assist in identifying a specific range of potential applications and guide the next phase of materials research/testing, which can aid decision-making before committing to extensive third party analysis. The coatings industry encompasses a wide range of uses and utilizing this set of assays can help ascertain any antibacterial, antifouling, and foul-releasing characteristics that are of particular value for commercial or industrial applications.

Supplementary Materials: The following supporting information can be downloaded at: https://www.mdpi.com/article/10.3390/coatings14070849/s1.

Author Contributions: Conceptualization, R.M.P., C.B.S. and J.C.F.; Methodology, R.M.P., C.B.S., X.T. and H.W.; Validation, R.M.P.; Investigation, R.M.P., M.J.F. and H.W.; Resources, X.T.; Writing—original draft, R.M.P. and C.B.S.; Writing—review & editing, X.T., H.W. and J.C.F.; Supervision, H.W. and J.C.F. All authors have read and agreed to the published version of the manuscript.

Funding: This research was funded by Bowling Green State University through startup funding to X.T.

Institutional Review Board Statement: Not applicable.

Informed Consent Statement: Not applicable.

Data Availability Statement: Data are contained within the article and supplementary materials.

Conflicts of Interest: The authors declare no conflict of interest.

Abbreviations

OD_{600}	Optical Density at 600 nm	CA	Contact Angle
SOC	Super Optimal Broth	RI	Radical Initiator
DMEM	Dulbecco'c Modified Eagle Medium	PAG	Photoacid Generator
FBS	Fetal Bovine Serum	ACN	Acetonitrile
MCE	Mixed Cellulose Ester	LB	Liquid Broth
Ra	Mean Roughness	BSL	Biological Saftey Lab
Rq	Root-mean-square roughness		

References

1. Krewski, D.; Andersen, M.E.; Tyshenko, M.G.; Krishnan, K.; Hartung, T.; Boekelheide, K.; Wambaugh, J.F.; Jones, D.; Whelan, M.; Thomas, R.; et al. *Toxicity Testing in the 21st Century: Progress in the Past Decade and Future Perspectives*; Springer: Berlin/Heidelberg, Germany, 2020; Volume 94, ISBN 0123456789.
2. Fischer, I.; Milton, C.; Wallace, H. Toxicity Testing Is Evolving! *Toxicol. Res.* **2020**, *9*, 67–80. [CrossRef]

3. Balouiri, M.; Sadiki, M.; Ibnsouda, S.K. Methods for in Vitro Evaluating Antimicrobial Activity: A Review. *J. Pharm. Anal.* **2016**, *6*, 71–79. [CrossRef] [PubMed]
4. Li, W.; Zhou, J.; Xu, Y. Study of the in Vitro Cytotoxicity Testing of Medical Devices. *Biomed. Rep.* **2015**, *3*, 617–620. [CrossRef]
5. Chellappa, M.; Anjaneyulu, U.; Manivasagam, G.; Vijayalakshmi, U. Preparation and Evaluation of the Cytotoxic Nature of TiO_2 Nanoparticles by Direct Contact Method. *Int. J. Nanomed.* **2015**, *10*, 31–41. [CrossRef]
6. Iverson, B.L.; Dervan, P.B. Rapid Screening of Biomedical Polymers by Two Methods of Tissue Cul. 7823–7830.
7. Piazza, V.; Gambardella, C.; Garaventa, F.; Massanisso, P.; Chiavarini, S.; Faimali, M. A New Approach to Testing Potential Leaching Toxicity of Fouling Release Coatings (FRCs). *Mar. Environ. Res.* **2018**, *141*, 305–312. [CrossRef]
8. Pantulap, U.; Unalan, I.; Zheng, K.; Boccaccini, A.R. Hydroxycarbonate Apatite Formation, Cytotoxicity, and Antibacterial Properties of Rubidium-Doped Mesoporous Bioactive Glass Nanoparticles. *J. Porous Mater.* **2024**, *31*, 685–696. [CrossRef]
9. Li, M.; Mondrinos, M.J.; Chen, X.; Gandhi, M.R.; Ko, F.K.; Lelkes, P.I. In Vitro and in Vivo (Cyto)Toxicity Assays Using PVC and LDPE as Model Materials. *J. Biomed. Mater. Res. A* **2006**, *79*, 175–182. [CrossRef]
10. HENSTEN-PETTERSEN, A. Comparison of the Methods Available for Assessing Cytotoxicity. *Int. Endod. J.* **1988**, *21*, 89–99. [CrossRef]
11. Gatto, C.; Ruzza, P.; Giurgola, L.; Honisch, C.; Rossi, O.; Romano, M.R.; Ragazzi, E.; D'Amato Tóthová, J. Comparison of Perfluorocarbon Liquids Cytotoxicity Tests: Direct Contact Versus the Test on Liquid Extracts. *ACS Omega* **2022**, *8*, 365–372. [CrossRef]
12. Srivastava, G.K.; Alonso-Alonso, M.L.; Fernandez-Bueno, I.; Garcia-Gutierrez, M.T.; Rull, F.; Medina, J.; Coco, R.M.; Pastor, J.C. Comparison between Direct Contact and Extract Exposure Methods for PFO Cytotoxicity Evaluation. *Sci. Rep.* **2018**, *8*, 1425. [CrossRef]
13. Liu, X.; Rodeheaver, D.P.; White, J.C.; Wright, A.M.; Walker, L.M.; Zhang, F.; Shannon, S. A Comparison of in Vitro Cytotoxicity Assays in Medical Device Regulatory Studies. *Regul. Toxicol. Pharmacol.* **2018**, *97*, 24–32. [CrossRef]
14. Wang, M.O.; Etheridge, J.M.; Thompson, J.A.; Vorwald, C.E.; Dean, D.; Fisher, J.P. Evaluation of the in Vitro Cytotoxicity of Cross-Linked Biomaterials. *Biomacromolecules* **2013**, *14*, 1321–1329. [CrossRef]
15. Li, M.; Neoh, K.G.; Xu, L.Q.; Wang, R.; Kang, E.T.; Lau, T.; Olszyna, D.P.; Chiong, E. Surface Modification of Silicone for Biomedical Applications Requiring Long-Term Antibacterial, Antifouling, and Hemocompatible Properties. *Langmuir* **2012**, *28*, 16408–16422. [CrossRef]
16. Alexandrescu, L.; Syverud, K.; Gatti, A.; Chinga-Carrasco, G. Cytotoxicity Tests of Cellulose Nanofibril-Based Structures. *Cellulose* **2013**, *20*, 1765–1775. [CrossRef]
17. Leonardo, M.R.; Da Silva, L.A.B.; Tanomaru Filho, M.; Bonifácio, K.C.; Ito, I.Y. In Vitro Evaluation of Antimicrobial Activity of Sealers and Pastes Used in Endodontics. *J. Endod.* **2000**, *26*, 391–394. [CrossRef]
18. Francolini, I.; Donelli, G. Prevention and Control of Biofilm-Based Medical-Device-Related Infections. *FEMS Immunol. Med. Microbiol.* **2010**, *59*, 227–238. [CrossRef]
19. Zare, M.; Ghomi, E.R.; Venkatraman, P.D.; Ramakrishna, S. Silicone-Based Biomaterials for Biomedical Applications: Antimicrobial Strategies and 3D Printing Technologies. *J. Appl. Polym. Sci.* **2021**, *138*, 50969. [CrossRef]
20. Lopez, A.I.; Kumar, A.; Planas, M.R.; Li, Y.; Nguyen, T.V.; Cai, C. Biofunctionalization of Silicone Polymers Using Poly(Amidoamine) Dendrimers and a Mannose Derivative for Prolonged Interference against Pathogen Colonization. *Biomaterials* **2011**, *32*, 4336–4346. [CrossRef]
21. Ciriminna, R.; Bright, F.V.; Pagliaro, M. Ecofriendly Antifouling Marine Coatings. *ACS Sustain. Chem. Eng.* **2015**, *3*, 559–565. [CrossRef]
22. Hu, P.; Xie, Q.; Ma, C.; Zhang, G. Silicone-Based Fouling-Release Coatings for Marine Antifouling. *Langmuir* **2020**, *36*, 2170–2183. [CrossRef]
23. Chambers, L.D.; Stokes, K.R.; Walsh, F.C.; Wood, R.J.K. Modern Approaches to Marine Antifouling Coatings. *Surf. Coat. Technol.* **2006**, *201*, 3642–3652. [CrossRef]
24. Sims, C.B.; Lenora, C.U.; Furgal, J.C. Hybrid Tri-Cure Organo-Silicon Coatings for Monument Preservation. *Coatings* **2022**, *12*, 1098. [CrossRef]
25. Omnirad 819. Material Safety Data Sheet. Available online: https://gato-docs.its.txstate.edu/jcr:65445a50-260d-47c8-a175-86b58b3fb23e/IGMR%20Resins%20USA%20Inc%20-%20Omnirad%20819.pdf (accessed on 3 July 2024).
26. Diphenyliodonium Hexafluorophosphate. Material Safety Data Sheet. Available online: https://www.fishersci.com/store/msds?partNumber=AC450910050&productDescription=DIPHENYLIODONIUM+HEXAFLU+5GR&vendorId=VN00032119&countryCode=US&language=en (accessed on 3 July 2024).
27. IGM Resins General Information Omnirad 819 Photoinitiator Product Data Sheet. Available online: https://www.penpoly.com/app/uploads/2021/07/Omnirad-819_tds.pdf (accessed on 3 July 2024).
28. Balan, L.; Melinte, V.; Buruiana, T.; Schneider, R.; Vidal, L. Controlling the Morphology of Gold Nanoparticles Synthesized Photochemically in a Polymer Matrix through Photonic Parameters. *Nanotechnology* **2012**, *23*, 415705. [CrossRef] [PubMed]
29. Uzunlar, E.; Schwartz, J.; Phillips, O.; Kohl, P.A. Decomposable and Template Polymers: Fundamentals and Applications. *J. Electron. Packag.* **2016**, *138*, 020802. [CrossRef]
30. Ghobashy, M.M. Ionizing Radiation-Induced Polymerization. In *Ionizing Radiation Effects and Applications*; InTech: Houston, TX, USA, 2018.

31. Dektar, J.L.; Hacker, N.P. Photochemistry of Diaryliodonium Salts. *J. Org. Chem.* **1990**, *55*, 639–647. [CrossRef]
32. Crivello, J.V. Diaryliodonium Salt Photoacid Generators. In *Wiley Blackwell 6*; Wiley Blackwell: Hoboken, NJ, USA, 2014; pp. 457–478. ISBN 9781118909911.
33. Hossain, M.L.; Hammer, K.; Lim, L.Y.; Hettiarachchi, D.; Locher, C. Optimisation of an Agar Overlay Assay for the Assessment of the Antimicrobial Activity of Topically Applied Semi-Solid Antiseptic Products Including Honey-Based Formulations. *J. Microbiol. Methods* **2022**, *202*, 106596. [CrossRef] [PubMed]
34. Sjollema, J.; Zaat, S.A.J.; Fontaine, V.; Ramstedt, M.; Luginbuehl, R.; Thevissen, K.; Li, J.; van der Mei, H.C.; Busscher, H.J. In Vitro Methods for the Evaluation of Antimicrobial Surface Designs. *Acta Biomater.* **2018**, *70*, 12–24. [CrossRef] [PubMed]
35. Ghamrawi, S.; Bouchara, J.P.; Corbin, A.; Rogalsky, S.; Tarasyuk, O.; Bardeau, J.F. Inhibition of Fungal Growth by Silicones Modified with Cationic Biocides. *Mater. Today Commun.* **2020**, *22*, 100716. [CrossRef]
36. Borowski, R.G.V.; Barros, M.P.; da Silva, D.B.; Lopes, N.P.; Zimmer, K.R.; Staats, C.C.; de Oliveira, C.B.; Giudice, E.; Gillet, R.; Macedo, A.J.; et al. Red Pepper Peptide Coatings Control Staphylococcus Epidermidis Adhesion and Biofilm Formation. *Int. J. Pharm.* **2020**, *574*, 118872. [CrossRef]
37. Lejars, M.; Margaillan, A.; Bressy, C. Fouling Release Coatings: A Nontoxic Alternative to Biocidal Antifouling Coatings. *Chem. Rev.* **2012**, *112*, 4347–4390. [CrossRef]
38. Yebra, D.M.; Kiil, S.; Dam-Johansen, K. Antifouling Technology—Past, Present and Future Steps towards Efficient and Environmentally Friendly Antifouling Coatings. *Prog. Org. Coat.* **2004**, *50*, 75–104. [CrossRef]
39. CELL CULTURE BASICS Handbook. Available online: https://www.vanderbilt.edu/viibre/CellCultureBasicsEU.pdf (accessed on 3 July 2024).
40. Piccinini, F.; Tesei, A.; Arienti, C.; Bevilacqua, A. Cell Counting and Viability Assessment of 2D and 3D Cell Cultures: Expected Reliability of the Trypan Blue Assay. *Biol. Proced. Online* **2017**, *19*, 8. [CrossRef]
41. Lebeau, P.F.; Chen, J.; Byun, J.H.; Platko, K.; Austin, R.C. The Trypan Blue Cellular Debris Assay: A Novel Low-Cost Method for the Rapid Quantification of Cell Death. *MethodsX* **2019**, *6*, 1174–1180. [CrossRef]
42. Kamiloglu, S.; Sari, G.; Ozdal, T.; Capanoglu, E. Guidelines for Cell Viability Assays. *Food Front.* **2020**, *1*, 332–349. [CrossRef]
43. Segeritz, C.P.; Vallier, L. *Cell Culture: Growing Cells as Model Systems In Vitro*; Elsevier: Amsterdam, The Netherlands, 2017; ISBN 9780128030783.
44. Buskens, P.; Wouters, M.; Rentrop, C.; Vroon, Z. A Brief Review of Environmentally Benign Antifouling and Foul-Release Coatings for Marine Applications. *J. Coat. Technol. Res.* **2013**, *10*, 29–36. [CrossRef]
45. Francolini, I.; Vuotto, C.; Piozzi, A.; Donelli, G. Antifouling and Antimicrobial Biomaterials: An Overview. *Apmis* **2017**, *125*, 392–417. [CrossRef]
46. Pistone, A.; Scolaro, C.; Visco, A. Mechanical Properties of Protective Coatings against Marine Fouling: A Review. *Polymers* **2021**, *13*, 173. [CrossRef]
47. Sfameni, S.; Rando, G.; Marchetta, A.; Scolaro, C.; Cappello, S.; Urzì, C.; Visco, A.; Plutino, M.R. Development of Eco-Friendly Hydrophobic and Fouling-Release Coatings for Blue-Growth Environmental Applications: Synthesis, Mechanical Characterization and Biological Activity. *Gels* **2022**, *8*, 528. [CrossRef]
48. Gipperth, L. The Legal Design of the International and European Union Ban on Tributyltin Antifouling Paint: Direct and Indirect Effects. *J. Environ. Manag.* **2009**, *90*, 86–95. [CrossRef]
49. McNeil, E.M. Antifouling: Regulation of Biocides in the UK before and after Brexit. *Mar. Policy* **2018**, *92*, 58–60. [CrossRef]
50. Rulison, C. Measure Surface Energy: A Tutorial Designed to Provide Basic Understanding of the Concept of Solid Surface Energy. Available online: https://www.sanyo-si.com/wp-content/uploads/kruss-tn306-en.pdf (accessed on 3 July 2024).
51. O'Toole, G.A. Microtiter Dish Biofilm Formation Assay. *J. Vis. Exp.* **2011**, *30*, e2437. [CrossRef]
52. Law, K.-Y. Definitions for Hydrophilicity, Hydrophobicity, and Superhydrophobicity: Getting the Basics Right. *J. Phys. Chem. Lett.* **2014**, *5*, 686–688. [CrossRef] [PubMed]
53. Erbil, H.Y. Contact Angle of Liquid Drops on Solids, 1st edition. *Surf. Chem.* **2006**, 308–337. [CrossRef]
54. Thomas, S.; Nochehdehi, A.R. *Handbook of Magnetic Hybrid Nanoalloys and Their Nanocomposites*; Springer: Berlin/Heidelberg, Germany, 2022; ISBN 9783030909482.
55. Ray, S.S.; Lee, H.K.; Kwon, Y.N. Review on Blueprint of Designing Anti-Wetting Polymeric Membrane Surfaces for Enhanced Membrane Distillation Performance. *Polymers* **2019**, *12*, 23. [CrossRef] [PubMed]
56. Candries, M.; Atlar, M.; Guerrero, A.; Anderson, C.D. Lower Frictional Resistance Characteristics of Foul Release Systems. In *Practical Design of Ships and Other Floating Structures*; Elsevier: Amsterdam, The Netherlands, 2001; pp. 517–523.
57. Sun, L.; Wang, N.; Wang, Y.; Chen, P.; Liu, H.; Cao, S.; Liu, Y.; Gao, C. Fabrication of an Anti-Fouling Coating Based on Epoxy Resin with a Double Antibacterial Effect via an in Situ Polymerization Strategy. *Prog. Org. Coat.* **2023**, *184*, 107837. [CrossRef]
58. Akuzov, D.; Vladkova, T.; Klöppel, A.; Brümmer, F. Composition antifouling coating: Effect of siloxane and fluorinated oil incorporation on marine biofilm formation. *J. Chem. Technol. Metall.* **2014**, *49*, 229–237.
59. Candries, M.; Atlar, M.; Anderson, C.; Candries, M.; Atlar, M.; Anderson, C.D. Low-Energy Surfaces on High-Speed Craft. *Proc. 2nd Euroconference High-Perform. Veh. HIPER* **2001**, *1*, 119–127.
60. Qiu, H.; Feng, K.; Gapeeva, A.; Meurisch, K.; Kaps, S.; Li, X.; Yu, L.; Mishra, Y.K.; Adelung, R.; Baum, M. Functional Polymer Materials for Modern Marine Biofouling Control. *Prog. Polym. Sci.* **2022**, *127*, 101516. [CrossRef]

61. Dataphyics Surface Tension Values of Some Common Test Liquids for Surface Energy Analysis Surface Tension Values of Some Common Test Liquids for Surface Energy Analysis. Available online: https://www.dataphysics-instruments.com/Downloads/Surface-Tensions-Energies.pdf (accessed on 3 July 2024).
62. Detty, M.R.; Ciriminna, R.; Bright, F.V.; Pagliaro, M. Environmentally Benign Sol-Gel Antifouling and Foul-Releasing Coatings. *Acc. Chem. Res.* **2014**, *47*, 678–687. [CrossRef]
63. Zhao, Q.; Liu, Y.; Wang, C.; Wang, S.; Müller-Steinhagen, H. Effect of Surface Free Energy on the Adhesion of Biofouling and Crystalline Fouling. *Chem. Eng. Sci.* **2005**, *60*, 4858–4865. [CrossRef]

coatings

MDPI

Article

A Comparative Study of the Corrosion Behavior of P110 Casing Steel in Simulated Concrete Liquid Containing Chloride and Annulus Fluid from an Oil Well

Yang Li [1,2,3], Zhongxu Cai [1,2,3], Lijuan Huang [1,2,3,*] and Ruiquan Liao [1,2,3]

1 Petroleum Engineering College, Yangtze University, Wuhan 430100, China; hiliyoung@163.com (Y.L.)
2 Key Laboratory of CNPC for Oil and Gas Production, Yangtze University Branch, Wuhan 430100, China
3 Key Laboratory of Drilling and Production Engineering for Oil and Gas, Wuhan 430100, China
* Correspondence: huanglijuan@yangtzeu.edu.cn

Abstract: The corrosion behavior of P110 casing steel in simulated concrete liquid and simulated annulus fluid was investigated to reveal the corrosion pattern and protective properties of corrosion products in the two environments. Potentiodynamic polarization curves, electrochemical impedance spectroscopy (EIS), Mott–Schottky tests, and electrochemical noise (EN) tests were used to study the corrosion behavior of P110 casing steel in simulated concrete liquid and simulated annulus fluid saturated with CO_2. Scanning electron microscopy (SEM) combined with Energy-Dispersive Spectrometer (EDS) mapping was used to characterize the corrosion morphology and elemental distribution of P110 casing steel. The results show that the corrosion resistance of P110 casing steel deteriorates with the increasing immersion days in the simulated annulus fluid, the impedance decreases gradually, and the corrosion-product film shows a loose and porous structure. In the simulated concrete liquid, under the condition of containing a low concentration of Cl^-, the protection of the corrosion products gradually increases with the extension of immersion days. With the increasing concentration of Cl^- and the extension of immersion days, the electrochemical noise resistance and charge transfer resistance of P110 steel decrease gradually, and the protective property of the corrosion-product film decreases, which is capable of forming steady pitting corrosion.

Keywords: P110 steel; corrosion behavior; annulus-fluid environment; concrete-liquid environment; electrochemical corrosion

1. Introduction

Corrosion failure of oil-casing steel in the downhole environment is one of the main factors affecting the safe production of oil and gas. Oil-casing tubing is a significant equipment in the process of oil and gas field development. On one hand, it can fix the formation and prevent collapse of the downhole environment. On the other hand, it can effectively isolate the penetration of the fluid in the wellbore and protect the underground environment as well. At present, a large number of oil and gas wells have entered the high water cut development stage, and coupled with the application of CO_2-enhanced oil and gas recovery technology, the working environment of oil-casing steel is more complicated. The outer wall of the casing steel provides an important service in concrete environments, which always contain high concentrations of Cl^- and high salinity, and is susceptible to local corrosion of the oil casing. The inner wall of the casing is mostly affected in annulus environments, where inorganic salts are mostly used as an aggravating agent, which can easily cause corrosion failure of the casing steel, and where the corrosive gas in the tubing can also invade the annular environment, further inducing casing failure [1].

Currently, some achievements have been made regarding the corrosion behavior of metals in simulated concrete-liquid environments, and the scope of this research has mainly been aimed at the corrosion behavior of simulated concrete liquids in high-pH-value environments. Shi et al. [2] investigated the corrosion behavior of low-alloy steel in a high-pH simulated concrete-pore solution using electrochemical tests, and their research found that the passivation film in low-alloy steel can significantly improve the pitting resistance of the metal, but the pitting resistance of low-alloy steel in an environment with a low concentration of Cl^- is lower than that of low-carbon steel. However, in a high-Cl^--concentration environment, the passivation film of low-alloy steel can dramatically reduce the pitting tendency of the metal. When the passivation film on the metal surface is locally destroyed and the cathode reaction is promoted, meanwhile, the anodic dissolution rate reaction on the metal surface is also accelerated. According to the study of Thangavel et al. [3], the initial concentration of Cl^- as well as O_2 diffusion needs to be considered when evaluating the corrosion behavior of concrete environments, and the ratio of Cl^- to OH^- individually cannot fully reflect the overall corrosion situation. The contamination of Cl^- in concrete environments is a common working condition in cementing processes in oil and gas fields, but due to the "size effect" of the real tube, it is difficult to evaluate the risk of corrosion failure using the concentration of Cl^- in a concrete environment [4], and the current research still needs to simulate real conditions in a laboratory to evaluate these corrosion conditions [5,6]. Huang et al. [7] investigated the corrosion behavior of P110 casing steel under alkaline-containing and oxidizing media conditions in oil- and gas-well environments, and their study showed that adjusting the ratio of initial oxide to NaOH could control the corrosion rate.

In the downhole environment of oil and gas fields, pitting corrosion and perforation of casing steel in the annulus fluid is another important factor causing casing failure [8]. The annulus fluid is relatively static, and corrosive gases such as CO_2 leaking from the pipe gap can easily gather in the annulus, accelerating the serious pitting corrosion of the inner wall of the casing. The existing studies were mainly conducted on the alloy composition of the casing steel, the temperature and pressure of the working environment, and corrosion failure forms [9–11]. Zeng et al. [12] found that in a CO_2 annulus environment, severe uniform corrosion would be induced in the saturated brine phase, and more serious pitting corrosion would occur in the simulated production solution. The study of Wei et al. [13,14] also pointed out that P110 steel exhibits mainly uniform corrosion failure in the liquid phase and mainly local corrosion in the CO_2 environment. Xu et al. [15] concluded that when the Cr content is less than 13%, the corrosion resistance of an alloy will increase with the increase in Cr content. In summary, certain progress has been made in material selection and corrosion failure forms of oil-casing steel. However, there are few reports on the mechanism of local corrosion and pitting corrosion formation, and there are few comparative analyses of the corrosion behavior of oil-casing steel in internal and external working environments.

In this study, the corrosion kinetics mechanisms of P110 casing steel in simulated annulus fluid saturated with CO_2 and a simulated concrete-liquid environment are analyzed, and the two kinds of corrosion environments are compared. Immersion experiments were used to study the electrochemical corrosion behavior and morphology of P110 casing steel after immersion for different lengths of time (in days), and EN and EIS techniques were used to investigate corrosion kinetics changes and local corrosion occurrence in the P110 casing steel after different durations of immersion. On this basis, the corrosion failure mechanism of P110 casing steel was combined with the results of Mott–Schottky tests and potentiodynamic polarization curves. Finally, by comparing the corrosion behavior of the P110 casing steel under the two environments, the high-risk working condition was obtained, which provides a theoretical basis for the protection of casing steel from corrosion.

2. Experiments and Materials

2.1. Experiment Material

The material used in this study is P110 casing steel, which was taken from a certified oil-well casing tube. An Optical Emission Spectrometer (OES) (GNR S3 Minilab, Milan, Italy) was used to test the material's element composition, and this element composition is shown in Table 1. The metallographic structure of P110 casing steel in its as-supplied state after polishing and etching with 4% nitrate alcohol is shown in Figure 1. The metallographic structure is mainly bainite and martensite, with a small amount of ferrite.

Table 1. Elemental composition of P110 casing steel.

Element	C	Si	Mn	P	S	Cr	Ni	V	Mo	Cu	Als	Fe
Mass fraction (%)	0.28	0.30	0.65	0.02	0.01	1.10	0.02	0.08	0.18	0.20	0.02	balance

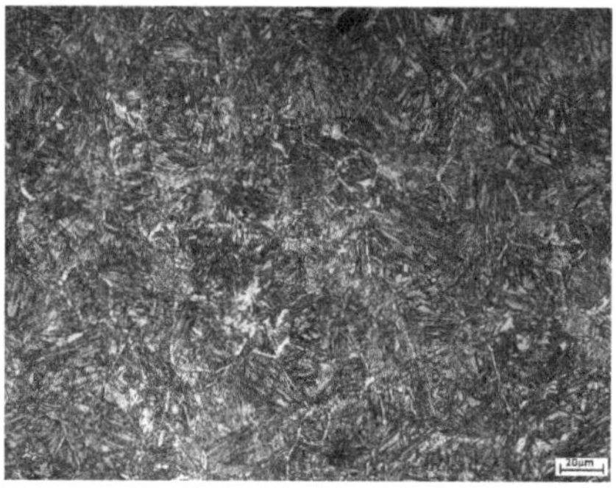

Figure 1. Metallographic structure of P110 casing steel.

2.2. Electrochemical Tests

The electrochemical working electrodes were P110 casing steel, the cutting size was 10 mm × 10 mm × 3 mm, and the electrode working area was 1 cm^2. Each specimen was welded with copper wires on the non-working surface, and then encapsulated with epoxy resin as a working electrode. Before the experiment, the working electrodes were polished step by step with metallographic silicon paper from $80^{\#}$ to $2000^{\#}$, wiped with anhydrous-ethanol, rinsed with deionized water, dried with cold air, and then stored. The electrochemical tests were displayed on a CS310M electrochemical workstation (CORRTEST, Wuhan, China). The tests adopted a three-electrode system with a platinum sheet as the counter electrode (CE) and a saturated calomel electrode (SCE) as the reference electrode. Open-circuit potential tests were performed at first, and electrochemical impedance spectroscopy (EIS) and potentiodynamic polarization curves were performed after the open-circuit potential was stabilized. The electrochemical impedance spectroscopy measurements were taken according to the AC sine-wave voltage from peak to peak at 10 mV at the open-circuit potential in the frequency range of 10^{-2} Hz to 10^5 Hz, and then an equivalent circuit was fitted to the electrochemical impedance spectroscopy. The potentiodynamic polarization curves were scanned from the cathode to the anode with a starting potential of −0.3 V (vs. OCP); the scanning was stopped at a current density of 10^{-2} A/m^2 and the scanning rate was 1 mV/s. The Mott–Schottky curves were tested at a frequency of 1 kHz, with a sine-wave amplitude of 5 mV. The Mott–Schottky curves' measured potentials spanned from high to low potential, ranging from −0.8 V to 0.5 V, with a potential interval of 50 mV. The electrochemical noise (EN) tests were measured using two identical P100 steel working electrodes and an SCE, with a sampling frequency of 5 Hz and a test time of 3600 s. To minimize the interference of external interferences, the EN measurement system was placed in a Faraday shielded box (Shanghai Xianren Instrument Co., Ltd., Shanghai,

China). After the electrochemical noise tests, the DC component of the original electrochemical noise was removed through polynomial fitting. All potentials herein are relative to the SCE in this paper.

Saturated $Ca(OH)_2$ supernatant was used as the base solution system to simulate the concrete environment [16,17], and the pH value of the base solution was 13. NaCl was added into the simulated concrete liquid, and the concentration of Cl^- was kept at 0.001, 0.05, 0.1, 0.3, 0.5 mol/L; then, deoxygenation with N_2 was performed for 3 h, and it was stored after sealing. The simulated annulus fluid was taken from a CO_2-driven oil-well environment with the following compositions: $NaHCO_3$ 2.7 g/L, NaCl 8.2 g/L, Na_2SO_4 0.33 g/L, $CaCl_2$ 1.1 g/L. The pH value was adjusted to 4 with 10% glacial acetic acid and 10% NaOH solution, and the N_2 was deoxidized for 3 h, and then saturated with CO_2 for 4 h [18]. The experimental container gaps or crevices were sealed with silicone rubber. The experimental simulation solutions were configured using analytically pure-grade (AR) reagents and deionized water. The experiment systems were immersed in the simulated solutions for 10 d. EIS and EN tests were carried out at 1 d, 3 d, 5 d, 7 d, and 10 d. Mott–Schottky curves and potentiodynamic polarization curves were measured at 10 d, respectively. The deoxygenation operation was performed in the experimental solutions by bubbling N_2 at the end of each experiment, and the simulated annulus fluid was saturated with CO_2. A water bath was used to provide a constant temperature environment to maintain the experimental temperature at 60 °C.

2.3. Surface Analysis

The P110 steel was made into 30 mm × 15 mm × 3 mm portions, and the test sample was soaked in a round hole with a diameter of 4 mm at 1 mm away from the edge of the sample. Before the experiment, the immersion specimens were polished step by step with metallographic silicon paper from 80# to 2000#, wiped with anhydrous ethanol, rinsed with deionized water, and the N_2 was blown dry for later use. The specimens were suspended with non-metallic wires and then fixed in a sealed container; the immersion experiment time was 10 d, and the experiment temperature was 60 °C. At the end of the experiment, the specimens were ultrasonically cleaned for 3 min, rinsed with deionized water, and purged with N_2. The corrosion morphology was observed using a ZEISS MELIN Compact scanning electron microscope (ZEISS, Oberkochen Germany), and the EDS mappings were performed using Oxford X-MAXN (Oxford Instruments, Oxford, UK) for elemental analysis.

3. Results

3.1. Potentiodynamic Polarization Measurements

The potentiodynamic polarization curves of P110 in simulated concrete liquid with different concentrations of Cl^- and simulated annulus fluid at 60 °C are shown in Figure 2. The potentiodynamic polarization curves containing passivation intervals were fitted to the passivation current density (i_p) and passivation potential (E_p), and the polarization curves in the activated state were fitted to the corrosion current density (i_{corr}) and corrosion potential (E_{corr}) using Tafel extrapolation; these fitting results are shown in Table 2. As can be seen from Figure 2, a protective passivation film structure was generated on the surface of the P110 casing steel after immersion for 10 d at five concentrations of Cl^- in the simulated concrete-liquid environment. In the simulated concrete-liquid environment, with the increase in Cl^- concentration, the polarization curve gradually shifted to the right, which indicates that the electrochemical reaction activity increased and promoted the corrosion of the P110 steel, and the passivation interval gradually became shorter, which indicates that the passivation film was less protective. Also, in the anodic region of the polarization curve, the anodic slope was changed, in which the anodic polarization curves went through the activation–passivation transition zone, and shifted to the stable passivation zone; finally, the passivation film was broken down because of the high anodic potential. Moreover, i_p was positively correlated with the Cl^- concentration: with the increasing Cl^- concentration, i_p increased, which indicates that the presence of Cl^- was detrimental to the structure of the passivation film of the P110 casing steel in the simulated concrete-liquid environment, and also indicates that the presence of Cl^- increases the corrosion tendency of N80 steel. In the simulated annulus-fluid environment, the P110 casing steel was activated, and the i_{corr} was significantly greater than that of the simulated concrete-liquid environment, in which no protective corrosion-product film was generated; also, the corrosion rate was significantly higher than that of the simulated concrete-liquid environment.

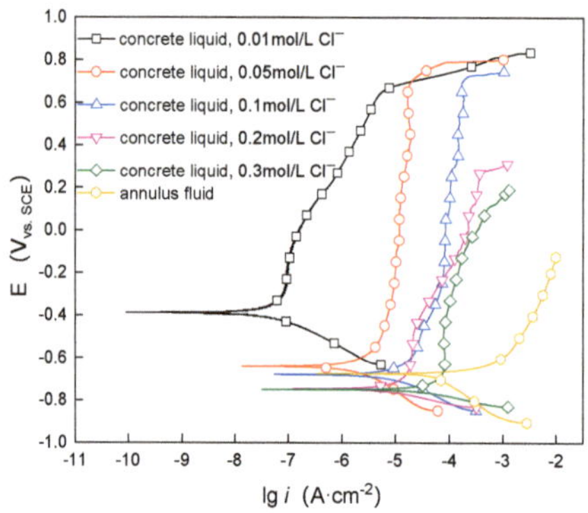

Figure 2. Potentiodynamic polarization curves of P110 after 10 d of immersion.

Table 2. Electrochemical fitting parameters for the polarization curves under different corrosion conditions.

Environmental Factors		i_p/i_{corr} A·cm^{-2}	E_p/E_{corr} V
Simulated concrete liquid Cl^- concentration (mol/L)	0.01	1.36×10^{-7}	−0.38
	0.05	3.86×10^{-6}	−0.57
	0.1	2.07×10^{-5}	−0.61
	0.2	2.93×10^{-5}	−0.63
	0.3	1.05×10^{-4}	−0.68
Simulated annulus fluid		2.21×10^{-4}	−0.71

3.2. Electrochemical Impedance Spectroscopy

Figure 3 shows the electrochemical impedance spectroscopy (EIS) results for P110 casing steel in simulated concrete-liquid solution and simulated annulus fluid at a test temperature of 60 °C. The EIS measurements were fitted using the equivalent circuits R(Q(R(QR))) and R(QR), as shown in Figure 4, and the fitting results are shown in Table 3, where R_s represents the solution resistance, Q_f is the passivated film capacitance constant phase angle element, Q_{dl} is the double-layer capacitance constant phase angle element, n is the dispersion coefficient, R_f is the passivated film resistance, and R_{ct} represents the charge transfer resistance [19,20].

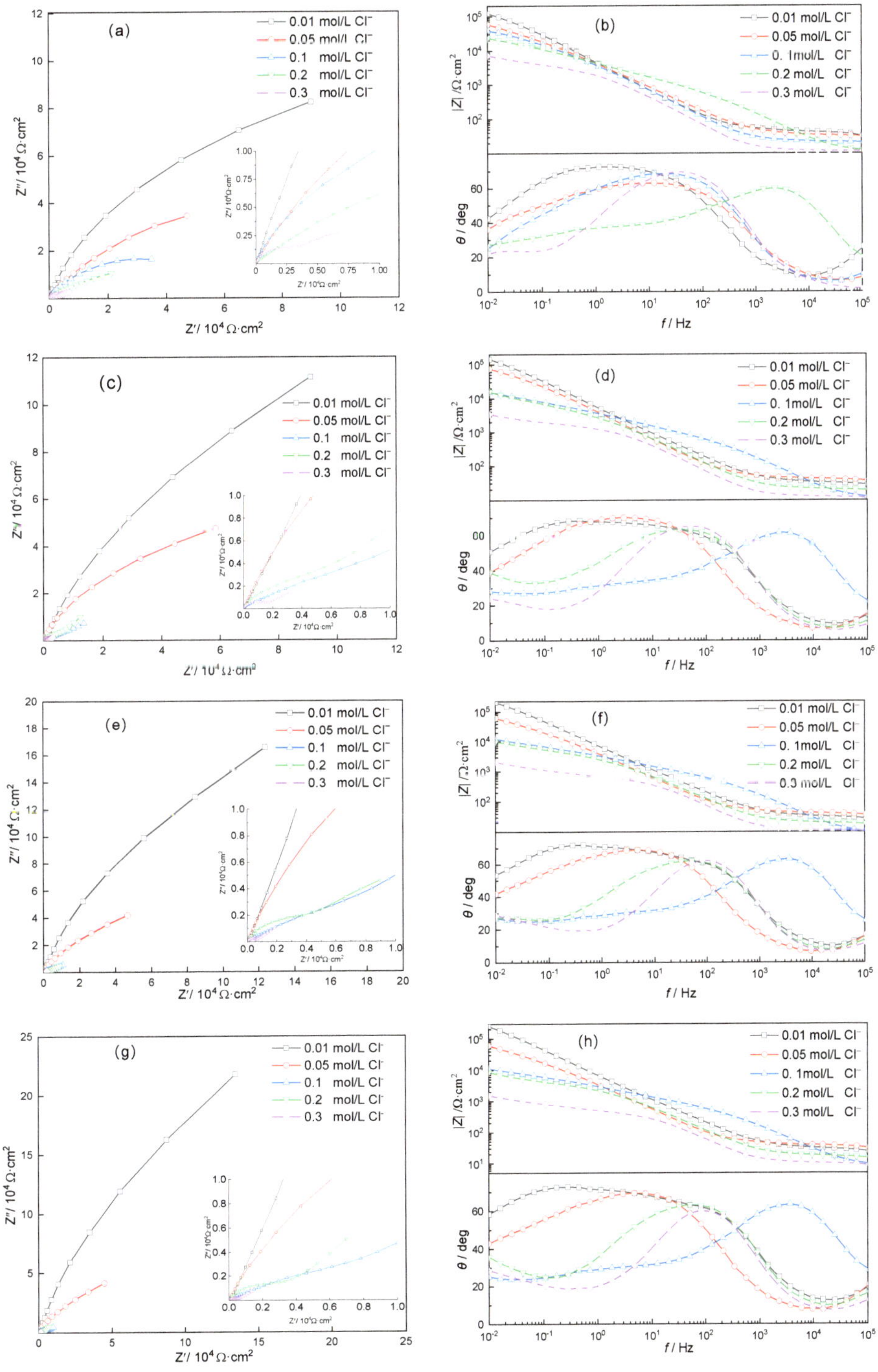

Figure 3. *Cont.*

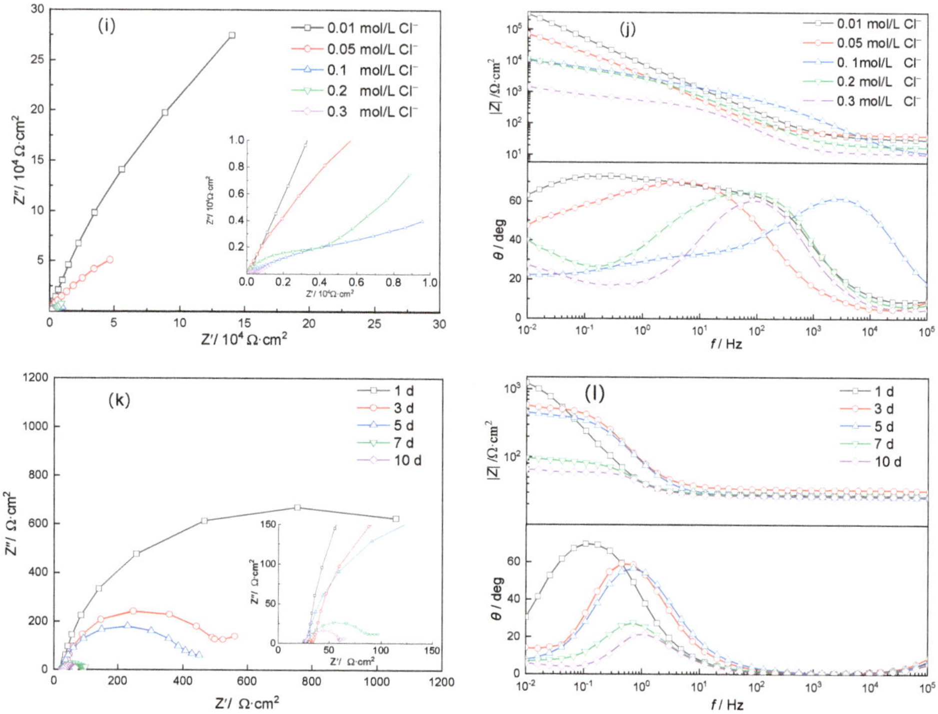

Figure 3. EIS results of P110 under different immersion times: (**a**,**c**,**e**,**g**,**i**) Nyquist diagrams after immersion for 1 d, 3 d, 5 d, 7 d, and 10 d in simulated concrete-liquid environments with different concentrations of Cl^-; (**b**,**d**,**f**,**h**,**j**) Bode diagrams after immersion for 1 d, 3 d, 5 d, 7 d, and 10 d in simulated concrete-liquid environments with different concentrations of Cl^-; (**k**,**l**) Nyquist and Bode diagrams after immersion for 1 d, 3 d, 5 d, 7 d, and 10 d in simulated annulus fluid.

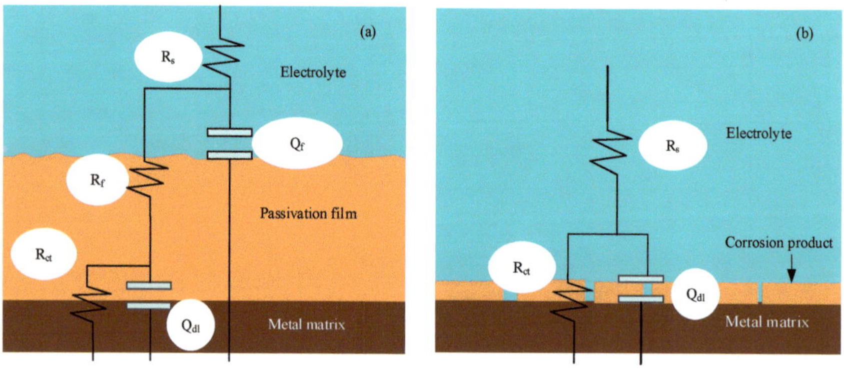

Figure 4. EIS equivalent circuit diagrams: (**a**) equivalent circuit diagram of EIS after 3–10 d of immersion in a simulated concrete-liquid environment; (**b**) equivalent circuit diagram of EIS after immersion for 1 d in a simulated concrete-liquid environment and simulated annulus fluid.

Table 3. EIS fitting results.

EIS Factors	Immersion Days	0.01 mol/L Cl^-	0.05 mol/L Cl^-	0.1 mol/L Cl^-	0.2 mol/L Cl^-	0.3 mol/L Cl^-	Annulus Fluid
R_s ($\Omega \cdot cm^{-2}$)	1 d	41.9	31.3	20.6	3.01	11.1	27.44
	3 d	5.27	41.53	10.51	19.7	6.5	33.27
	5 d	11.9	37.81	8.70	19.36	9.78	26.17
	7 d	10.38	38.71	7.57	16.7	9.85	26.75
	10 d	22.48	42.75	10.32	17.05	10.7	26.21
$Q_f \times 10^{-5}$ ($\Omega^{-1} \cdot cm^{-2} \cdot s^n$)	1 d	/	/	/	/	/	/
	3 d	8.28×10^{-2}	5.83	2.97×10^{-1}	7.04	6.78	/
	5 d	1.02×10^{-1}	2.84	3.07×10^{-1}	5.43	15.8	/
	7 d	1.09×10^{-1}	6.53	3.49×10^{-1}	5.58	7.56	/
	10 d	2.11×10^{-1}	6.35	2.87×10^{-1}	5.22	6.99	/
n1	1 d	/	/	/	/	/	/
	3 d	0.73	0.80	0.85	0.75	0.84	/
	5 d	0.77	0.43	0.85	0.79	0.71	/
	7 d	0.78	0.79	0.84	0.76	0.84	/
	10 d	0.79	0.80	0.86	0.78	0.85	/
R_f ($\Omega \cdot cm^{-2}$)	1 d	/	/	/	/	/	/
	3 d	26.88	47.11	13.43	17.47	5.39	/
	5 d	21.89	18.55	350.6	264.0	13.77	/
	7 d	21.69	46.84	347.5	295.3	6.07	/
	10 d	11.69	45.62	281.1	8.92	2.66	/
$Q_{dl} \times 10^{-5}$ ($\Omega^{-1} \cdot cm^{-2} \cdot s^n$)	1 d	4.88	6.29	6.55	6.04	9.63	611.3
	3 d	4.42	4.83	1.44×10^{-1}	2.56	13.9	228.1
	5 d	3.45	4.93	16.3	34.3	19.5	244.5
	7 d	3.13	8.06	17.21	60.7	25.9	815.6
	10 d	2.67	5.61	15.5	43.5	28.6	711.4
n2	1 d	0.80	0.71	0.76	0.55	0.78	0.94
	3 d	0.76	0.83	0.89	0.47	0.40	0.93
	5 d	0.78	0.83	0.36	0.37	0.42	0.91
	7 d	0.79	0.69	0.37	0.50	0.41	0.83
	10 d	0.78	0.66	0.38	0.51	0.43	0.91
R_{ct} ($\Omega \cdot cm^{-2}$)	1 d	2.17×10^5	8.14×10^4	4.26×10^4	2.46×10^4	5413.0	1494.0
	3 d	4.37×10^5	1.49×10^5	7.72×10^4	4.98×10^4	1823.0	533.6
	5 d	7.41×10^5	2.36×10^5	3.56×10^4	1.31×10^4	1352.0	415.9
	7 d	1.33×10^6	3.03×10^5	3.02×10^4	6435	1021	67.24
	10 d	5.11×10^6	2.34×10^5	2.03×10^4	3769	505.4	36.09

Figure 3a,c,e,g,i,k show the Nyquist diagrams of P110 for different days of immersion, and Figure 3b,d,f,h,j,l show the Bode diagrams of P110 for different days of immersion. From the Nyquist diagrams, it can be seen that the EIS diagrams of P110 are all characterized by a single capacitive semicircular, indicating that the corrosion process was mainly controlled by electrochemical reactions. Under the same number of immersion days, but with increases in the concentration of Cl^- in the simulated concrete-liquid environment, the capacitive semicircular radius in the Nyquist plot gradually decreases, and the impedance modulus in the Bode plot also decreases, which indicates that the corrosion resistance of the P110 casing steel decreased with the increase in the concentration of Cl^-. Also, the impedance modulus in the simulated annulus fluid is smaller than that in the simulated concrete-liquid environment after the same number of days of immersion, which indicates that the corrosion degree was higher (i.e., the P110 corroded to a greater extent), which is consistent with the results of the polarization curves.

By analyzing the EIS diagrams of P110 casing steel in the two environments, it can be seen that in the simulated concrete-liquid environment, when the immersion time of the P110 steel was in the range of 3 d to 10 d, the frequency–phase angle diagrams of the

Bode plots in the middle and low frequency regions have constant phase angle intervals, and the phase angle values are larger than other conditions, which indicates that in these environments, a passivation film was generated by the P110 steel; at this time, there were defects in the passivation film on the surface of the metal, and corrosive ions could be transported to the metal substrate through these defects. Therefore, the components Q_f and R_f correlated to the corrosion film layer were considered in the equivalent circuit to characterize the changes in the film layer of the corrosion products, and the equivalent-circuit diagram of Figure 4a was used for the EIS fitting in such conditions [18,19]. After 1 d of immersion, the Nyquist diagram is dominated by a single capacitive semicircle, indicating that the electrochemical reaction was mainly controlled by charge transfer. In the annulus-fluid environment, the highest value of the phase angle in the Bode plot of P110 gradually shifts to the high-frequency region as the number of days of immersion increases, indicating that the electrochemical reaction activity increased, the charge transfer rate accelerated, and the corrosion became more severe with the increase in the number of days of immersion; so, the equivalent circuit diagram of Figure 4b was used for the fitting in the two environments mentioned above.

From the EIS results, it can be seen that in the simulated concrete-liquid environment, when the concentration of Cl^- was 0.01 mol/L, the value of R_{ct} increased gradually with the increase in immersion time, which indicates that the corrosion products of the P110 casing steel had protective properties under a low concentration of Cl^-, the protective properties increased gradually, the corrosion tendency decreased, and the generation of the protective corrosion products was greater than the dissolution rate of the products. The protective properties of the corrosion products increased with the extension of the immersion time. When the concentration of Cl^- was 0.05 to 0.2 mol/L, the value of R_{ct} increased and then decreased with the prolongation of the immersion time; the R_{ct} reached the peak value at 7 d of immersion with the 0.05 mol/L concentration of Cl^-, and the R_{ct} reached the peak value at 3 d of immersion in 0.1 mol/L and 0.2 mol/L Cl^-, which meant that the corrosion rate decreased and then increased during the immersion process. This indicates that the corrosion rate decreased and then increased during the immersion process, the charge transfer rate increased, and the degree of corrosion increased, which was related to the weakening of the protective property of the corrosion products. When the concentration of Cl^- reached 0.3 mol/L, the value of R_{ct} decreased with the increase in immersion time, indicating that the corrosion products generated in this case were not protective. In the annulus fluid, the Nyquist diagram of P110 showed an eccentric arc, and the immersion time was prolonged; the radius of the capacitive arc was gradually reduced, the impedance modulus value decreased, and the decline was most drastic after 7 d of immersion, which indicates that the corrosion resistance of P110 casing steel gradually decreased with the prolongation of immersion time, and there was a tendency towards localized corrosion.

3.3. Mott–Schottky Measurements

The results of the Mott–Schottky curves of P110 casing steel in the simulated concrete-liquid environment are shown in Figure 5. These tests mainly illustrated the semiconductor nature of the passivation film generated by the P110 casing steel in conditions of different concentrations of Cl^-, and discussed the effect of structural changes in the passivation film on the corrosion behavior in Cl^- containing environments. According to the Mott–Schottky theory, there existed a relationship between the space charge layer capacitance C_{sc} and the electrode potential E, as shown in Equation (1):

$$\frac{1}{C_{SC}^2} = \frac{2}{\varepsilon\varepsilon_0 e N_q}(E - E_{fb} - \frac{kT}{e}) \quad (1)$$

where N_q is the carrier concentration, ε is the semiconductor permittivity, ε_0 is the vacuum permittivity, E_{fb} is the flat-band point potential, k is the Boltzmann constant, q is the elementary charge, and T is the absolute temperature [21]. It can be seen that in plotting between E and C_{sc}^{-2}, the two should be linear when the Mott–Schottky relation is satisfied. When the slope of this linear relationship is positive, the potential interval presents n-type semiconductor properties, and when the slope of the straight line is negative, it presents p-type semiconductor properties, and the slope of the straight line on the potential axis can be used to find out E_{fb} [22,23].

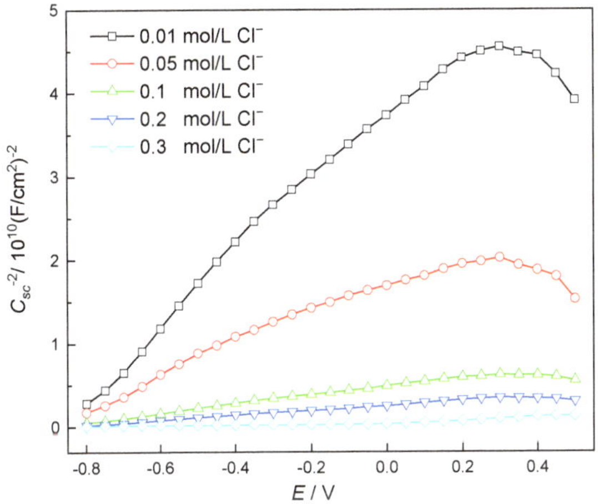

Figure 5. Mott–Schottky results of P110 casing steel in simulated concrete liquid after 10 d of immersion.

As can be seen in Figure 5, P110 casing steel showed a linear relationship in the range of −0.8 V to 0.2 V with a positive slope under the five concentrations of Cl^-, showing an n-type semiconductor nature, and the slope was related to the donor density (N_D), which indicates that the carriers were mainly supplied by electrons [24]. The relative permittivity of the iron oxide was chosen to be taken as 12, and the donor density (N_D) and E_{fb} on the surface of the P110 casing steel were fitted according to the test results of the Mott–Schottky curves in combination with Equation (1); the results of this fitting are shown in Table 4.

Table 4. Mott–Schottky curves' fitting results.

Concentrations of Cl^-/(mol/L)	N_D/cm^{-3}	E_{fb}/V
0.01	2.802×10^{20}	−0.90
0.05	5.685×10^{20}	−0.89
0.1	2.201×10^{21}	−0.88
0.2	3.914×10^{21}	−0.85
0.3	6.462×10^{21}	−0.23

From Table 4, when the concentration of Cl^- was 0.01 mol/L, the N_D value was the smallest, indicating that at this time, the passivation film of the P110 steel was the most dense. The charge transfer was slow when the concentration of Cl^- was increased to more than 0.1 mol/L, the sizing density increased by one order of magnitude, and with the increase in the concentration of Cl^-, the N_D continued to increase, indicating that at this time, the defects in the passivation film on the surface of the P110 steel increased, with the charge transfer accelerating and the passivation film conductivity increasing. This led to the protective properties being decreased.

3.4. Electrochemical Noise Measurements

3.4.1. Time-Domain Analysis

Figure 6 shows the time-domain spectra of EN of P110 casing steel in the simulated concrete-liquid environment with different immersion days for removing the DC drift. The corrosion state of P110 casing steel under different immersion times can be evaluated based on the variation in the time-domain spectra [25,26]. When the concentration of Cl^- was 0.01 mol/L, a small number of transient peaks appeared in the P110 casing steel at an immersion time of 1 d (Figure 6a), the potential noise was synchronized with the trend of the current noise, the transient peaks of the current and potential showed a rapid rise followed by a rapid fall, and a passivation film was gradually formed at this stage. When immersed for 5 d (Figure 6d), a larger transient peak appeared, the current and potential fluctuated in the opposite direction, the potential decreased and the current then increased, and the changes in the time domain were synchronized, but in the opposite direction. The overall trend of the electrochemical potential noise (EPN) was upward, and the EPN peak rapidly decreased and then slowly increased, which indicates that the passivation film of P110 casing steel in the simulated concrete-liquid environment containing Cl^- was in the process of pitting nucleation and repair. The electrochemical current noise (ECN) then rose rapidly and then decreased slowly—the reason for such a phenomenon being related to the expansion of metastable pitting in the passivation membrane breakage—and the noise current decreased after the passivation film was broken. The passivation membrane repair rate was greater than the dissolution rate of aggressive ions. During the immersion time of 10 d (Figure 6g), the potential decreased and then rose slowly, but at this time, there was no noise current flow, and there were fewer transient peaks in the time domain, indicating that at this time a steady-state passivation film had been generated, and the erosive dissolution of Cl^- did not constitute an effect on the structure of the passivation film. In the P110 casing steel immersed in a simulated concrete-liquid environment with a concentration of Cl^- at 0.1 mol/L, from 1 d to 5 d of immersion (Figure 6b,e), the current potential presented alternating synchronous anisotropic transient peaks, and this process was related to the repair of the surface generated passivation film and the generation of sub-stable pitting corrosion. The repair of the passivation film and the dissolving action of Cl^- exhibited a dynamic interaction process, and Cl^- competitive adsorption occurred on the active sites of the steel matrix. The passivation film was also repaired at this stage, indicating that the P110 casing steel had a tendency towards localized corrosion at this time. When the immersion time was 10 d (Figure 6h), the transient peak fluctuations and the quantities of current and potential increased significantly, which indicates that steady-state pitting corrosion had been generated at this time. Also, the potential and current presented a synchronous trend in the same direction, indicating that this time was mainly affected by Faraday current. The transient peak was mainly the current peak formed by pitting corrosion, the electrode surface was mainly for the solubilization of Cl^-, and the protective properties of the passivation film disappeared. When the P110 casing steel was in a simulated concrete-liquid environment with a concentration of Cl^- at 0.3 mol/L, simultaneous isotropic fluctuation transient peaks of EPN and ECN appeared at the early stages of immersion (Figure 6c), which indicates that the metastable pitting stage had been entered at this time, with an enrichment of Cl^- on the surface, an increase in the local electrical conductivity of the passivation film, local dissolution of the passivation film, and a reduction in the protective properties. When the immersion time was 5 d to 10 d (Figure 6f,i), the number of potential transient peaks was higher, and the synchronization of the current and potential disappeared; combined with the results of the EN resistance calculation, it can be seen that the electrode surface was in a state of active dissolution, the electrode surface was mainly undergoing an anodic dissolution reaction, and the electrode surface entered into the period of localized corrosion. In the annulus-fluid environment of P110 casing steel, when the immersion time was 1 d, there were synchronous potential and current peaks and isotropic fluctuations, and the electrode surface entered the pitting-corrosion formation period. When the immersion time was 5 d to 10 d, the number of transient peaks of current and potential increased, the electrode surface underwent localized corrosion, and the amplitude of the transient peaks of ECN and EPN were significantly higher than that of the simulated concrete-liquid environment, which indicates that the electron transfer rate was faster in the annulus-fluid environment, and electrochemical reactions were more likely to happen [27].

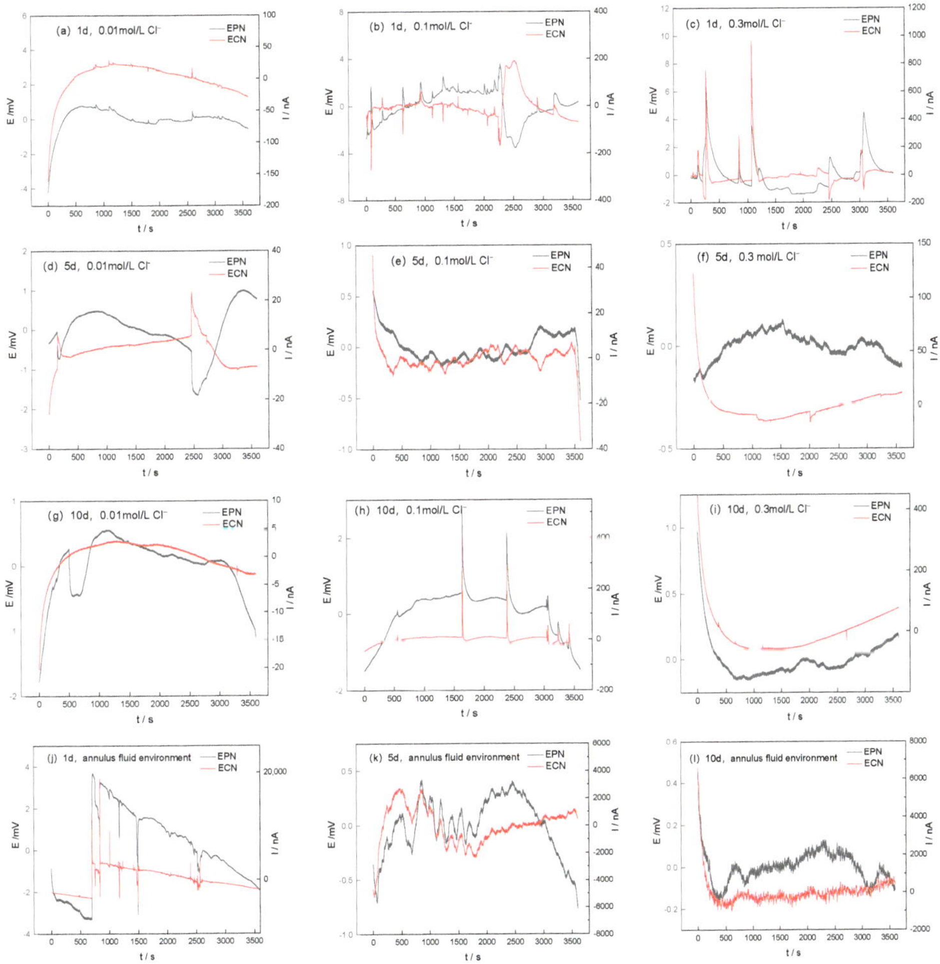

Figure 6. Time-domain spectra of removing DC drift of P110 steel after soaking for different amounts of time: (**a–i**) immersion times of 1 d, 5 d, and 10 d in the simulated concrete-liquid environment, with concentrations of Cl^- of 0.01 mol/L, 0.1 mol/L, and 0.3 mol/L; (**j–l**) immersion times of 1 d, 5 d, 10 d in the simulated annular environment.

In time-domain analyses of electrochemical noise, statistical analysis is often used to quantitatively describe the degree of corrosion, and the electrochemical noise resistance R_n is calculated using Formula (2) [28]:

$$R_n = S_v / S_I \tag{2}$$

where S_V is the standard deviation of potential, V; S_I is the standard deviation of current, A/cm^2; R_n is the noise resistance, $\Omega \cdot cm^2$. The R_n is negatively correlated with the corrosion rate. The larger the R_n, the smaller the corrosion degree and the better the protection of the passivation film [29]. From the fitting results in Table 5, it can be seen that when the concentration of Cl^- was 0.01 mol/L, R_n increased with the extension of the immersion time, which indicates that the corrosion resistance of P110 casing steel gradually increased in this environment, and the passivation film protection was strengthened. When the concentration of Cl^- was 0.1 mol/L, with the extension of immersion time, R_n was basically in the same order of magnitude, and the fluctuation was relatively stable, which was related to the formation and development of pitting corrosion in the P110 casing steel in this period; in such a concentration, passivation film dissolution and repair occurred basically in a stage of dynamic equilibrium. Deposition and dissolution of the corrosion products were mainly carried out on the surface of the P110 casing steel at this stage. When the concentration of Cl^- was 0.3 mol/L, R_n decreased by one order of magnitude with the prolongation of the immersion time, indicating that the corrosion resistance of P110 was reduced in this environment, and locally stabilized pitting had been generated. In the annulus-fluid environment, the R_n value of P110 gradually decreased with the extension of the immersion time, which implies that localized corrosion had occurred in the simulated annulus fluid, and the R_n value was significantly lower than that in the simulated concrete-liquid environment, which suggests that the corrosion of P110 was more intense in the simulated annulus-fluid environment, which is in agreement with the results of the time-domain analysis and the electrochemical tests.

Table 5. Calculation results of R_n in the EN time domain.

Environmental Factors		R_n $\Omega \cdot cm^2$
Simulated concrete liquid	1 d, 0.01 mol/L Cl^-	2.55×10^4
	1 d, 0. 1 mol/L Cl^-	2.33×10^4
	1 d, 0.3 mol/L Cl^-	1.42×10^4
	5 d, 0.01 mol/L Cl^-	1.05×10^5
	5 d, 0.1 mol/L Cl^-	2.63×10^4
	5 d, 0.3 mol/L Cl^-	1.22×10^4
	10 d, 0.01 mol/L Cl^-	1.43×10^5
	10 d, 0.1 mol/L Cl^-	3.06×10^4
	10 d, 0.3 mol/L Cl^-	2.16×10^3
Annulus fluid	1 d	9.34×10^2
	5 d	1.90×10^2
	10 d	1.01×10^2

3.4.2. Frequency-Domain Analysis

The power spectral density (PSD) of the EN was obtained from frequency-domain analysis using the fast Fourier transform (FFT), and the following relationship was found between the PSD and the frequency of the electrochemical noise:

$$\log \mathrm{PSD} = A_i + S_i \log f \quad (3)$$

where PSD is the power spectral density, $A^2 \cdot Hz^{-1}$; A_i represents the noise intensity; S_i is the diagonal part of the power spectral density curve; and f is the frequency, Hz. The PSD spectra were obtained through FFT processing of the EN under different conditions, and the PSD spectra of the ECN are shown in Figure 7.

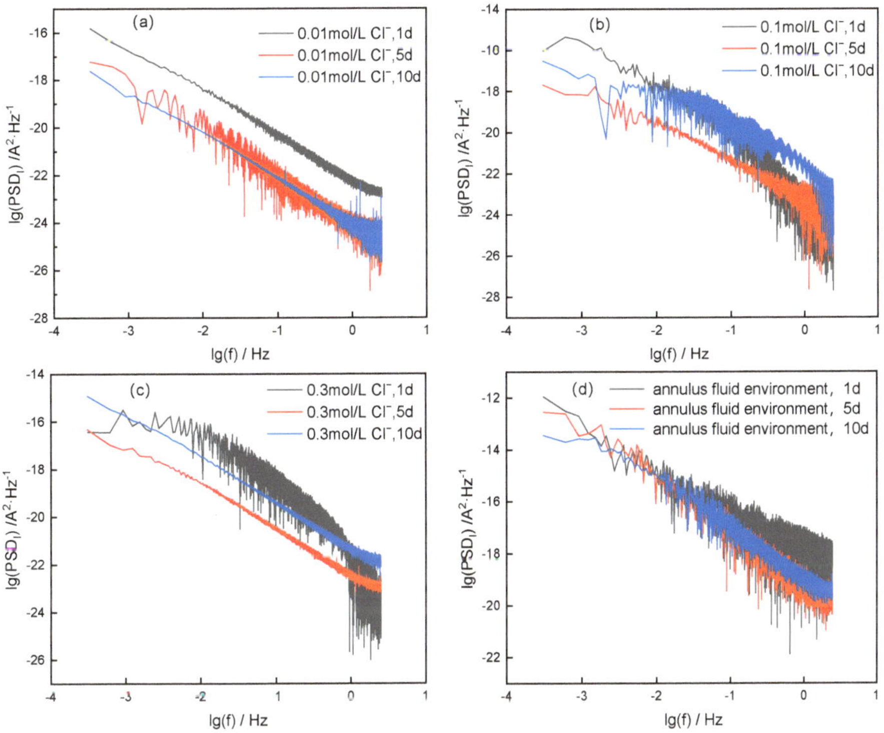

Figure 7. ECN frequency-domain PSD spectra of P110 steel after 1 d, 5 d, and 10 d of immersion: (**a**) 0.01 mol/L Cl^-, (**b**) 0.1 mol/L Cl^-, (**c**) 0.3 mol/L Cl^-, (**d**) annulus-fluid environment.

When the frequency in the PSD spectra is less than a 10^{-2} interval, there is a level of white noise (W_I); the higher the level of W_I in the PSD spectra, the more intense the corrosion behavior becomes. At a concentration of Cl^- at 0.01 mol/L, the W_I level gradually shifted to the negative direction, which indicates that the corrosion rate gradually decreased [30]. When the concentrations of Cl^- were 0.1 and 0.3 mol/L, the W_I values decreased and then increased, and this process was related to the generation of the passivation film and the competitive adsorption of Cl^-. After 5 d of immersion, the protective property of the passivation film decreased and the corrosion rate increased. In the annulus-fluid environment, the W_I level was two orders of magnitude higher than that in the simulated concrete-liquid environment, indicating that the corrosion rate of P110 in the annulus fluid was significantly higher than that in the simulated concrete liquid. The order of magnitude of the W_I did not change significantly after different days of immersion and there was no protective corrosion product generated in the simulated annulus-fluid environment.

3.5. Corrosion Morphology Analysis of P110-Casing-Steel Specimens

The SEM corrosion morphology of P110 casing steel in different environments is shown in Figure 8. As can be seen from the figures, the surface of P110 casing steel in the simulated concrete-liquid environment generated a corrosion-product film, and the corrosion products on the surface of P110 steel presented a double-layer film structure. When the concentration of Cl^- was 0.01 and 0.1 mol/L, the surface of the P110 casing steel formed a relatively dense corrosion product: when observed under low magnification (500×), the corrosion product's structure was regular, and it could cover the metal substrate more completely; under high magnification (5000×), the corrosion product was looser in the outer layer and more dense in the inner layer, and the collective outer layer of the corrosion product was almost complete, which indicates that the corrosion products generated in this state were more protective than those generated in the other simulated concrete-liquid environments. This indicated that the corrosion products generated in this state were better protected, and the passivation film was denser, hindering the further corrosion of aggressive ions. When the concentration of Cl^- was 0.1 mol/L, breakage of the corrosion products was visible under the high-magnification field of view, and local defects that were not covered by the corrosion products were visible. When the concentration of Cl^- was 0.3 mol/L, some corrosion products were visible under low magnification, and the distribution of corrosion products was discontinuous, with defects; also, the corrosion products in the inner and outer layers under high magnification showed a loose and porous structure, and the base metal matrix was visible at the uncovered place of the corrosion products. At this time, Cl^- could be adsorbed to the specific active site to induce steel corrosion of the P110 casing steel, so the corrosion products generated under this condition had poor protective properties. As the corrosion products generated in this condition were less protective and could not hinder the contact between the aggressive medium and the metal substrate, localized pitting corrosion was likely to occur. In the simulated annulus fluid, under low magnification, the corrosion products on the surface of the P110 casing steel were unevenly distributed, no regular corrosion product structure could be seen, and local pitting defects were visible on the metal substrate; under high magnification, the corrosion-product film on the P110 casing steel formed a loose and porous structure, and the corrosion products showed a stepped irregular distribution, which was likely related to the CO_2 corrosion.

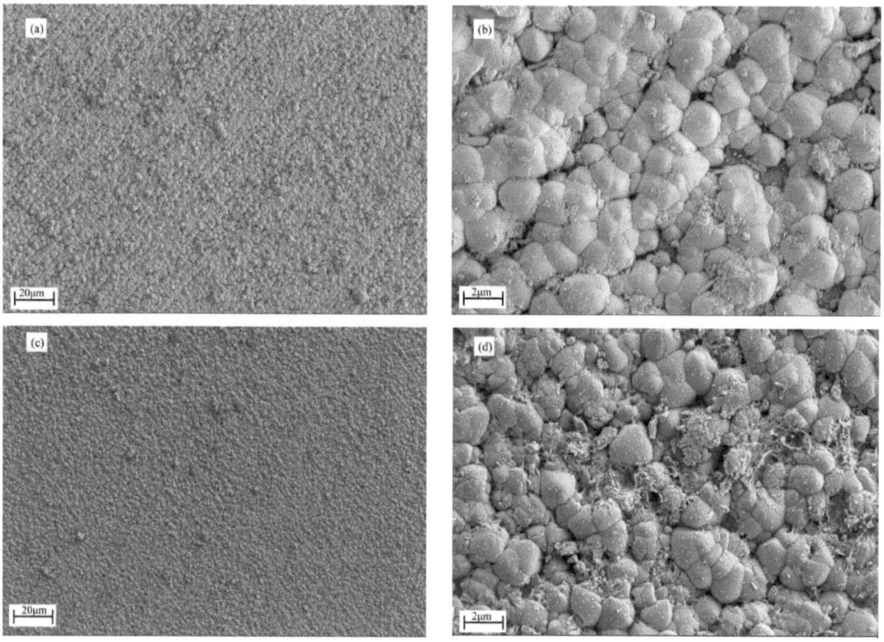

Figure 8. *Cont.*

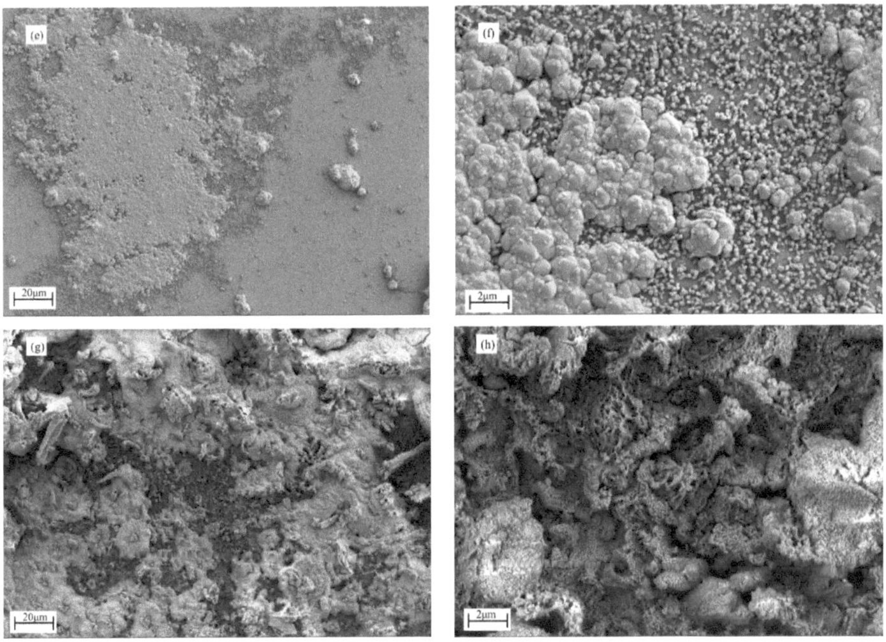

Figure 8. SEM microscopic morphology of P110 casing steel after 10 d of immersion in different environments: (**a**,**b**) 0.01 mol/L Cl^-, (**c**,**d**) 0.1 mol/L Cl^-, (**e**,**f**) 0.3 mol/L Cl^-, (**g**,**h**) annulus fluid.

Figure 9 shows the EDS mapping results of P110 casing steel after 10 d of immersion. From this figure, it can be seen that the corrosion products on the surface of P110 casing steel in the simulated concrete-liquid environment mainly contained Fe, O, and Ca elements, and in the simulated annulus-fluid environment mainly contained Fe, O, C, and a small number of Cl elements. The Ag element in the EDS mapping originated from a silver-spraying treatment that was applied in order to enhance the conductivity of the substrate. When the concentration of Cl^- was 0.01 and 0.1 mol/L, Fe, O, and Ca elements were uniformly distributed on the surface of the metal substrate; at this time, the corrosion products were more homogeneous, and the main corrosion products were iron oxides and a calcium-based scale layer. When the concentration of Cl^- was 0.3 mol/L, Ca and O elements were mainly covered in the outer layer of the metal substrate, and Fe and O elements were mainly in the inner layer, which indicates that the corrosion products were mainly a calcium-containing scale layer in the outer layer and iron oxides in the inner layer. In the simulated annulus fluid, the corrosion products mainly contained Fe, C, O, and Cl elements, which was related to the CO_2 and high concentration of Cl^- in the working environment; these corrosion products might have been composed of iron–carbon oxides, iron oxides, and part of the scale material.

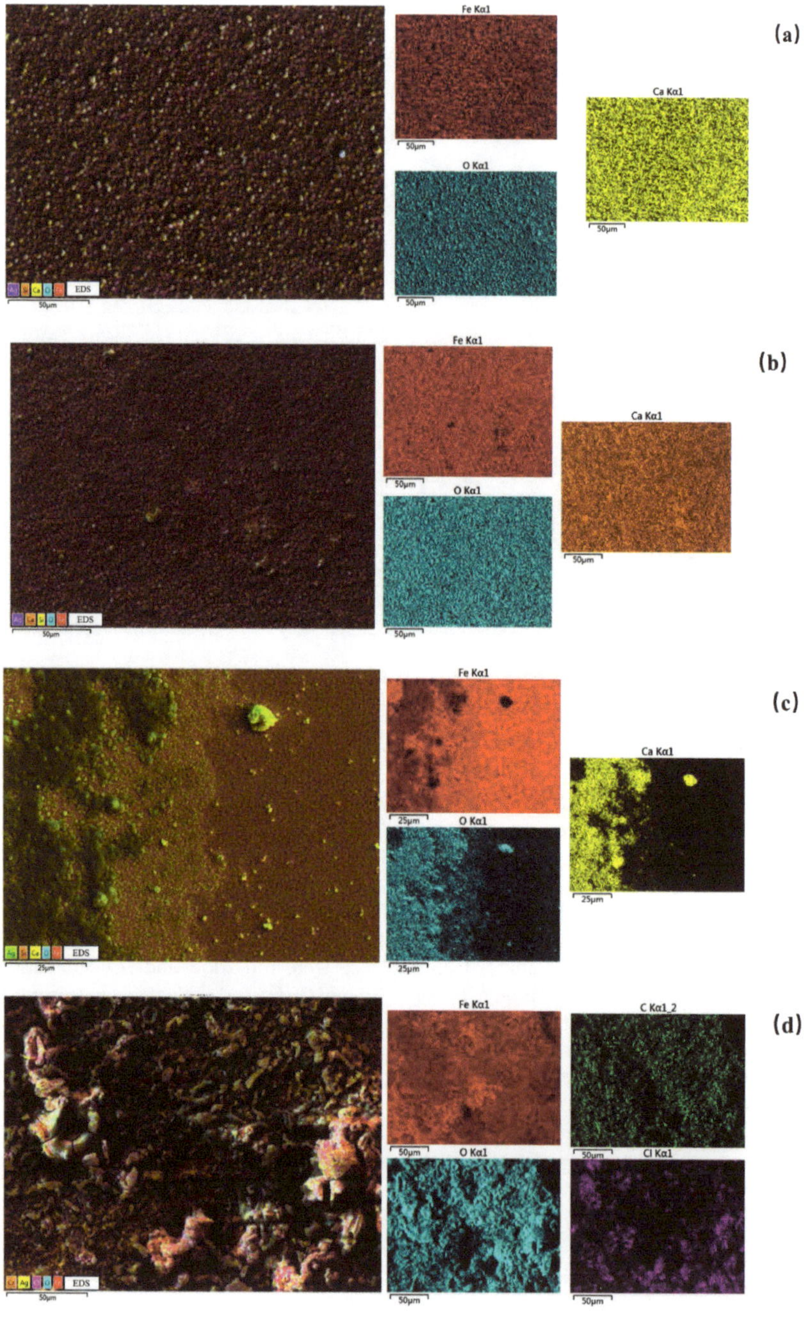

Figure 9. EDS mapping of P110 steel after immersion for 10 d in different environments: (**a**) 0.01 mol/L Cl^-, (**b**) 0.1 mol/L Cl^-, (**c**) 0.3 mol/L Cl^-, (**d**) annulus fluid.

3.6. Corrosion Mechanism Analysis

P110 casing steel had a tendency towards local corrosion in the simulated annulus fluid and in the simulated concrete liquid, where the anodic reaction was mainly the dissolution reaction of iron. For the simulated concrete-liquid environment, the corrosion mechanism is shown in Figure 10a, and the cathodic reaction that occurred in the strongly alkaline environment was mainly oxygen absorption corrosion. In the simulated concrete-liquid environment containing Cl^-, the self-repair of the passivation film and the dissolution process of Cl^- alternated with each other, and the passivation film was in a dynamic process of destruction and self-repair, forming the passivation film of P110 casing steel mainly by following the process described below:

$$3Fe + 8OH^- - 8e^- \rightarrow Fe_3O_4 + 4H_2O \tag{4}$$

$$2Fe + 6OH^- - 6e^- \rightarrow Fe_2O_3 + 3H_2O \tag{5}$$

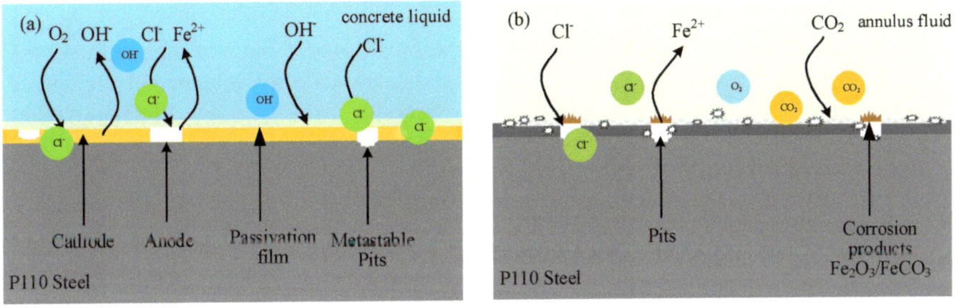

Figure 10. Corrosion mechanism diagram of P110 steel: (**a**) simulated concrete-liquid environment; (**b**) simulated annulus-fluid environment.

When the concentration of Cl^- was 0.01 mol/L, a dense passivation film was formed on the surface of the P110 steel at a high concentration of OH^-. During the immersion process, R_{ct} and R_n gradually increased, while W_I levels gradually decreased, and the electrochemical reaction was inhibited. At this time, the passivation film formed was the most compact and had the best protection, and the protection of the passivation film gradually increased with the extension of immersion time. When the concentration of Cl^- was 0.05~0.2 mol/L, competitive adsorption of the Cl^- and OH^- occurred; then, at this concentration, the R_{ct} increased and then decreased with the prolongation of the immersion time. The effective protection time of the passivation film was shortened with the increase in the concentration of Cl^-. In this period, Cl^- and OH^- were competitively adsorbed on the active sites on the surface of the steel substrate, and the initial passivation film repair process was greater than the effect of the local dissolution process of Cl^-; at this time, the EN test results showed metastable pitting corrosion. When the concentration of Cl^- increased to 0.3 mol/L, R_{ct} and R_n decreased with the extension of immersion time, and the transient peak of EPN increased. The corrosion current density reached its maximum at 10 d of immersion. At this time, the passive film on the surface of the P110 steel exhibited local complete rupture, exposing the metal matrix, and local corrosion occurred.

The minimum transient of the electrochemical potential corresponds to the initiation of the passivation film repair from the metastable pitting. Comparing the EN results of P110 casing steel in the simulated concrete-liquid environments at 10 d of immersion, it can be seen that there was no current flowing on the surface of P110 steel after the potential decreased under the condition of the 0.01 mol/L concentration of Cl^-, whereas in the environments containing Cl^- at 0.1 and 0.3 mol/L, fluctuations in the electrochemical potential noise could cause fluctuations in the passivation film of the P110 steel. This is because the P110 casing steel's passivation film was denser in low concentrations of Cl^-: due to the cathodic reaction in the simulated concrete liquid environment being mainly in the progress of an oxygen reduction reaction, the electrons generated by the dissolution process of iron that occurred in the anodic reaction were consumed in the cathodic reaction, and the total current of the whole system remained unchanged; the electrons consumed by the cathodic reaction were not enough to offset the electrons generated by the anodic reaction, and the excess charge was used to conduct the recharging of the passivation film; the role of the capacitance of the passivation-film electrode began to appear, and there was no current noise fluctuation on the electrode surface during this process. In an environment with a high concentration of Cl^-, the ability of Cl^- to penetrate the passivation film was stronger, and when Cl^- continued to accumulate at the defects of the passivation film, the passivation film underwent dissolution rupture, which promoted the dissolution of the P110 substrate and the formation of pitting corrosion. The charge of the cathodic reaction was also consumed by the charging process of the passivation film, which was a Faraday–cathodic charge in two parts. When the passivation film could not completely cover the metal substrate, the capacitive effect of the passivation film was weakened, the Faraday current appeared at the electrode interface, and the electrode current fluctuated synchronously. According to the Point Defect Model (PDM) theory [31], as the donor density N_d increases, the oxygen vacancy concentration on the metal surface increases, which promotes the adsorption of aggressive ions such as Cl^-, resulting in a severe degree of damage to the passivation film. With the extension of immersion time, the passivation film becomes thin due to the effects of hydration and dissolution; moreover, the tendency of the passivation film to rupture increases.

The main electrochemical reactions in the simulated annulus fluid were as follows:

$$Fe + H_2O + CO_2 \rightarrow FeCO_3 + H_2 \tag{6}$$

$$Fe^{2+} + O_2 + 2OH^- \rightarrow Fe_2O_3 + H_2O \tag{7}$$

For the simulated annulus-fluid environment, the corrosion mechanism of this process is shown in Figure 10b. The corrosion form of P110 casing steel in such an environment was mainly pitting corrosion, and the corrosion product could not cover the metal substrate completely. In this P110 steel, with the prolongation of the immersion time, the corrosion resistance was gradually reduced, and local pitting occurred. The corrosion product of the outer layer exhibited a loose and porous structure. On one hand, for the metal substrate with the loose and porous corrosion-product film working in the corrosive medium, these corrosion products could not effectively protect the metal matrix. This accelerated the occurrence of local corrosion. On the other hand, these porous corrosion products were also conducive to the adsorption of corrosive environmental factors such as Cl^-, O_2, and CO_2. In this environment, various corrosive media cooperated to promote the occurrence of pitting corrosion.

4. Conclusions

In this paper, the corrosion behavior of P110 casing steel in internal and external environments was studied in simulated concrete-liquid and simulated annulus-fluid environments. The electrochemical kinetics process of the corrosion behavior of P110 casing steel was investigated using potentiodynamic polarization curves, EIS, EN, and Mott–Schottky tests. The corrosion morphology and elemental distribution of the P110 casing steel under the two environments were analyzed through SEM and EDS mapping, and the results of this study can provide technical references for the corrosion and protection of casing steel. The following conclusions were obtained:

(1) P110 casing steel is prone to local corrosion in both simulated concrete liquid containing chloride and simulated annular fluid. With the extension of immersion time, the corrosion procedure of P110 steel mainly includes cracking and repair of the passivation film, the formation of metastable pitting corrosion, and the occurrence of steady-state pitting corrosion. In a simulated concrete liquid with a low concentration of Cl^-, the repair effect of the passivation film is greater than the dissolution effect of aggressive ions, and the capacitance effect of the dense passivation film hinders the

metastable pitting into a steady pitting formation process. With increases in the concentration of Cl^-, the protection of the passivation film on the metal matrix becomes worse. With the extension of immersion time, the trend of Cl^- penetrating the passivation film increases, the donor density N_D increases, the defects in the passivation film increase, and the self-healing ability of the passivation film is decreased.

(2) P110 steel cannot form a protective corrosion-product film in a simulated annular environment containing CO_2. With the extension of immersion time, the trend of pitting corrosion increases significantly, and the corrosion resistance of the P110 steel decreases. In this environment containing CO_2, loose and porous corrosion products are formed on the surface of P110 steel. These corrosion products have no protective effect and will promote the occurrence of local corrosion of the metal matrix.

(3) According to the research results of this paper, the corrosion degree of P110 steel in an annular environment is much greater than that in a simulated concrete-liquid environment, and the risk of corrosion failure in the annular environment inside the casing is higher in oil and gas production processes. Annular environments and concrete environments with high chloride contents are important technical nodes for investigating corrosion management in oil fields.

Author Contributions: Conceptualization, Y.L. and Z.C.; methodology, Y.L.; validation, Y.L., Z.C. and L.H.; formal analysis, Y.L. and L.H.; writing—original draft preparation, Y.L.; writing—review and editing, Y.L.; visualization, L.H.; supervision, R.L.; project administration, R.L. All authors have read and agreed to the published version of the manuscript.

Funding: This research received no external funding.

Institutional Review Board Statement: Not applicable.

Informed Consent Statement: Not applicable.

Data Availability Statement: Data are contained within the article.

Conflicts of Interest: The authors declare no conflicts of interest.

References

1. Liu, Z.; Li, H.; Jia, Z.; Du, C.; Li, X. Failure analysis of P110 steel tubing in low-temperature annular environment of CO_2 flooding wells. *Eng. Fail. Anal.* **2016**, *60*, 296–306. [CrossRef]
2. Shi, X.; Anh Nguyen, T.; Kumar, P.; Liu, Y. A phenomenological model for the chloride threshold of pitting corrosion of steel in simulated concrete pore solutions. *Anti-Corros. Methods Mater.* **2011**, *58*, 179–189. [CrossRef]
3. Thangavel, K.; Rengaswamy, N. Relationship between chloride/hydroxide ratio and corrosion rate of steel in concrete. *Cem. Concr. Compos.* **1998**, *20*, 283–292. [CrossRef]
4. Tran, V.Q.; Soive, A.; Baroghel-Bouny, V. Modelisation of chloride reactive transport in concrete including thermodynamic equilibrium, kinetic control and surface complexation. *Cem. Concr. Res.* **2018**, *110*, 70–85. [CrossRef]
5. Covelo, A.; Diaz, B.; Freire, L.; Nóvoa, X.R.; Perez, M.C. Microstructural changes in a cementitious membrane due to the application of a DC electric field. *J. Environ. Sci. Health Part A* **2008**, *43*, 985–993. [CrossRef]
6. Angst, U.M. Predicting the time to corrosion initiation in reinforced concrete structures exposed to chlorides. *Cem. Concr. Res.* **2019**, *115*, 559–567. [CrossRef]
7. Huang, S.; Yang, Y.; Li, Z.; Liu, Y.; Su, D. Corrosion behavior and mechanism of P110 casing steel in alkaline-activated persulfate-based preflush fluid. *Eng. Fail. Anal.* **2023**, *152*, 107482. [CrossRef]
8. Abubakar, S.A.; Mori, S.; Sumner, J. A Review of Factors Affecting SCC Initiation and Propagation in Pipeline Carbon Steels. *Metals* **2022**, *12*, 1397. [CrossRef]
9. Gao, X.; Zhang, D.; Lu, Y.; Fan, Z.; Du, L.; Yuan, G.; Qiu, C.; Kang, J. CO_2 Corrosion Behavior of High-Strength and Toughness V140 Steel for Oil Country Tubular Goods. *J. Mater. Eng. Perform.* **2020**, *29*, 8451–8460. [CrossRef]
10. Elgaddafi, R.; Ahmed, R.; Osisanya, S. Modeling and experimental study on the effects of temperature on the corrosion of API carbon steel in CO_2-Saturated environment. *J. Pet. Sci. Eng.* **2021**, *196*, 107816. [CrossRef]
11. Hua, Y.; Yue, X.; Liu, H.; Zhao, Y.; Wen, Z.; Wang, Y.; Zhang, T.; Zhang, L.; Sun, J.; Neville, A. The evolution and characterisation of the corrosion scales formed on 3Cr steel in CO_2-containing conditions relevant to geothermal energy production. *Corros. Sci.* **2021**, *183*, 109342. [CrossRef]
12. Zeng, D.; Dong, B.; Zhang, S.; Yi, Y.; Huang, Z.; Tian, G.; Yu, H.; Sun, Y. Annular corrosion risk analysis of gas injection in CO_2 flooding and development of oil-based annulus protection fluid. *J. Pet. Sci. Eng.* **2022**, *208*, 109526. [CrossRef]

13. Wei, L.; Gao, K.; Li, Q. Corrosion of low alloy steel containing 0.5% chromium in supercritical CO_2-saturated brine and water-saturated supercritical CO_2 environments. *Appl. Surf. Sci.* **2018**, *440*, 524–534. [CrossRef]
14. Wei, L.; Gao, K. Understanding the general and localized corrosion mechanisms of Cr-containing steels in supercritical CO_2-saturated aqueous environments. *J. Alloys Compd.* **2019**, *792*, 328–340. [CrossRef]
15. Xu, L.; Xu, X.; Yin, C.; Qiao, L. CO_2 corrosion behavior of 1% Cr–13% Cr steel in relation to Cr content changes. *Mater. Res. Express* **2019**, *6*, 096512. [CrossRef]
16. Peng, Y.; Lin, Y.; Xia, R.; Dai, Z.; Zhang, W.; Liu, W. Electrochemical Investigation of Chloride Ion-Induced Breakdown of Passive Film on P110 Casing Steel Surface in Simulated Pore Solution: Behavior and Critical Value Determination. *Metals* **2024**, *14*, 93. [CrossRef]
17. Liu, G.; Zhang, Y.; Ni, Z.; Huang, R. Corrosion behavior of steel submitted to chloride and sulphate ions in simulated concrete pore solution. *Constr. Build. Mater.* **2016**, *115*, 1–5. [CrossRef]
18. Zhang, H.; Li, Y.; Wang, S.; Zhang, J.; Deng, T.; Zhang, X.; Zhang, Z. A Study on the CO_2 Corrosion Behavior of P110 Steel in High-Density Cement. *Int. J. Electrochem. Sci.* **2022**, *17*, 22023. [CrossRef]
19. Liu, G.; Zhang, Y.; Wu, M.; Huang, R. Study of depassivation of carbon steel in simulated concrete pore solution using different equivalent circuits. *Constr. Build. Mater.* **2017**, *157*, 357–362. [CrossRef]
20. Eliyan, F.F.; Mahdi, E.-S.; Alfantazi, A. Electrochemical evaluation of the corrosion behaviour of API-X100 pipeline steel in aerated bicarbonate solutions. *Corros. Sci.* **2012**, *58*, 181–191. [CrossRef]
21. Zhu, G.; Li, Y.; Hou, B.; Zhang, Q.; Zhang, G. Corrosion behavior of 13Cr stainless steel under stress and crevice in high pressure CO_2/O_2 environment. *J. Mater. Sci. Technol.* **2021**, *88*, 79–89. [CrossRef]
22. Feng, Z.; Cheng, X.; Dong, C.; Xu, L.; Li, X. Passivity of 316L stainless steel in borate buffer solution studied by Mott–Schottky analysis, atomic absorption spectrometry and X-ray photoelectron spectroscopy. *Corros. Sci.* **2010**, *52*, 3646–3653. [CrossRef]
23. Belo, M.D.C.; Hakiki, N.; Ferreira, M. Semiconducting properties of passive films formed on nickel–base alloys type Alloy 600: Influence of the alloying elements. *Electrochim. Acta* **1999**, *44*, 2473–2481. [CrossRef]
24. Macdonald, D.D. The history of the Point Defect Model for the passive state: A brief review of film growth aspects. *Electrochim. Acta* **2011**, *56*, 1761–1772. [CrossRef]
25. Xia, D.-H.; Song, S.-Z.; Behnamian, Y. Detection of corrosion degradation using electrochemical noise (EN): Review of signal processing methods for identifying corrosion forms. *Corros. Eng. Sci. Technol.* **2016**, *51*, 527–544. [CrossRef]
26. Al-Mazeedi, H.; Cottis, R. A practical evaluation of electrochemical noise parameters as indicators of corrosion type. *Electrochim. Acta* **2004**, *49*, 2787–2793. [CrossRef]
27. Dong, Z.H.; Shi, W.; Guo, X.P. Initiation and repassivation of pitting corrosion of carbon steel in carbonated concrete pore solution. *Corros. Sci.* **2011**, *53*, 1322–1330. [CrossRef]
28. Mansfeld, F.; Sun, Z. Localization index obtained from electrochemical noise analysis. *Corrosion* **1999**, *55*, 915–918. [CrossRef]
29. Obot, I.B.; Onyeachu, I.B.; Zeino, A.; Umoren, S.A. Electrochemical noise (EN) technique: Review of recent practical applications to corrosion electrochemistry research. *J. Adhes. Sci. Technol.* **2019**, *33*, 1453–1496. [CrossRef]
30. Markhali, B.; Naderi, R.; Mahdavian, M. Characterization of corrosion inhibition performance of azole compounds through power spectral density of electrochemical noise. *J. Electroanal. Chem.* **2014**, *714*, 56–62. [CrossRef]
31. Macdonald, D.D. The point defect model for the passive state. *J. Electrochem. Soc.* **1992**, *139*, 3434. [CrossRef]

coatings

Review

Application Research of Ultrasonic-Guided Wave Technology in Pipeline Corrosion Defect Detection: A Review

Feng Lyu [1], Xinyue Zhou [1], Zheng Ding [1], Xinglong Qiao [2] and Dan Song [2,*]

[1] State Grid Wuxi Power Supply Company, Wuxi 214000, China
[2] College of Materials Science and Engineering, Hohai University, Nanjing 211100, China
* Correspondence: songdancharls@hhu.edu.cn

Abstract: This paper presents research on the application of ultrasonic-guided wave technology in corrosion defect identification, expounds the relevant ultrasonic-guided wave theories and the principle of ultrasonic-guided wave non-destructive testing of pipelines, and discusses the Lamb wave and shear horizontal wave mode selection that is commonly used in ultrasonic-guided wave corrosion detection. Furthermore, research progress in the field of ultrasonic-guided wave non-destructive testing (NDT) technology, i.e., regarding transducers, structural health monitoring, convolutional neural networks, machine learning, and other fields, is reviewed. Finally, the future prospects of ultrasonic-guided wave NDT technology are discussed.

Keywords: ultrasonic-guided waves; pipeline; non-destructive testing; structural health monitoring; corrosion defect identification

1. Introduction

As vital components of industrial transportation, pipeline structures find extensive applications in fields such as petroleum, chemical engineering, aerospace, nuclear power, and more. During their service life, pipelines are susceptible to corrosion and degradation due to environmental and other factors. Detecting corrosion and pitting defects is crucial, as these can occur at various locations within pipelines and vessels, constituting a significant portion of maintenance expenditures. These issues can lead to pipeline cracking, leakage, reduced operational lifespan, and, in severe cases, catastrophic accidents resulting in significant economic losses [1–7]. Therefore, it is crucial to detect corrosion and pitting defects for the safety of the pipeline industry. It is necessary to conduct regular pipeline inspections and to provide maintenance and repair of pipeline structures, ensure the safe operation and use of pipelines, maximize the lifespan of pipelines, and pursue the most significant economic and social benefits [8,9].

Non-destructive testing (NDT) technology is a method used to detect various defects in components without causing damage to the inspected object or affecting the performance of the testing components. It provides information on the size, location, quantity, and severity of defects, allowing for the assessment of the component's condition. NDT technology has widespread applications across multiple fields and is crucial in product quality inspections and equipment examinations. It makes it possible to identify the initial damage stage in structures, prevent structural failure, and reduce economic losses; it therefore serves as a valuable defect assessment method. Among the advantages of NDT are its remote controllability, ease of operation, and low operating costs [10–14].

In infrastructure maintenance, NDT for pipelines has become an essential practice. Standard techniques for pipeline defect identification include ultrasonic testing (UT), eddy current testing (ECT), radiographic testing (RT), penetrant testing (PT), and magnetic particle testing (MT) [15]. UT may be applied to various materials, including metals, non-metals, and composite materials. It offers advantages such as a broad inspection range, significant detection depth, precise defect localization, high sensitivity, low cost, lack of pollution, and ease of on-site use. However, UT is a point-by-point inspection method, resulting in relatively slow inspection speeds.

Eddy current testing is widely used in various industrial fields. The procedure makes it possible to perform rapid and automated inspections. However, this technique can only be employed for surface defect detection in electrically conductive materials. For seamless steel pipes such as those commonly used in pipelines, eddy current testing is too fast; as such, more precision is needed for individual pipe inspections. Magnetic particle testing is known for its high sensitivity and ability to visually display defects' location and shape. However, this approach can only be used to detect surface defects in ferromagnetic materials. Radiographic testing may be applied to various materials, providing intuitive inspection results. However, it can be detrimental to the health of inspectors due to radiation exposure and is associated with high inspection costs and low efficiency. Traditional inspection methods involve point-by-point examinations in pressure pipelines, steam pipelines, gas pipelines, and similar structures, requiring significant work and demonstrating low efficiency. Metal defect features, such as general corrosion, result in an attenuation in the energy of a propagating wave due to the rough corroded surface, reducing the detection range. In contrast, defect sizes can reduce the detectable range for identification [15–17].

Ultrasonic-guided waves propagate along the walls of pipes over long distances, and their acoustic field covers the entire wall thickness of a pipe or plate. They possess high energy levels and experience minimal self-attenuation over extended distances. With strong penetrating capabilities, these waves can detect defects on the external surface and within the internal structure of components. This capability significantly enhances inspection efficiency and reduces the workload. As the theory of ultrasonic-guided waves has been more deeply researched, the application areas of ultrasonic-guided wave testing technology have gradually expanded, primarily in non-destructive testing and structural health monitoring.

Ultrasonic-guided wave technology is applied to the non-destructive testing of materials such as pipelines, steel rails, high-voltage transmission lines, plates, composite materials, and more. It is also used for damage detection and assessments of various structural components in engineering structures, including honeycomb structures, aircraft fuselage structures or components, and infrastructure [18–27]. For pipeline inspection, the utilization of the multimodal and dispersive characteristics of a guided wave is essential. Considering factors such as component properties, dispersion curves, and attenuation, it is crucial to excite the appropriate guided wave modes to detect various types and levels of defects. In the case of plate materials, ultrasonic-guided wave testing has proven to be an effective means of inspecting critical components in metal plates. This involves selecting suitable excitation methods to stimulate single-mode-guided waves. Analyzing the signals from the defect location provides information about the position, size, and shape of the defect [28].

Given the significance of pipeline corrosion detection and the diversity of research that has been carried out on ultrasonic-guided waves, there is currently only a limited number of comprehensive reviews of research on pipeline corrosion detection based on guided waves. Recognizing the necessity of such a review, this article provides a comprehensive overview of the research applications of ultrasonic-guided waves in detecting corrosion defects in pipelines. It primarily elucidates the relevant theories of pipeline ultrasonic-guided waves and discusses the modal selection of commonly used Lamb waves and shear horizontal waves in ultrasonic-guided wave corrosion detection. Additionally, the article explores the research progress in ultrasonic-guided wave non-destructive testing technology for structural health monitoring and the combination of ultrasonic-guided waves with fiber optic acoustics. The second section introduces the basic theory and technical characteristics of ultrasonic-guided wave detection, while the third section delves into the in-depth research on pipeline ultrasonic-guided wave detection technology, its development, and related transducer studies. The fourth section focuses on the application of ultrasonic-guided wave detection technology in structural health monitoring.

2. Ultrasonic-Guided Wave Fundamentals

2.1. Concept of Ultrasonic-Guided Waves

Ultrasonic-guided waves pertain to either supersonic or subsonic mechanical or elastic waves that propagate within a medium parallel to its boundary surfaces, exemplified by pipes and plates. The propagation state of UGW is influenced by the geometric shape of the waveguide. UGW comprises mechanical vibrations generated beneath the excitation source and subsequently propagated through the waveguide [27,28]. At the interfaces of elastic media, UGW can induce reflection, refraction, and waveform transformation. During propagation, UGW carries distinctive information about both the excitation source and the material. This unique characteristic, in conjunction with alterations in sound wave intensity, establishes physical properties such as the size and location of internal defects within the component [28].

Ultrasonic-guided waves emerge in elastic media when a particle is subjected to force, inducing neighboring particles to experience a distinct force and oscillate around their equilibrium positions. This sequential propagation results in the formation of mechanical waves. As ultrasonic waves traverse a medium of a particular size, they traverse discontinuous cross-sections, leading to multiple reflections, refraction, mode conversion, and interference. This culminates in the creation of plate-like ultrasonic-guided waves, commonly referred to as Lamb waves [29–32]. Classical surface wave propagation encompasses surface waves, Lamb waves, and Stonely waves. Strictly speaking, Lamb waves propagate in uniformly isotropic plates devoid of traction forces. Nevertheless, this term has been broadened to encompass diverse structures, including plates, multilayer boards, rods, pipes, and beyond, where Lamb wave-type propagation takes place. In Lamb waves, the components of the wave vector can be parallel or perpendicular to the vertical plane of particle vibration, in contrast to horizontal shear waves. As Lamb waves travel along a structure, particle motion takes place solely in the horizontal plane, perpendicular to the wave vector [32–34].

2.2. Group Velocity and Phase Velocity of Guided Waves

In guided wave theory, the fundamental concepts of group velocity (c_g) and phase velocity (c_p) play a crucial role. Owing to the dispersion characteristics, guided waves manifest multiple frequencies as they propagate. Group velocity denotes the speed at which an entire pulse wave propagates and encompasses a family of waves with comparable frequencies. This is the speed at which the energy of the waveform travels. In the realm of guided waves, it elucidates the movement of a guided wave pulse through the medium. Conversely, phase velocity pertains to the speed at which a fixed phase point on a guided wave travels in the direction of propagation. In simple terms, it signifies the speed at which a point on the wave with a constant phase moves through the medium. However, in the majority of cases, especially when dealing with guided waves in waveguides, the term "propagation velocity" typically pertains to the group velocity, providing an overall description of how guided waves traverse the medium [28,35]. The formulas for calculating group and phase velocities are typically employed based on known frequencies.

2.3. The Excitation and Attenuation of Guided Waves

The guided wave testing of pipelines offers broad coverage for long-distance pipelines, but its detectable range is constrained by various attenuation factors, encompassing energy dispersion and absorption [16]. In elastic materials, wavefront dispersion is a prevalent phenomenon. In contrast, viscoelastic materials not only demonstrate wavefront dispersion but also manifest energy absorption phenomena [36].

The attenuation of guided waves denotes the phenomenon where the energy of the waves progressively diminishes as they propagate through a medium with an increasing propagation distance. The principal factors contributing to the attenuation of guided waves include diffusive attenuation, scattering attenuation, and absorption attenuation. Diffusive attenuation arises as guided waves propagate over distances, leading to the continuous expansion of non-planar wavefronts, resulting in a reduction of acoustic energy per unit area as the distance increases. Scattering attenuation: Scattering attenuation stems from non-uniformities in the material, resulting in variations in the acoustic impedance of the medium. This induces the scattering of guided waves, generating thermal energy as the waves persist in their propagation. Absorption attenuation: Absorption attenuation takes place as guided waves propagate through a medium due to the viscous properties of the material, leading to internal friction between particles and the conversion of some acoustic energy into thermal energy [37]. When performing guided wave testing of pipeline structures, scattering and diffusive attenuation are typically the more significant factors to consider.

2.4. Dispersion and Multimodal Characteristics of Guided Waves

Leveraging plate and shell theory, Gazis investigated the propagation of guided waves in infinite hollow cylindrical structures. He acquired comprehensive analytical solutions for both axisymmetric- and non-axisymmetric-guided wave modes within an infinite hollow cylindrical structure. He deduced frequency equations for diverse guided wave modes in hollow cylindrical structures, illustrated dispersion curves depicting the propagation of guided waves in hollow pipes, and ascertained cutoff frequencies for distinct modes. His contributions have laid a theoretical groundwork for subsequent studies on guided waves in pipelines [38,39]. Wave dispersion pertains to the alteration of group velocity and phase velocity with frequency during the propagation of guided waves. Dispersion arises from both geometric and physical factors. When ultrasonic waves propagate in waveguides like plates and pipes, the velocity of the waves varies with frequency owing to the geometric dimensions of the material structure, resulting in geometric dispersion.

Conversely, physical dispersion arises from the alterations in the physical properties of the waveguide material itself. When guided waves traverse a pipeline, the group velocity and phase velocity fluctuate with the product of frequency and thickness (frequency-thickness product). The structural dimensions and frequency predominantly influence the propagation of guided waves within the pipeline. Owing to the impact of dispersion characteristics, guided waves with distinct frequencies exhibit varying propagation speeds. Consequently, guided wave echoes with identical characteristics reach the transducer at different times. As a result, the echo waveform progressively broadens in the time domain as the propagation distance increases, giving rise to spatial energy dispersion. Subsequently, this influences signal sensitivity and the signal-to-noise ratio, affecting the inspection outcomes. Owing to the curvature of pipes and dispersion characteristics, wave propagation in such structures becomes intricate and encompasses an infinite array of possible wave modes.

Meitzler categorized modal-guided waves in hollow cylindrical pipes into three classes: longitudinal modes L (n, m), torsional modes T (n, m), and flexural modes F (n, m). Here, m symbolizes the azimuthal harmonic number, and n signifies the order of the guided wave mode. When $n = 0$ denotes an axisymmetric mode, $n > 0$ indicates a non-axisymmetric mode [40]. The multimodal nature of guided waves implies that various modes of guided waves propagate concurrently at the same frequency. In pipes, for instance, within the longitudinal modal-guided waves, the vibration displacement predominantly occurs in the axial and radial directions. Conversely, the torsional modal-guided waves mainly entail circumferential motion. The flexural modal-guided waves display concurrent axial, circumferential, and radial displacement and are deemed non-axisymmetric modes. Rose developed a methodology for solving dispersion curves of hollow cylindrical-guided waves, considering the geometry and material properties of pipes. This method illustrates the correlation between the nature of these wave modes and their frequencies. Figure 1 presents an example of a dispersion curve for a 3-inch Schedule 40 steel pipe [29,41].

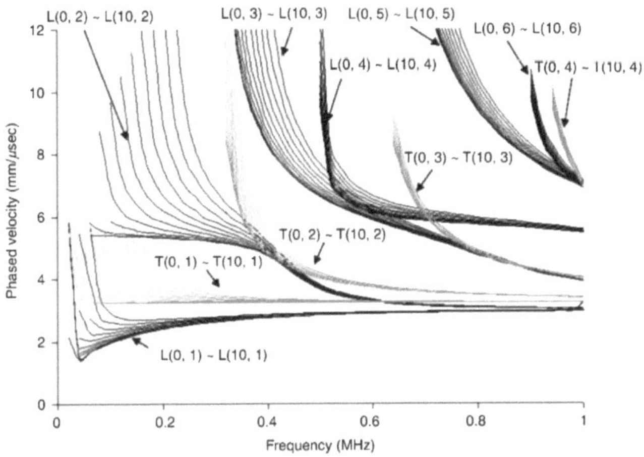

Figure 1. Dispersion curve phase velocity diagram [30].

Dispersion and multimodality stand out as pivotal traits in the propagation of guided waves within pipelines. Frequency-domain dispersion results in spectrum broadening as frequency components, initially with the same phase shift when passing through the waveg uide, causing diminished frequency resolution [42]. In testing scenarios, the multimodal characteristic of guided waves can result in mode interference, diminishing signal sensitivity and the signal-to-noise ratio, adversely affecting accurate detection. Nevertheless, if the multimodal characteristics are efficiently harnessed, they can augment the detection process. Different modes display diverse sensitivity to various defect types. Utilizing multimodal characteristics enables the collection of more detection information, thereby enhancing the accuracy of the detection results. In engineering applications such as pipeline inspection, waveguides can accommodate multiple guided wave modes, each possessing distinct energy and excitability characteristics. Hence, relying on the multimodal and dispersion characteristics of ultrasonic-guided waves, it is crucial to select suitable guided wave modes and frequencies based on specific inspection requirements and dispersion properties to attain more sensitive and accurate detection results. In practical inspections, factors like instrument constraints and coupling conditions can impact precise mode control. Therefore, in most cases, it is recommended to choose excitation frequencies with fewer modes on the dispersion curve to minimize the occurrence of multimodal effects and mitigate the impact of dispersion on the inspection process.

The utilization of ultrasonic-guided wave technology in pipeline inspection stems from the exploration of both axisymmetric- and non-axisymmetric-guided wave modes [43]. Ultrasonic-guided waves showcase multimodal characteristics, requiring deliberations on mode selection in guided wave inspections. Both axisymmetric and non-axisymmetric mode-guided waves find application in long-distance pipeline inspections. The most extensively investigated and commercially established applications of ultrasonic-guided wave detection technology encompass the axisymmetric longitudinal mode L (0, 2) and torsional mode T (0, 1) guided waves. Nevertheless, employing multimodal excitations amplifies the complexity and challenges in analyzing reflected signals [44–46].

In practical applications, the axisymmetric longitudinal mode L (0, 2) is widely adopted due to its ease of excitation and absence of dispersion within a specific frequency range, among other favorable characteristics. When the L (0, 2) mode is introduced into a pipeline, it often simultaneously excites the low-order longitudinal mode L (0, 1) guided wave, which is commonly used in pipeline inspections [29,47]. The predominant motion direction of guided wave pipeline particles in the L (0, 2) mode is axial displacement. In this mode, both group velocity and phase velocity show limited changes with frequency across a broad frequency range, yielding a stable and readily identifiable waveform. Furthermore, the L (0, 2) mode undergoes comparatively low energy attenuation during propagation within the pipeline, rendering it suitable for long-distance transmission. Moreover, the L (0, 2) mode is the fastest propagating mode and demonstrates equal sensitivity to internal and external surface defects, as well as defects along the thickness direction [39,48–50]. While the L (0, 1) mode frequently coexists with the L (0, 2) mode, its slower group velocity can be eliminated in the time domain or suppressed by incorporating a ring into the transducer or adjusting the length of the transducer unit.

Researchers have comprehensively examined and applied the torsional mode T (0, 1) as well. Based on dispersion curves, it is observable that T (0, 1) exhibits non-dispersive advantages across the complete frequency spectrum. If the excitation signal frequency stays below the cutoff frequency of other torsional modes, such as T (0, 2), they do not become excited [30,51–54]. In the T (0, 1) mode, the primary motion direction of pipeline particles is circumferential displacement. T (0, 1) guided waves display non-dispersive characteristics throughout all frequency ranges. Additionally, longitudinal waves are prone to circumferential defects in pipelines, whereas torsional waves exhibit greater sensitivity to axial defects [19,48,53]. T (0, 1) is more sensitive to pipeline axial cracks than L (0, 2). In addition to axisymmetric modes, in some cases, non-axisymmetric modes are also employed for damage detection in pipelines after appropriate mode tuning. These non-axisymmetric modes are sensitive to different types of defects, and therefore, the choice of excitation method and frequency should be tailored to the type of defect. Longitudinal guided waves, characterized by fast propagation speed and long detection distances, offer advantages in the inspection of long-distance pipelines. However, the selection of an appropriate excitation frequency should also take into account the material of the pipeline. The cutoff frequencies for each guided wave mode provide effective separation, especially in the low to mid-frequency range. Choosing an excitation frequency below the cutoff frequency of higher-order modes makes it possible to excite a single mode effectively [55–57].

2.5. Ultrasonic-Guided Wave Non-Destructive Testing Technology Principle and Technical Characteristics

Ultrasonic-guided wave non-destructive testing predominantly employs low-frequency ultrasound to comprehensively screen pipeline damage by capturing scattered waves from defects. As guided waves encounter discontinuous cross-sections, like cracks and voids, during propagation, they experience reflection, transmission, and mode conversion at the boundaries. This leads to changes in acoustic impedance, modifying the received signals. These signals encapsulate pertinent information regarding the size and location of defects. By processing the received echo signals and calculating the axial position of defects based on wave velocity and signal time delay, it becomes feasible to ascertain the component's overall or local health status [37,55,58]. The schematic diagram of ultrasonic-guided wave non-destructive testing technology for pipelines is depicted in Figure 2. The principle of an ultrasonic-guided wave detection system is fundamentally centered on generating pulse signals of a specific frequency using a pulse transmitter. After amplification by a power amplifier, the signals are fed into guided wave transducers. During the propagation of ultrasonic-guided waves in pipelines, the transducers convert the signals into electrical signals upon encountering defects. Subsequently, these signals undergo conditioning and amplification and are input into a computer for further processing to obtain results.

Ultrasonic-guided wave non-destructive testing technology is characterized by single-point excitation, long propagation distances with minimal energy attenuation, high detection efficiency, and the capability to achieve complete cross-sectional inspections. It is particularly beneficial for the non-destructive testing of large-scale and long-distance structures such as pipelines, plates, and railway tracks, significantly improving the efficiency of inspections. Ultrasonic-guided waves find wide-ranging applications in the non-destructive testing of various material structures, with the detection echo signals containing nearly all the information about the inspected structures. Its primary advantages and characteristics are as follows [25,27,51,59–62]:

(1) Single-point excitation allows for extensive and long-distance structural inspections. By inducing vibrations in particles throughout the specimen, it achieves 100% full coverage inspection of the entire cross-section of the tested component. This technology can inspect the component's inner and outer surfaces, revealing internal and external defects.
(2) Ultrasonic-guided wave exhibits minimal energy attenuation during propagation. Low-frequency elastic waves can propagate over tens of meters within the component with minimal attenuation, enabling long-distance inspections.
(3) Ultrasonic-guided wave possesses multimodal characteristics. Controlling the modes and frequencies can significantly enhance the precision of inspections, providing high redundancy and multidimensional inspection information.
(4) It has high inspection efficiency with relatively low costs. It eliminates the need for probes, making it suitable for inspecting buried, pressurized, coated, and transmission pipelines, among other components. Installing a single transducer at an appropriate location can conveniently conduct inspections for complex structures.

UGW shows promising potential in detecting various defects in the pipeline network at a long enough distance. However, for traditional guided wave detection, only the low-frequency guided waves are usually used, and the low-frequency guided waves have less attenuation. While low-frequency guided waves may not be very sensitive to small defects, existing guided wave techniques often struggle to detect the responses generated by local small defects or in cases where the wall is smoothly thinned. This limitation can lead to the oversight and under-detection of defects, highlighting a challenge in current guided wave inspection capabilities [63–66].

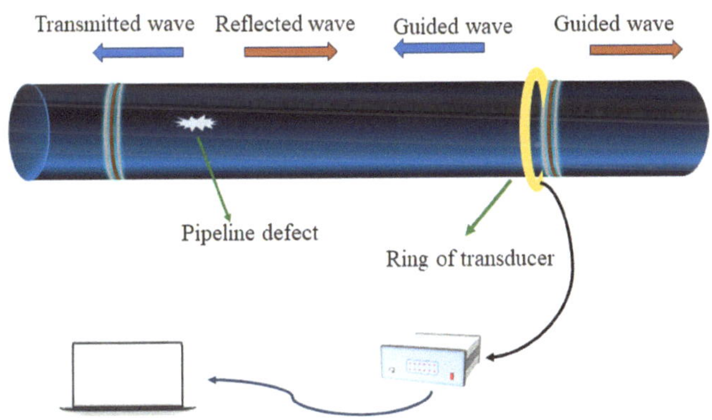

Figure 2. Schematic diagram of ultrasonic-guided wave non-destructive testing technology for pipelines.

3. Pipeline Structures Ultrasonic-Guided Wave Non-Destructive Testing Technology

3.1. Development and Application Status of Pipeline Ultrasonic-Guided Wave Non-Destructive Testing Technology

The investigation into guided ultrasonic waves in pipelines has its origins in the early 20th century, marked by Rayleigh and Lamb's initiation of the study on the propagation theory of elastic waves in infinite media. They forged a connection between plate wave numbers and frequencies [67,68]. Within the theoretical exploration of hollow cylindrical structures, Love delved into investigating the propagation characteristics of guided waves within hollow cylindrical pipes [69]. Gazis extensively explored the propagation of guided waves in infinitely long hollow cylinders employing plate shell theory. This led to the derivation of comprehensive analytical solutions for both axisymmetric and non-axisymmetric mode-guided waves within infinite hollow cylindrical structures. Gazis formulated frequency equations for diverse guided wave modes, charted dispersion curves, and analytically determined the cutoff frequencies for various modes through meticulous data analysis. These contributions lay the essential theoretical groundwork for subsequent research on guided waves in pipelines [38,39,70,71]. Rose et al. conducted theoretical research on axisymmetric and non-axisymmetric surface waves in hollow cylinders at low-order modes, elucidating the influence of amplitude parameters on propagation characteristics and explaining the methods for solving the dispersion curves of guided waves in hollow cylinders [41]. Silk, Bainton, and others studied guided ultrasonic waves excited in thin-walled metal pipes using piezoelectric ultrasound probes to investigate their modes [72]. Lowe, Cawley, and others researched defects such as cracks, holes, and corrosion in pipelines and found that when the L (0, 2) mode-guided waves propagate in pipelines and encounter non-axisymmetric damaged defects, some of the L (0, 2) mode waves undergo mode conversion, generating F (1, 3) and F (2, 3) mode waves. This highlights the significance of mode conversion at defect locations for pipeline inspection [73–75]. Brook verified the effectiveness of L (0, 2) mode-guided waves for detecting pipeline defects [76]. Demma, Cawley, and others systematically analyzed the effects of the pipe size, defect size, guided wave mode, and frequency on the reflectivity of notches. They studied the relationship between the damage reflectance coefficient and the type of damage for T (0, 1) mode-guided waves at different frequencies. Imperial College London developed a detection system for controlling corrosion in long pipelines using guided ultrasonic waves [19,77]. Sharma used guided ultrasonic waves for non-destructive testing to monitor invisible corrosion in concrete [78]. Farhidzadeh et al. proposed a non-referenced guided wave method for diagnosing corrosion in prestressed steel strands. This method combines dispersion curves, continuous wavelet transforms, and velocity measurements to quantitatively assess the corrosion damage in prestressed steel strands [79].

Extensive research has been conducted on the application of ultrasonic-guided waves to coatings. Pan et al. explored the mode selection challenge of guided waves in elastic steel plates with substantial coatings. Through an investigation into the influence of thick elastic coatings and coating damping on dispersion and mode shapes, they identified the introduction of new modes primarily associated with the coating. It was found that within the range of $10 \leq \omega l \leq 13$, the modes S1, A1, and A0 of bare steel plates are most suitable for the ultrasonic detection of stress corrosion cracks in gas pipelines with thick elastic coatings. This can be applied to the ultrasonic testing of stress corrosion cracks in thin-walled gas pipelines with coal tar protective layers [80]. Ostiguy et al. introduced a technique based on guided waves to characterize the coating thickness. They employed the S0 mode, which demonstrates linear behavior at low frequencies and is particularly suitable for thin coatings. The variation in time-of-flight (ToF) of the S0 mode between two reference measurements was carefully measured. The actual coating thickness, estimated to be as low as 10 μm, can be precisely determined, underscoring the considerable potential of this method [81]. Zhang et al. introduced a guided wave-based technique for characterizing the bonding zone state of coatings. They employed FFT analysis of different bonding zone states to demonstrate its effectiveness in characterizing the bonding zone state of water-based coatings [17]. Duan et al. combined the traditional piezoresistive effect with the tunneling effect to fabricate a graphene-based nanocomposite material sensor capable of responding to ultra-low dynamic deformations. This sensor exhibits a response frequency range of up to 1 MHz, providing sensitive and rapid responses to broadband ultrasonic waves. With a high signal-to-noise ratio, it holds promising applications in structural health monitoring [82]. Pinilla et al. applied an efficient Gauss–Lobatto–Legendre (GLL) collocation method to establish a scale boundary finite-element method (SBFEM). They investigated the influence of viscoelastic coatings on the dispersion, attenuation, and wave structure in elastic–viscoelastic (steel–asphalt) double-layer structures. The study provided insights into the relationship between the displacement amplitudes of the free plate and coated plate under wave action [83]. Schmitz et al. applied machine learning and deep learning techniques to non-destructively characterize the thickness and uniformity of coatings in layered systems using dispersion curve analysis. Further, for coatings with non-uniform thickness, a convolutional neural network architecture was employed for classification through deep learning. The assessment and testing of the method were carried out [84].

Research into ultrasonic-guided wave non-destructive testing techniques for pipeline corrosion defects has been increasingly widespread and in-depth. For instance, Zhu et al. conducted numerical studies to assess the generation and reflection of guided elastic waves in hollow cylindrical structures with corrosion defects for non-destructive evaluation (NDE) of pipeline-like structures. They employed finite-element simulations to model time-delayed pitch-catch annular arrays (TDPRAs) and analyzed axisymmetric- and non-axisymmetric-guided wave reflections. The study validated the correlation between mode conversion (the L (0, 1) and L (0, 2) modes) of guided waves with the corrosion depth, axial, and circumferential extent within hollow cylindrical bodies [85,86]. Shivaraj et al. employed high-order cylindrical-guided ultrasonic waves to detect and quantify hidden pitting corrosion in concealed gap areas (between the pipelines and supports). They designed a manually operated pipe crawler and examined experimental parameters such as guided wave mode dispersion, particle displacement, and wavelength to address this issue, providing a device for securing wedges [87]. Demma et al. systematically analyzed the impact of the pipeline dimensions, defect sizes, guided wave modes, and frequencies on defect reflections. They identified the maximum and minimum values of the reflection coefficients within different axial ranges and employed them for defect quantification. An extrapolation formula was proposed and assessed to extend these results to other pipeline dimensions [77].

Moustafa et al. introduced a novel approach based on guided ultrasonic waves and fractal analysis for monitoring the corrosion evolution path in post-tensioning systems. They utilized permanently attached piezoelectric transducers on rebars to transmit and receive guided waves and proposed an outlier detection algorithm to enhance sensitivity to corrosion damage [88]. Løvstad et al. investigated reflections of the T (0, 1) mode from a cluster of pits, evaluating the relationship between the maximum pit depth and maximum reflection coefficient. They extensively studied the issue of obtaining low maximum reflection coefficients from clusters with deep maximum pit depths, aiding in estimating the maximum depth of problematic corrosion profiles [89]. Cawley et al. reviewed the application of ultrasonic-guided waves in the long-range detection of large structures [90]. They measured the attenuation characteristics of the fundamental torsional guided wave mode under different inspection frequencies for pipes with three standard anti-corrosion coatings, assessing the degree of corrosion in coated pipes [91]. Toppe et al. utilized ultrasonic-guided wave technology for corrosion and erosion monitoring in pipelines [92].

Yuan et al. introduced a non-destructive testing method to address the challenge of wall thinning due to corrosion in the bending region of an L-shaped pipe. They applied the time reversal focused guided wave (FGW) technique to evaluate the wall thickness reduction caused by corrosion in the L-shaped bent pipe. The SAFE-PML method was utilized to compute the dispersion and transmission characteristics of the L-shaped bent fiber Bragg grating. Through the analysis of the attenuation, wave structures, and energy concentration coefficients of the FGW modes in these bends, they investigated the interactions with wall-thinning defects in the bending region [93]. Zhu et al. integrated numerical simulations with the boundary element method (BEM) for experimental investigations into hidden corrosion detection using ultrasonic waves. They examined corrosion simulation specimens created through machine cutting and electrochemical processing of naturally corroded samples. By simulating the transmission and reflection of guided waves upon entering the corroded region, they devised a method to evaluate the hidden corrosion depth using guided waves, employing hybrid boundary element calculations [94]. Sargent et al. measured the amplitude of ultrasonic Lamb waves reflected from single 6 mm diameter flat-bottom holes near welds in 20 mm thick butt-welded steel plates. They simulated corrosion defects within the heat-affected zone. They assessed the applicability of using the fundamental S0 Lamb wave to remotely detect tank corrosion filled with liquid [95]. Pouyan Khalili et al. systematically analyzed the detection performance of each method for sharp and gradient defects, as well as their sensitivity to attenuation coatings, liquid loads, surface roughness, and features other than defects. They proposed the use of the A1 reflection mode for severe, sharp, pitting-type defects and the SH1 reflection mode for remote-guided wave reflection in the case of large area thinning and shallow gradient defects [25]. In recent years, many studies have applied deep learning to guided ultrasonic wave non-destructive testing. Xu et al. proposed a guided wave convolutional neural network for diagnosing fatigue cracks and evaluating 19 crack grades [96]. To address the diffraction limit and dispersion issues of guided waves, Song et al. proposed a subwavelength defect non-contact super-resolution guided wave array imaging method based on multiscale deep learning [97]. Wang et al. introduced a rapid guided wave imaging method based on convolutional neural networks (CNN) for quantitative evaluation of corrosion damage [98].

3.2. Key Components of Pipeline Ultrasonic-Guided Wave Detection Technology-Transducer

Transducers primarily transform electrical energy into acoustic energy and vice versa. They function as waveguide transmitters, converting electrical energy into acoustic energy, and as receivers, converting acoustic energy into electrical energy. The transducers utilized to generate ultrasonic-guided waves are primarily classified into piezoelectric, magnetostrictive, electromagnetic acoustic transducer (EMAT), and laser ultrasonic types based on their transduction mechanisms. Inadequate control over the dispersion and multimodal characteristics of ultrasonic-guided waves can result in mode conversion, waveform broadening, and a diminished signal-to-noise ratio. Consequently, the precise choice of transducers for both exciting and receiving guided waves plays a critical role in the ultrasonic-guided wave testing process. Frequently employed ultrasonic-guided wave instruments are predominantly grounded in piezoelectric and magnetostrictive principles. Piezoelectric ultrasonic-guided wave instruments are predominantly multi-channel systems, facilitating multimodal detection and approximate circumferential localization. Nonetheless, they necessitate distinct transducers tailored to different test objects, which might incur expenses. Magnetostrictive ultrasonic-guided wave instruments typically manifest as single-channel systems with steadfast structures. They can be adapted to the pipeline and facilitate an accurate circumferential localization of defects, rendering them convenient for detection and extensively adopted.

3.2.1. Piezoelectric Transducers

Piezoelectric transducers are widely used, and PZT can be fabricated into various shapes with modal selectivity [99]. Commonly used piezoelectric-based transducers include lead zirconate titanate (PZT) and PVDF (polyvinylidene fluoride film) transducers. In addition, there are EMATs, such as Lorentz force EMATs and magnetostrictive EMATs [31,100]. PVDF transducers have several advantages, including being lightweight, flexible, and have the ability to customize electrode shapes. Current research on PZTs has focused on developing transducers for specific guided wave mode excitation and investigating the effect of the coupling layer on detection [101]. Hay and Rose proposed a flexible PVDF multimodal comb-like transducer, improving coupling reliability in pipeline monitoring [102]. Additionally, piezoelectric spray comb-like sensors have shown excellent performance in measuring the state of cracks in high-temperature pipeline materials, making them potentially useful in the health monitoring of steam pipelines. Although PVDF exhibits high flexibility and weak piezoelectric properties, its usage as an exciter and receiver is limited due to constraints such as the inability to embed PVDF thin films into composite structures. [103,104].

3.2.2. Electromagnetic Ultrasonic Transducer

EMAT is a device with the capability to generate and detect ultrasonic waves in metals [105]. EMATs possess the advantage of not necessitating extensive surface preparation of the specimen. They can be readily configured for mode selection, generating families of SH-wave modes under non-contact conditions and demonstrating tolerance to surface conditions. These merits render them a swift screening tool for pipeline inspection [106–109]. Nevertheless, they do have certain limitations, including the low energy of emitted ultrasonic waves, a low signal-to-noise ratio, bulky equipment, dependency of induced energy on the proximity of the probe to the tested object (typically within 1 mm in practical applications and material testing), and low energy conversion efficiency [110–112].

PPM EMAT transducers are commonly used to excite torsional mode-guided waves in pipelines, and the width of the permanent magnet determines the excitation wavelength [31,100]. Compared to the excitation based on the Lorentz force principle, guided wave excitation based on magnetostrictive effects typically generates higher power [85]. Table 1 is a part of the list of GW detection techniques for pipeline corrosion through transducers, mainly for different transducers, guided wave modes, and a brief overview of the detection defects. There are many options for transducers used in pipeline corrosion. Here is only a part of the list, but we must continue to summarize and improve.

Table 1. GW detection technology of pipeline corrosion based on partial transducer.

Different Types of GW Transducers	Guided Wave Mode and Type	A Brief Introduction of the Research Content	Defect Type
Piezoelectric transducers (PZT)	L (0, 1) and L (0, 2) F (1, 1), F (1, 2), and F (1, 3)	(1) Quantitative characterization of defects was performed (2) FE simulation is used to show the complexity of the result of the defect reflection signal received from the overlap between the reflections of the defect edges with different characteristics [49].	Corrosion and notch-type defects
	Higher-order circumferential	The 2D FE model using ABAQUS® verified that the system can detect defects as small as 1.5 mm in diameter and 25% of the penetrating wall thickness [87].	(1) Defects in the pipe support (2) Pitting corrosion
Phased array transducers (PA)	Circumferential GWs	(1) Higher-order mode cluster (HOMC)-GW (2) Identifying defect sizes through high-frequency mode recognition. (3) Improving small defect imaging capability through cross-sectional (CS) views of the pipeline [113].	(1) axial cracks (2) Pinhole-like corrosion
	L (0, 1) and L (0, 2)	Three-dimensional and two-dimensional finite-element simulation analysis was employed to study the reflection of guided waves (GWs) from non-axisymmetric and axisymmetric corrosion defects [86].	Non-axisymmetric and axisymmetric corrosion defects
	T (m, 1) and L (m, 2)	(1) Performing circumferential scans at the defect location to determine the circumference of the defect, enabling circumferential and axial positioning, as well as defect detection. (2) By applying different delay inputs and amplitudes to the phased array (PA), the energy of the GW is concentrated to cover the entire pipeline. This enhances resolution and penetration. (3) Ring-shaped distributed PA used for focusing techniques [114].	(1) volumetric through-wall hole (2) Volumetric elliptical corrosion

3.3. *Selection of Guided Wave Modes for Assessment of Pipeline Corrosion Defect Identification*

Ultrasonic-guided waves have found extensive application in defect detection through the monitoring of reflected and transmitted signals, showcasing significant promise in evaluating the distributed corrosion across extensive pipeline distances. Conventional low-frequency guided wave testing is frequently utilized due to its capacity to offer comprehensive coverage from a singular sensor location. Nevertheless, the testing of ultrasonic-guided waves poses challenges when dealing with extremely shallow and sharp defects, owing to the constrained sensitivity of low-frequency guided waves to such characteristics. In the propagation of guided waves within an elastic medium in a hollow cylindrical structure, we commonly simulate this phenomenon by employing the displacement described in the Navier equation, as illustrated in the following Equation (1) [29,115].

$$(\lambda + \mu)\nabla\nabla\cdot u + \mu\nabla^2 u = \rho\frac{\partial^2 u}{\partial t^2} \tag{1}$$

where μ and λ are the Lamé constants, ρ is the density, t is time, ∇^2 is the Laplace operator, and u is the displacement vector.

Following guided wave excitation and propagation, signal processing becomes imperative. Conventional methods for processing ultrasonic-guided wave signals encompass the wavenumber-domain, time-domain, and frequency-domain techniques [116,117]. For time-domain analysis, the Hilbert transform is commonly used for processing ultrasonic-guided wave data signals, as represented by the following Equation (2) [118]:

$$\hat{X}(t) = \frac{1}{\pi}\int_{-\infty}^{+\infty} \frac{X(t)}{t-\tau}d\tau \tag{2}$$

where $\hat{X}(t)$ is the Hilbert transform of the signal $X(t)$.

c primarily acquires information from ultrasonic-guided waves through the Fourier transform for signal analysis in the frequency domain. The Fourier transform is given by the following Equation (3):

$$u(\omega) = \int_{-\infty}^{+\infty} e^{-j\omega t}u(t)dt \tag{3}$$

where $u(\omega)$ represents the signal in the frequency domain, $u(t)$ is the original time-domain signal, ω is the angular frequency, and j is the imaginary unit.

When employing ultrasonic waves for the detection of corrosion defects in pipelines, directly measuring wall loss is not a precise method. Auyeung et al. identified the instantaneous damage resulting from corrosion by leveraging additional current density I_d and the material characteristics of the pipeline. The cumulative corrosion damage (CCD) in the wall thickness of the metal pipe over time index n_i is measured by the following Equation (4) [91,119,120]:

$$CCD(n_i) = \sum_{n=0}^{n_i} \frac{A_W}{k\cdot F\rho_{Fe}} I_d(n) \tag{4}$$

where A_W represents the steel atomic weight, F is the Faraday's constant, and ρ_{Fe} is the density of iron. For the Fe/Fe^{2+} reaction, the rate constant (k) is equal to 2, taking into account the anticipated current level.

With the progress of guided wave theory research, high-order guided wave modes are commonly employed. These modes propagate along the circumferential direction of the pipeline, rendering them more sensitive to minor variations in the wall thickness. Nevertheless, contemporary high-order guided wave techniques predominantly rely on non-dispersive shear modes or HOMC waves, primarily attuned to surface defects. Hence, there is a necessity to investigate additional modes that can augment resolution and sensitivity to local corrosive-type defects.

Khalili et al. conducted a comprehensive study on applying Lamb waves and shear horizontal waves to detect defects in pipelines, including corrosion. Their research systematically examined various modes, such as the S0 mode, SH0 and SH1 modes, M-skip, and high-order mode clusters (A1 mode at approximately 18 MHz-mm) [25,121,122]. Additionally, they investigated the possibility of exciting low-dispersion single-mode Lamb waves at around 20 MHz-mm thickness and the SH1 mode at approximately 3 MHz-mm thickness. Figure 3 shows the phase velocity dispersion curves of guided waves in steel plates. Table 2 contains relevant information about some commonly used Lamb wave and shear horizontal wave modes in pipeline defect detection.

Table 2. Lamb wave and shear horizontal wave modes in pipeline defect detection.

Phase Velocity Dispersion Curves of Guided Waves in Steel Plate		Frequency-Thickness	Key Information	Partial Scope of Application	Defect Type	See for Details
Lamb modes	S0	1 MHZ-mm 1.5 MHZ-mm	(1) It exhibits ideal characteristics in the 1 MHz to millimeter range and shows minimal dispersion at low frequencies. (2) Provides approximate constant sensitivity to defects at different depths.	(1) Provides approximate constant sensitivity to defects at different depths. (2) Pipe supports (3) S0 mode in transmission used for guided wave tomography configuration in CUPS monitoring.	(1) Defects of different depths (2) Pitting	Khalili et al. [25]
	A1	20 MHZ-mm	(1) Non-dispersive nature (2) Low surface motion at high-frequency-thickness products (3) It is very similar to HOMC (4) Sensitivity to small defects	(1) Pure mode generation at high-frequency-thickness products (2) T-joints	severe, sharp, pitting-type defects	Balasubramaniam et al. [123,124] and Khalili et al. [121]
Shear horizontal (SH) modes	SH0	2.5 MHz-mm	Not affected by non-viscous liquid loads on the structure. Non-dispersive inspection can be conducted in the fundamental SH0 mode.	For pipes with relatively thin walls (<15 mm), EMATs are employed to excite SH0 and SH1 modes.	(1) Corrosion detection in inaccessible regions (2) Axial and circumferential corrosion defects	Cawley et al. [33–37]
	SH1	3 MHz-mm	The mode exhibits dispersion characteristics around 3 MHz·mm and demonstrates a mode shape with high surface energy.	Advantageous for short-range detection and very shallow crack-like defects.	SH1 mode to very shallow (<10% thickness loss) crack-like defects	Howard et al. [125]

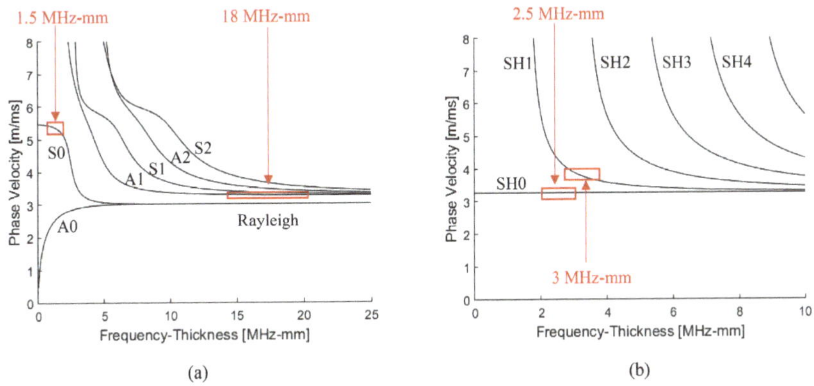

Figure 3. Guided wave phase velocity dispersion curves in a steel plate: (**a**) Lamb modes and (**b**) shear horizontal (SH) modes [25].

While these techniques can yield satisfactory results, they are not without limitations. The critical selection of an appropriate inspection mode is imperative for successfully assessing the corrosion levels. Mode selection criteria should encompass considerations such as mode excitability and receptivity, leakage loss, attenuation, dispersion, and velocity. Certain studies have indicated that conventional inspection techniques primarily utilize PZT transducers for exciting asymmetric wave modes, while methods involving shear mode excitations necessitate the use of EMATs. Conversely, symmetric modes are typically less susceptible to liquid loading, ensuring extended inspection distances [126,127]. Advanced technologies incorporating higher frequencies and selective higher-order symmetric mode generation hold the potential to enhance defect detectability and inspection coverage.

Cirtautas et al. [7] conducted a comprehensive study on the utilization of high-order symmetric modes for corrosion detection, underscoring the significance of evaluating various factors associated with higher mode selection, wave excitability, detectability, defect sensitivity, and leakage loss. Figure 4 represents the normalized displacement distributions of the A0-A3 and S0-S3 modes estimated at 1 MHz. The findings suggest that the out-of-plane displacement of the S3 mode peaks at depths of 1 mm and −1 mm relative to the surface and mid-thickness of the plate, respectively. Conversely, all symmetric modes demonstrate near-zero out-of-plane displacement at the mid-thickness of the plate, distinguishing them from the asymmetric modes. The S2 mode possesses a robust in-plane component at the plate's surface, leading to comparatively low leakage loss, rendering it more apt for detecting shallow surface defects. Nonetheless, the S2 mode displays a relatively low group velocity at 1 MHz, complicating the analysis of incoming signals. Moreover, the leakage loss of this mode becomes unacceptable above 1 MHz. Cirtautas employed high-order symmetric modes for the corrosion assessment, employing techniques like phased array or angle-beam excitation to selectively induce chosen wave modes into the structure, thereby minimizing the generation of unnecessary waves. The study illustrated that the S3 mode is prone to generating local and distributed defects associated with corrosion in metal structures, rendering it suitable for swift screening around pipelines.

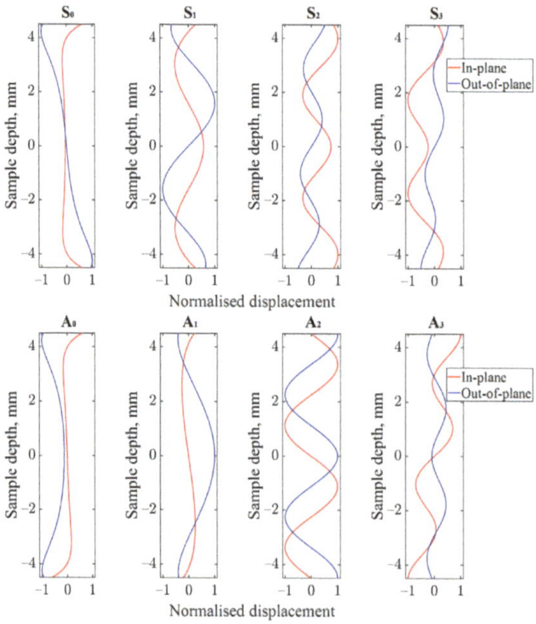

Figure 4. Normalized displacement of wave modes in a 9-millimeter steel plate at 1 MHZ [7].

Moreover, Raišutis introduced a highly efficient technique relying on fundamental UGW spiral waves to assess the integrity of steel pipe walls. By conducting finite-element

(FE) modeling and experimental investigations, the generation and reception mechanisms of UGW spiral waves were systematically examined. This research substantiated the efficacy of spiral UGW in the detection and characterization of defects [128]. Figure 5 illustrates simulated B-scan images featuring circular defects of diverse diameters and depths. The results depicted in Figure 5b,d,f,h indicate that the phase delay values of the signal amplitude peaks for spiral waves are elevated in the defect area.

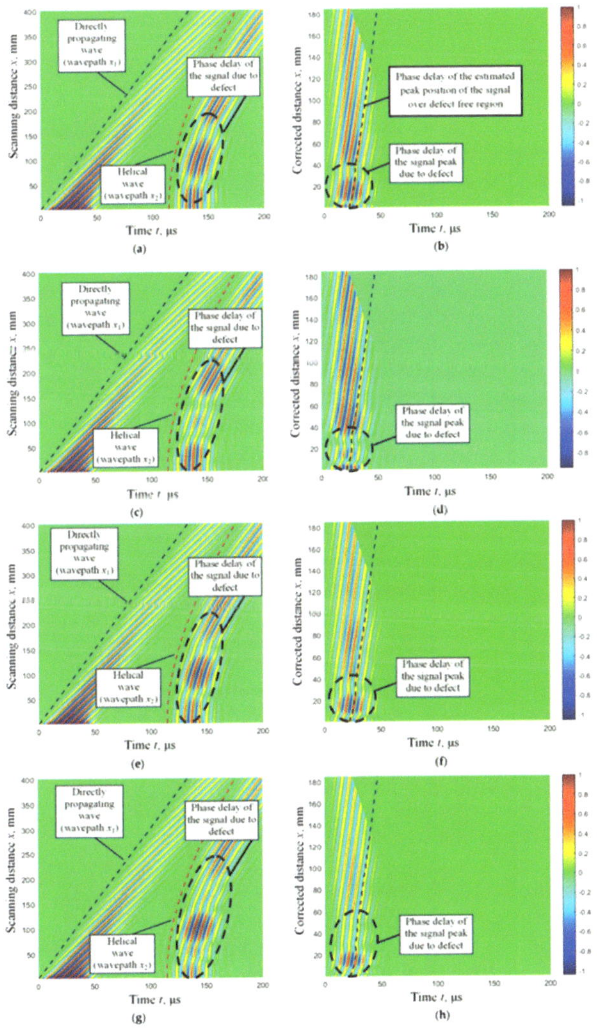

Figure 5. Simulated B-scan images capturing the propagation of direct waves and spiral waves in the presence of circular defects with varying diameters and depths are presented. The B-scan images portray both direct waves and spiral waves (**a**,**c**,**e**,**g**) alongside specifically chosen spiral waves and corrected spiral waves (**b**,**d**,**f**,**h**) for defects of different dimensions: (**a**,**b**)—30 mm (depth 25%), (**c**,**d**)—30 mm (75%), (**e**,**f**)—40 mm (25%), and (**g**,**h**)—60 mm (25%). The dashed lines indicate the phase delay values of the signal amplitude peaks for spiral waves [128].

4. Structural Health Monitoring Based on Pipeline Ultrasonic-Guided Wave

Some characteristic differences between NDT and structural health monitoring (SHM) are that NDT is an offline assessment that uses time-based maintenance, discovers the presence of damage, has more cost and labor, and the test system is independent of the test bed and does not require a baseline. In contrast, SHM is an online assessment that determines the suitability and service life of equipment through online evaluation, with lower costs and less labor involved. A test system integrated into the test bed requires baselines and has more requirements for algorithms, energy harvesting, data transmission, and processing while maintaining safety standards [32].

The SHM methods encompass data collection from various sensors installed on structures, interpreting the results to achieve reliable and cost-effective diagnostics under diverse operating conditions. Technologies for health assessment, including guided wave ultrasonics, X-rays, infrared thermography, and eddy current, have proven effective for non-destructive evaluations of structures. However, these technologies find quicker and more convenient applications in inspecting extensive structures such as pipelines, ships, and aerospace components [35]. Guided waves can propagate over large areas with minimal attenuation and energy loss [32,90]. This method can be employed for the integrity assessment of large continuous pipelines. Guided waves have the potential for application within curved structures, making them suitable for detecting various shapes and geometries over longer distances. Due to their excellent remote diagnostic capabilities, this method can effectively detect cracks, material losses, and fatigue defects in both isotropic and anisotropic structures. Ultrasonic-guided waves are one of the most commonly used techniques in current non-destructive testing and structural health monitoring of structural components [129].

Guided wave-based SHM uses non-destructive reference signals as baselines and extracts the defect signals by subtracting them from the baseline. As a result, it is more sensitive to small defects, enabling the timely detection of maintenance needs and predicting the remaining lifespan based on the condition of the pipeline [130]. The components of the complete pipeline SHM process and the requirements for an ideal pipeline SHM system are illustrated in Figure 6 [131,132].

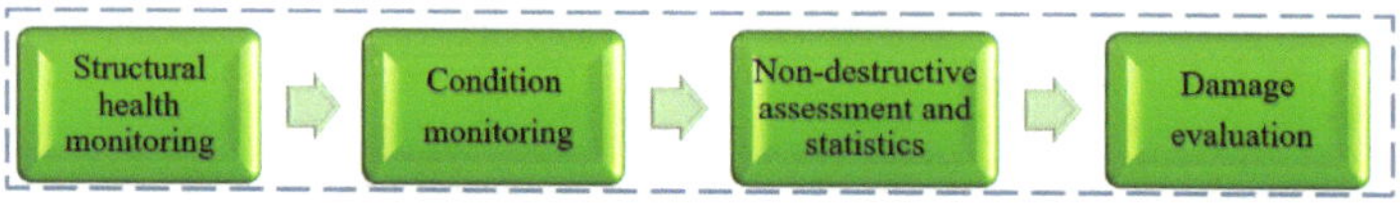

Figure 6. Complete process of pipeline SHM.

Contrary to the limitations associated with other defect diagnostic techniques, guided waves have lower operational barriers across various materials and environmental conditions. The flexibility of both contact and non-contact methods enhances their applicability. They are widely used in magnetic/non-magnetic, metal/non-magnetic, and isotropic/anisotropic structures of various shapes. SHM systems used for pipes buried in soil, insulation, exhaust and air conditioning pipes, railroad tracks, marine pipelines, etc., are frequently used to study defects in complex, sensitive, dynamically loaded structures. SHM systems collect data from sensors permanently installed on structures and diagnose damage by processing the collected data. Typically, damage detection is achieved by comparing the acquired data during the structure's healthy state with reference data. Automatic damage detection is triggered when the data deviates from the reference data, prompting an alarm.

However, the detection of UGW is highly sensitive to the presence of damage and is also affected by various factors, primarily changes in the environmental and operational conditions (EOCs). These EOCs can significantly impact the reference data, leading to false alarms [133]. Although there is a growing number of publications on employing SHM to detect damage in fully operational structures, many proposed SHM methods have been developed under controlled laboratory or simulated conditions, neglecting the influence of EOCs [134–137]. Mountassir et al. have systematically investigated the effect of EOCs on guided-wave-based SHM, with a particular focus on the impact of EOCs. The overview delineates various methods and strategies for reducing false alarms in SHM caused by EOCs. Each method is comprehensively described, identifying its optimal application and elucidating how to effectively implement it under natural conditions [27]. Methods tailored to specific environments should be chosen for SHM to mitigate the effects of EOCs, thereby enhancing the accuracy of SHM monitoring and ensuring the safety and stability of engineering applications.

Moreover, guided-wave non-destructive testing has the potential to significantly reduce the number of sensors required for structural monitoring. Therefore, the excitation transducer is frequently utilized as a measurement sensor in guided ultrasonic wave sensing to measure the backscattered ultrasonic waves. However, this installation scheme is limited by the information that can be extracted, as it is constrained by the measurement of excitation positions. This limitation may be particularly significant when detecting defects in long-distance and remote areas. An alternative solution has been proposed, utilizing distributed optical fiber sensors as measurement sensors, which can be strategically placed at multiple points along the pipeline to monitor damage [138,139]. Zhang et al. proposed a pipeline health monitoring method based on quasi-distributed fiber optic sensors, integrating simulated guided wave results with simulations of DAS/quasi-DAS systems. This work exploits the spatiotemporal data matrix in a fully distributed sensor system to generate a simulated dataset, holding significant potential for applications in pipeline SHM and defect identification using DAS data. The integration of physics-based simulated datasets with DAS systems enables real-time data analysis for detecting and identifying pipeline defects, thereby significantly enhancing the accuracy of SHM [139].

Various studies indicate that guided waves can be utilized for inspecting underwater structures, such as pipelines, coatings, or other difficult-to-reach structures, demonstrating excellent sensitivity. However, due to the dispersive nature of guided waves, mode conversion phenomena at defect locations, and variations in the wave structure of different mode waves at different frequencies, classifying and determining the size of defects detected by guided waves remains a challenge to be addressed. In recent years, the interaction of horizontal SH waves with various cracks and corrosion boundaries in structures has highlighted the significant potential of guided waves in addressing issues related to structural defect characterization and measurement [16].

5. Prospects of Ultrasonic-Guided Wave Detection Technology

Ultrasonic-guided waves exhibit substantial promise in identifying diverse defects within pipeline networks over considerable distances. Nevertheless, conventional guided wave detection often relies on low-frequency guided waves because of their reduced attenuation. While guided waves propagate, the signal amplitude gradually diminishes with increasing propagation distance. Consequently, the reflection wave packet from early defects becomes heavily contaminated or may even be submerged in noise, posing challenges in effectively identifying defect characteristics [140]. While low-frequency guided waves may not be very sensitive to small defects, in cases of locally small defects or wall thinning with smooth surfaces, existing guided wave technologies often struggle to detect the responses generated by defects. This limitation can lead to missed detections and new leaks, presenting challenges in current guided wave detection [63–66].

In contrast to the limitations associated with alternative defect diagnostic techniques, guided waves boast low operational barriers, making them applicable in diverse material and environmental conditions. The flexibility of employing contact or non-contact methods further broadens their range of utilization. They find extensive application in structures of various shapes, whether magnetic or non-magnetic, metal or composite, isotropic or anisotropic. In piping applications, such as those involving pipes buried in soil, insulation, exhaust and air conditioning pipes, railroad tracks, marine pipelines, etc., guided waves are often employed to investigate defects in complex, sensitive, and dynamically loaded structures.

Nevertheless, the intricate propagation characteristics of ultrasonic-guided waves within in-service pipeline engineering structures present formidable challenges for developing precise and efficient methods of signal feature extraction. Current methods for detecting damage using ultrasonic-guided waves in pipelines exhibit shortcomings in terms of their recognition efficiency and reliability, thus falling short of meeting the engineering requirements. With the advancement of machine learning (ML), numerous researchers are incorporating it into the diagnostic technology of ultrasonic-guided wave SHM in practical engineering structures. In SHM systems employing guided waves, data are collected from sensors permanently mounted on the structure, and the damage is diagnosed through the processing of the collected data. Typically, damage detection involves comparing the reference data acquired during the healthy state of the structure. An alarm is activated to ensure the automatic detection of damage when the data deviates from the reference data.

ML technology has shown significant advantages in the realm of SHM based on guided waves in recent years. Nevertheless, ML-based guided wave diagnostic techniques encounter crucial challenges, including dependable in-service data collection/management, ML models with generalization and real-time monitoring capabilities, integration of data from multiple frequency bands for passive collision and active damage-guided wave monitoring, in-service health assessment, and lifespan prediction, among others [117]. Further research is needed on the propagation model of guided waves, particularly in areas like data analysis and simulation data. The extension capability of machine learning models in guided wave structural health monitoring (SHM) also requires improvement, taking into account the sensitivity of guided waves to factors like structural types, noise, and temperature variations, as mentioned earlier. In large-scale engineering structures such as pipelines, these models might yield inaccurate prediction results. Presently, there is limited research on the correlation between machine learning-based damage detection and in-service structures, as well as environmental changes.

In addition, for ultrasonic-guided wave detection in large structures such as pipelines, as the guided waves propagate, the signal amplitude gradually attenuates with increasing propagation distance. This phenomenon leads to the reflection wave packet from early defects being severely contaminated or even submerged in noise, making it challenging to effectively identify defect features. In the future, to detect defects at an early stage and ensure structural safety, further research should be conducted on suppressing background noise and enhancing the reflection wave packet from defects.

6. Conclusions

This paper provides a review of the application of ultrasonic-guided wave technology in pipeline defect assessment, discussing the principles and applications of ultrasonic-guided wave-related theories and ultrasonic-guided wave non-destructive testing technology.

1. It elaborates on the standard modal selections of Lamb waves and shear horizontal waves commonly used in corrosion detection using ultrasonic-guided wave non-destructive testing technology.
2. The study focused on the research progress of non-destructive testing techniques for pipeline ultrasonic-guided waves. It highlighted the applications of ultrasonic-guided waves in coating research, relevant studies on transducers, and advancements in structural health monitoring. The paper elucidated the applications of ultrasonic-guided waves in conjunction with convolutional neural networks and machine learning for signal processing and visualizing the measurement results in related research areas.
3. The paper also introduces relevant research advancements. The comprehensive literature survey indicates that ultrasonic-guided wave technology is considered an ideal technical approach for pipeline defect detection and safety assessment, providing convenience to ensure the safety and stability of pipeline operation and maintenance while also achieving cost savings.

Currently, research into ultrasonic-guided wave theory is continuously deepening, and ultrasonic-guided wave detection technology is increasingly integrating and innovating with computer and simulation technologies. This has led to the gradual maturation of ultrasonic-guided wave detection technology, allowing it to serve in various fields.

In the future, research on ultrasonic-guided wave detection technology for pipelines should persist in concentrating on multimodal characteristics and optimizing interference modes. This can be accomplished by continuously improving the neural network intelligent algorithms and conducting simulation experiments. To quantitatively characterize defects in components, the precise feature recognition ability of defect signals needs improvement. The accuracy of multi-type defect detection should be heightened to minimize errors. Continuous efforts to optimize structural health monitoring based on ultrasonic-guided waves will further reduce the likelihood of false positives and omissions. It is hoped that the personnel involved in pipeline inspection and technicians unfamiliar with this technology can gain a better understanding of it through this manuscript, thereby providing assistance in enhancing the accuracy and convenience of engineering pipeline inspection. Ultimately, these efforts will contribute to the growing maturity of ultrasonic-guided wave inspection technology, laying a solid foundation for maintaining engineering safety and ensuring stable industrial operations.

Author Contributions: Formal analysis, F.L. and X.Z.; investigation, F.L. and X.Z.; resources, F.L. and X.Z.; writing—original draft preparation, Z.D. and X.Q.; writing—survey and editing, Z.D. and X.Q.; visualization, Z.D. and X.Q.; supervision, D.S.; project administration, D.S.; funding acquisition, D.S. All authors have read and agreed to the published version of the manuscript.

Funding: This research was funded by the State Grid Jiangsu Electric Power Co., Ltd. Technology Project, grant number J2023009.

Institutional Review Board Statement: Not applicable.

Informed Consent Statement: Not applicable.

Data Availability Statement: Data are included in the article and the corresponding references.

Conflicts of Interest: Authors Feng Lyu, Xinyue Zhou and Zheng Ding was employed by the State Grid Wuxi Power Supply Company. The remaining authors declare that the research was conducted in the absence of any commercial or financial relation-ships that could be construed as a potential conflict of interest.

References

1. Papadakis, G.A. Major Hazard Pipelines: A Comparative Study of Onshore Transmission Accidents. *J. Loss Prev. Process Ind.* **1999**, *12*, 91–107. [CrossRef]
2. Papadakis, G.A.; Porter, S.; Wettig, J. EU Initiative on the Control of Major Accident Hazards Arising from Pipelines. *J. Loss Prev. Process Ind.* **1999**, *12*, 85–90. [CrossRef]
3. Smith, L. Control of Corrosion in Oil and Gas Production Tubing. *British Corrosion Journal* **1999**, *34*, 247–253. [CrossRef]
4. Cawley, P.; Cegla, F.; Stone, M. Corrosion Monitoring Strategies—Choice between Area and Point Measurements. *J. Nondestruct. Eval.* **2013**, *32*, 156–163. [CrossRef]
5. Wood, M.H.; Arellano, A.V.; Van Wijk, L. Corrosion Related Accidents in Petroleum Refineries. *Eur. Comm. Jt. Res. Cent. Rep. No. EUR* **2013**, *26331*. [CrossRef]
6. El-Sherik, A.M. *Trends in Oil and Gas Corrosion Research and Technologies: Production and Transmission*; Woodhead Publishing: Cambridge, UK, 2017; ISBN 0-08-101219-5.
7. Cirtautas, D.; Samaitis, V.; Mažeika, L.; Raišutis, R.; Žukauskas, E. Selection of Higher Order Lamb Wave Mode for Assessment of Pipeline Corrosion. *Metals* **2022**, *12*, 503. [CrossRef]
8. Marcantonio, V.; Monarca, D.; Colantoni, A.; Cecchini, M. Ultrasonic Waves for Materials Evaluation in Fatigue, Thermal and Corrosion Damage: A Review. *Mech. Syst. Signal Process.* **2019**, *120*, 32–42. [CrossRef]
9. Broda, D.; Staszewski, W.J.; Martowicz, A.; Uhl, T.; Silberschmidt, V.V. Modelling of Nonlinear Crack–Wave Interactions for Damage Detection Based on Ultrasound—A Review. *J. Sound Vib.* **2014**, *333*, 1097–1118. [CrossRef]
10. Márquez, F.P.G.; Muñoz, J.M.C. A Pattern Recognition and Data Analysis Method for Maintenance Management. *Int. J. Syst. Sci.* **2012**, *43*, 1014–1028. [CrossRef]
11. García, F.P.; Pedregal, D.J.; Roberts, C. Time Series Methods Applied to Failure Prediction and Detection. *Reliab. Eng. Syst. Saf.* **2010**, *95*, 698–703. [CrossRef]

12. Pedregal, D.J.; García, F.P.; Roberts, C. An Algorithmic Approach for Maintenance Management Based on Advanced State Space Systems and Harmonic Regressions. *Ann. Oper. Res.* **2009**, *166*, 109–124. [CrossRef]
13. Gupta, M.; Khan, M.A.; Butola, R.; Singari, R.M. Advances in Applications of Non-Destructive Testing (NDT): A Review. *Adv. Mater. Process. Technol.* **2022**, *8*, 2286–2307. [CrossRef]
14. Wang, Z.D.; Gu, Y.; Wang, Y.S. A Review of Three Magnetic NDT Technologies. *J. Magn. Magn. Mater.* **2012**, *324*, 382–388. [CrossRef]
15. Yun, H.; Zhang, W. Damage Detection Based on the Propagation of Longitudinal Guided Wave in a Bimetal Composite Pipe. *Theor. Appl. Mech. Lett.* **2011**, *1*, 021004. [CrossRef]
16. Olisa, S.C.; Khan, M.A.; Starr, A. Review of Current Guided Wave Ultrasonic Testing (GWUT) Limitations and Future Directions. *Sensors* **2021**, *21*, 811. [CrossRef]
17. Zhang, J.; Cho, Y.; Kim, J.; Malikov, A.K.U.; Kim, Y.H.; Yi, J.-H. Nondestructive Inspection of Underwater Coating Layers Using Ultrasonic Lamb Waves. *Coatings* **2023**, *13*, 728. [CrossRef]
18. Alleyne, D.N. Rapid, Long Range Inspection of Chemical Plant Pipework Using Guided Waves. *AIP Conf. Proc.* **2001**, *557*, 180–187.
19. Demma, A.; Cawley, P.; Lowe, M.; Roosenbrand, A.G. The Reflection of the Fundamental Torsional Mode from Cracks and Notches in Pipes. *J. Acoust. Soc. Am.* **2003**, *114*, 611–625. [CrossRef]
20. Beard, M.D. Development of a Guided Wave Inspection Technique for Rock Bolts. *AIP Conf. Proc.* **2002**, *615*, 1318–1325.
21. Monnier, T. Lamb Waves-Based Impact Damage Monitoring of a Stiffened Aircraft Panel Using Piezoelectric Transducers. *J. Intell. Mater. Syst. Struct.* **2006**, *17*, 411–421. [CrossRef]
22. Dalton, R.P.; Cawley, P.; Lowe, M.J.S. The Potential of Guided Waves for Monitoring Large Areas of Metallic Aircraft Fuselage Structure. *J. Nondestruct. Eval.* **2001**, *20*, 29–46. [CrossRef]
23. Thwaites, S.; Clark, N.H. Non-Destructive Testing of Honeycomb Sandwich Structures Using Elastic Waves. *J. Sound Vib.* **1995**, *187*, 253–269. [CrossRef]
24. Castaings, M. The Propagation of Guided Waves in Composite, Sandwich-like Structures and Their Use for NDT. *AIP Conf. Proc.* **2001**, *557*, 999–1006.
25. Khalili, P.; Cawley, P. The Choice of Ultrasonic Inspection Method for the Detection of Corrosion at Inaccessible Locations. *NDT E Int.* **2018**, *99*, 80–92. [CrossRef]
26. Andruschak, N.; Saletes, I.; Filleter, T.; Sinclair, A. An NDT Guided Wave Technique for the Identification of Corrosion Defects at Support Locations. *NDT E Int.* **2015**, *75*, 72–79. [CrossRef]
27. El Mountassir, M.; Yaacoubi, S.; Dahmene, F. Reducing False Alarms in Guided Waves Structural Health Monitoring of Pipelines: Review Synthesis and Debate. *Int. J. Press. Vessel. Pip.* **2020**, *188*, 104210. [CrossRef]
28. He, J.; Zhou, C.; Yang, L.; Sun, X. Research on Pipeline Damage Imaging Technology Based on Ultrasonic Guided Waves. *Shock Vib.* **2019**, *2019*, 1470761. [CrossRef]
29. Rose, J.L. *Ultrasonic Waves in Solid Media*; Cambridge University Press: Cambridge, UK, 2014.
30. Guan, R.; Lu, Y.; Duan, W.; Wang, X. Guided Waves for Damage Identification in Pipeline Structures: A Review. *Struct Control Health Monit* **2017**, *24*, e2007. [CrossRef]
31. Ling, E.H.; Abdul Rahim, R.H. A Review on Ultrasonic Guided Wave Technology. *Aust. J. Mech. Eng.* **2020**, *18*, 32–44. [CrossRef]
32. Rose, J.L. A Baseline and Vision of Ultrasonic Guided Wave Inspection Potential. *J. Press. Vessel Technol.* **2002**, *124*, 273–282. [CrossRef]
33. Xu, Z.-D.; Liu, M.; Wu, Z.; Zeng, X. Energy Damage Detection Strategy Based on Strain Responses for Long-Span Bridge Structures. *J. Bridge Eng.* **2011**, *16*, 644–652. [CrossRef]
34. Xu, Z.-D.; Wu, K.-Y. Damage Detection for Space Truss Structures Based on Strain Mode under Ambient Excitation. *J. Eng. Mech.* **2012**, *138*, 1215–1223. [CrossRef]
35. Abbas, M.; Shafiee, M. Structural Health Monitoring (SHM) and Determination of Surface Defects in Large Metallic Structures Using Ultrasonic Guided Waves. *Sensors* **2018**, *18*, 3958. [CrossRef] [PubMed]
36. Barshinger, J.N.; Rose, J.L. Guided Wave Propagation in an Elastic Hollow Cylinder Coated with a Viscoelastic Material. *IEEE Trans. Ultrason. Ferroelectr. Freq. Control* **2004**, *51*, 1547–1556. [CrossRef]
37. Zang, X.; Xu, Z.-D.; Lu, H.; Zhu, C.; Zhang, Z. Ultrasonic Guided Wave Techniques and Applications in Pipeline Defect Detection: A Review. *Int. J. Press. Vessel. Pip.* **2023**, *206*, 105033. [CrossRef]
38. Gazis, D.C. Exact Analysis of the Plane-Strain Vibrations of Thick-Walled Hollow Cylinders. *J. Acoust. Soc. Am.* **1958**, *30*, 786–794. [CrossRef]
39. Gazis, D.C. Three-dimensional Investigation of the Propagation of Waves in Hollow Circular Cylinders. I. Analytical Foundation. *J. Acoust. Soc. Am.* **1959**, *31*, 568–573. [CrossRef]
40. Meitzler, A.H. Mode Coupling Occurring in the Propagation of Elastic Pulses in Wires. *J. Acoust. Soc. Am.* **1961**, *33*, 435–445. [CrossRef]
41. Rose, J.L. *Ultrasonic Waves in Solid Media*; Acoustical Society of America: Melville, NY, USA, 2000; ISBN 0521640431.
42. Wilcox, P.; Lowe, M.; Cawley, P. The Effect of Dispersion on Long-Range Inspection Using Ultrasonic Guided Waves. *NDT E Int.* **2001**, *34*, 1–9. [CrossRef]

43. Shin, H.J.; Rose, J.L. Guided Waves by Axisymmetric and Non-Axisymmetric Surface Loading on Hollow Cylinders. *Ultrasonics* **1999**, *37*, 355–363. [CrossRef]
44. Koduru, J.P.; Momeni, S.; Rose, J.L. Phased Annular Array Transducers for Omnidirectional Guided Wave Mode Control in Isotropic Plate like Structures. *Smart Mater. Struct.* **2013**, *22*, 125022. [CrossRef]
45. Niu, X.; Tee, K.F.; Marques, H.R. Enhancement of Unidirectional Excitation of Guided Torsional T (0, 1) Mode by Linear Superposition of Multiple Rings of Transducers. *Appl. Acoust.* **2020**, *168*, 107411. [CrossRef]
46. Wang, X.; Gao, H.; Zhao, K.; Wang, C. Time-Frequency Characteristics of Longitudinal Modes in Symmetric Mode Conversion for Defect Characterization in Guided Waves-Based Pipeline Inspection. *NDT E Int.* **2021**, *122*, 102490. [CrossRef]
47. Michaels, J.E.; Michaels, T.E. Guided Wave Signal Processing and Image Fusion for in Situ Damage Localization in Plates. *Wave Motion* **2007**, *44*, 482–492. [CrossRef]
48. Alleyne, D.N.; Lowe, M.J.S.; Cawley, P. The Reflection of Guided Waves from Circumferential Notches in Pipes. *J. Appl. Mech.* **1998**, *65*, 635–641. [CrossRef]
49. Wang, X.; Peter, W.T.; Mechefske, C.K.; Hua, M. Experimental Investigation of Reflection in Guided Wave-Based Inspection for the Characterization of Pipeline Defects. *NDT E Int.* **2010**, *43*, 365–374. [CrossRef]
50. Verma, B.; Mishra, T.K.; Balasubramaniam, K.; Rajagopal, P. Interaction of Low-Frequency Axisymmetric Ultrasonic Guided Waves with Bends in Pipes of Arbitrary Bend Angle and General Bend Radius. *Ultrasonics* **2014**, *54*, 801–808. [CrossRef]
51. Lowe, M.J.; Alleyne, D.N.; Cawley, P. Defect Detection in Pipes Using Guided Waves. *Ultrasonics* **1998**, *36*, 147–154. [CrossRef]
52. Carandente, R.; Lovstad, A.; Cawley, P. The Influence of Sharp Edges in Corrosion Profiles on the Reflection of Guided Waves. *NDT E Int.* **2012**, *52*, 57–68. [CrossRef]
53. Løvstad, A.; Cawley, P. The Reflection of the Fundamental Torsional Guided Wave from Multiple Circular Holes in Pipes. *NDT E Int.* **2011**, *44*, 553–562. [CrossRef]
54. Zhang, L.; Gavigan, B.J.; Rose, J.L. High Frequency Guided Wave Natural Focusing Pipe Inspection with Frequency and Angle Tuning. *J. Pressure Vessel Technol.* **2006**, *128*, 433–438. [CrossRef]
55. Rose, J.L. Standing on the Shoulders of Giants: An Example of Guided Wave Inspection. *Mater. Eval* **2002**, *60*, 53–59.
56. Cawley, P.; Chua, C. Crack Growth Monitoring Using Fundamental Shear Horizontal Guided Waves. *Struct. Health Monit.* **2020**, *19*, 1311–1322.
57. Chen, S.; Yu, Q.; Xu, H.; Yang, Q.; Chen, Z. Investigation of Pipelines Defect Localization for Fusion Reactor by Using T (0, 1) Mode Ultrasonic Guided Waves. *Fusion Eng. Des.* **2023**, *195*, 113937. [CrossRef]
58. Auld, B.A. Application of Microwave Concepts to the Theory of Acoustic Fields and Waves in Solids. *IEEE Trans. Microw. Theory Tech.* **1969**, *17*, 800–811. [CrossRef]
59. Xu, Q.; Zhang, L.; Liang, W. Acoustic Detection Technology for Gas Pipeline Leakage. *Process Saf. Environ. Prot.* **2013**, *91*, 253–261. [CrossRef]
60. Nienwenhui, J.H.; Neumann, J.J.; Greve, D.W.; Oppenheim, I.J. Generation and Detection of Guided Waves Using PZT Wafer Transducers. *IEEE Trans. Ultrason. Ferroelectr. Freq. Control* **2005**, *52*, 2103–2111. [CrossRef] [PubMed]
61. Qi, M.; Zhou, S.; Ni, J.; Li, Y. Investigation on Ultrasonic Guided Waves Propagation in Elbow Pipe. *Int. J. Press. Vessel. Pip.* **2016**, *139*, 250–255. [CrossRef]
62. Howard, R.; Cegla, F. Detectability of Corrosion Damage with Circumferential Guided Waves in Reflection and Transmission. *NDT E Int.* **2017**, *91*, 108–119. [CrossRef]
63. Wong, B.; McCann, J.A. Failure Detection Methods for Pipeline Networks: From Acoustic Sensing to Cyber-Physical Systems. *Sensors* **2021**, *21*, 4959. [CrossRef]
64. Kain, V.; Roychowdhury, S.; Ahmedabadi, P.; Barua, D.K. Flow Accelerated Corrosion: Experience from Examination of Components from Nuclear Power Plants. *Eng. Fail. Anal.* **2011**, *18*, 2028–2041. [CrossRef]
65. Catton, P.; Mudge, P.; Balachandran, W. Advances in Defect Characterisation Using Long-Range Ultrasonic Testing of Pipes. *Insight-Non-Destr. Test. Cond. Monit.* **2008**, *50*, 480–484. [CrossRef]
66. Ida, N.; Meyendorf, N. *Handbook of Advanced Nondestructive Evaluation*; Springer International Publishing: Cham, Switzerland, 2019; Volume 10.
67. Yoo, B.; Na, S.-M.; Flatau, A.B.; Pines, D.J. Directional Magnetostrictive Patch Transducer Based on Galfenol's Anisotropic Magnetostriction Feature. *Smart Mater. Struct.* **2014**, *23*, 095035. [CrossRef]
68. Rayleigh, L. Lord on the Free Vibrations of an Infinite Plate of Homogeneous Isotropic Elastic Matter. *Proc. Lond. Math. Soc.* **1888**, *1*, 225–237. [CrossRef]
69. Love, A.E.H. Mathematical Theory of Elasticity. Available online: https://hal.science/hal-01307751/document (accessed on 16 January 2024).
70. Gazis, D.C. Three-dimensional Investigation of the Propagation of Waves in Hollow Circular Cylinders. II. Numerical Results. *J. Acoust. Soc. Am.* **1959**, *31*, 573–578. [CrossRef]
71. Armenàkas, A.E.; Gazis, D.C.; Herrmann, G. *Free Vibrations of Circular Cylindrical Shells*; Elsevier: Amsterdam, The Netherlands, 2013; ISBN 1-4831-5861-6.
72. Silk, M.G.; Bainton, K.F. The Propagation in Metal Tubing of Ultrasonic Wave Modes Equivalent to Lamb Waves. *Ultrasonics* **1979**, *17*, 11–19. [CrossRef]

73. Lowe, M.J.S.; Alleyne, D.N.; Cawley, P. The Mode Conversion of a Guided Wave by a Part-Circumferential Notch in a Pipe. *J. Appl. Mech.* **1998**, *65*, 649–656. [CrossRef]
74. Zheng, M.; Lu, C.; Chen, G.; Men, P. Modeling Three-Dimensional Ultrasonic Guided Wave Propagation and Scattering in Circular Cylindrical Structures Using Finite Element Approach. *Phys. Procedia* **2011**, *22*, 112–118. [CrossRef]
75. Carandente, R.; Ma, J.; Cawley, P. The Scattering of the Fundamental Torsional Mode from Axi-Symmetric Defects with Varying Depth Profile in Pipes. *J. Acoust. Soc. Am.* **2010**, *127*, 3440–3448. [CrossRef]
76. Brook, M.V.; Ngoc, T.D.; Eder, J.E. Ultrasonic Inspection of Steam Generator Tubing by Cylindrical Guided Waves. In *Review of Progress in Quantitative Nondestructive Evaluation*; Springer: Berlin/Heidelberg, Germany, 1990; pp. 243–249.
77. Demma, A.; Cawley, P.; Lowe, M.; Roosenbrand, A.G.; Pavlakovic, B. The Reflection of Guided Waves from Notches in Pipes: A Guide for Interpreting Corrosion Measurements. *NDT E Int.* **2004**, *37*, 167–180. [CrossRef]
78. Sharma, A.; Sharma, S.; Sharma, S.; Mukherjee, A. Monitoring Invisible Corrosion in Concrete Using a Combination of Wave Propagation Techniques. *Cem. Concr. Compos.* **2018**, *90*, 89–99. [CrossRef]
79. Farhidzadeh, A.; Salamone, S. Reference-Free Corrosion Damage Diagnosis in Steel Strands Using Guided Ultrasonic Waves. *Ultrasonics* **2015**, *57*, 198–208. [CrossRef] [PubMed]
80. Pan, E.; Rogers, J.; Datta, S.K.; Shah, A.H. Mode Selection of Guided Waves for Ultrasonic Inspection of Gas Pipelines with Thick Coating. *Mech. Mater.* **1999**, *31*, 165–174. [CrossRef]
81. Ostiguy, P.-C.; Quaegebeur, N.; Masson, P. Non-Destructive Evaluation of Coating Thickness Using Guided Waves. *NDT E Int.* **2015**, *76*, 17–25. [CrossRef]
82. Duan, F.; Liao, Y.; Zeng, Z.; Jin, H.; Zhou, L.; Zhang, Z.; Su, Z. Graphene-Based Nanocomposite Strain Sensor Response to Ultrasonic Guided Waves. *Compos. Sci. Technol.* **2019**, *174*, 42–49. [CrossRef]
83. Pinilla, C.A.G.; Quiroga, J.E.; Ballesteros, D.Y.P.; Cañas, C.A.T.; Minoli, C.A.A. Effect of Viscoelastic Coating on Lamb Wave Propagation in Plates. *Procedia Struct. Integr.* **2024**, *52*, 20–27. [CrossRef]
84. Schmitz, M.; Kim, J.-Y.; Jacobs, L.J. Machine and Deep Learning for Coating Thickness Prediction Using Lamb Waves. *Wave Motion* **2023**, *120*, 103137. [CrossRef]
85. Ghavamian, A.; Mustapha, F.; Baharudin, B.H.T.; Yidris, N. Detection, Localisation and Assessment of Defects in Pipes Using Guided Wave Techniques: A Review. *Sensors* **2018**, *18*, 4470. [CrossRef]
86. Zhu, W. An FEM Simulation for Guided Elastic Wave Generation and Reflection in Hollow Cylinders with Corrosion Defects. *J. Press. Vessel Technol.* **2002**, *124*, 108–117. [CrossRef]
87. Shivaraj, K.; Balasubramaniam, K.; Krishnamurthy, C.V.; Wadhwan, R. Ultrasonic Circumferential Guided Wave for Pitting-Type Corrosion Imaging at Inaccessible Pipe-Support Locations. *J. Pressure Vessel Technol.* **2008**, *130*, 021502. [CrossRef]
88. Moustafa, A.; Niri, E.D.; Farhidzadeh, A.; Salamone, S. Corrosion Monitoring of Post-tensioned Concrete Structures Using Fractal Analysis of Guided Ultrasonic Waves. *Struct. Control Health Monit.* **2014**, *21*, 438–448. [CrossRef]
89. Løvstad, A. *Detection of Localised Corrosion in Pipes Using Guided Waves*; Norwegian University of Science and Technology: Trondheim, Norway, 2012.
90. Cawley, P.; Lowe, M.J.S.; Alleyne, D.N.; Pavlakovic, B.; Wilcox, P. Practical Long Range Guided Wave Inspection-Applications to Pipes and Rail. *Mater. Eval.* **2003**, *61*, 66–74.
91. Cobb, A.C.; Kwun, H.; Caseres, L.; Janega, G. Torsional Guided Wave Attenuation in Piping from Coating, Temperature, and Large-Area Corrosion. *NDT E Int.* **2012**, *47*, 163–170. [CrossRef]
92. Instanes, G.; Toppe, M.; Lakshminarayan, B.; Nagy, P.B. Corrosion and Erosion Monitoring of Pipes by an Ultrasonic Guided Wave Method. In *Advanced Ultrasonic Methods for Material and Structure Inspection*; Wiley: Hoboken, NJ, USA, 2007; pp. 115–157. [CrossRef]
93. Yuan, X.; Li, W.; Deng, M. Quantitative Assessment of Corrosion-Induced Wall Thinning in L-Shaped Bends Using Ultrasonic Feature Guided Waves. *Thin-Walled Struct.* **2023**, *196*, 111493. [CrossRef]
94. Zhu, W.; Rose, J.L.; Barshinger, J.N.; Agarwala, V.S. Ultrasonic Guided Wave NDT for Hidden Corrosion Detection. *J. Res. Nondestruct. Eval.* **1998**, *10*, 205–225. [CrossRef]
95. Sargent, J.P. Corrosion Detection in Welds and Heat-Affected Zones Using Ultrasonic Lamb Waves. *Insight-Non-Destr. Test. Cond. Monit.* **2006**, *48*, 160–167. [CrossRef]
96. Xu, L.; Yuan, S.; Chen, J.; Ren, Y. Guided Wave-Convolutional Neural Network Based Fatigue Crack Diagnosis of Aircraft Structures. *Sensors* **2019**, *19*, 3567. [CrossRef]
97. Song, H.; Yang, Y. Noncontact Super-Resolution Guided Wave Array Imaging of Subwavelength Defects Using a Multiscale Deep Learning Approach. *Struct. Health Monit.* **2021**, *20*, 1904–1923. [CrossRef]
98. Wang, X.; Lin, M.; Li, J.; Tong, J.; Huang, X.; Liang, L.; Fan, Z.; Liu, Y. Ultrasonic Guided Wave Imaging with Deep Learning: Applications in Corrosion Mapping. *Mech. Syst. Signal Process.* **2022**, *169*, 108761. [CrossRef]
99. Yu, H.; Quan, Q.; Tian, X.; Li, H. Optimization and Analysis of a U-Shaped Linear Piezoelectric Ultrasonic Motor Using Longitudinal Transducers. *Sensors* **2018**, *18*, 809. [CrossRef]
100. Miao, H.; Li, F. Shear Horizontal Wave Transducers for Structural Health Monitoring and Nondestructive Testing: A Review. *Ultrasonics* **2021**, *114*, 106355. [CrossRef] [PubMed]
101. Sun, Z.; Rocha, B.; Wu, K.-T.; Mrad, N. A Methodological Review of Piezoelectric Based Acoustic Wave Generation and Detection Techniques for Structural Health Monitoring. *Int. J. Aerosp. Eng.* **2013**, *2013*, 928627. [CrossRef]

102. Hay, T.R.; Rose, J.L. Flexible PVDF Comb Transducers for Excitation of Axisymmetric Guided Waves in Pipe. *Sens. Actuators A Phys.* **2002**, *100*, 18–23. [CrossRef]
103. Raghavan, A. Guided-Wave Structural Health Monitoring. Ph.D. Thesis, University Avenue, Ann Arbor, MI, USA, 2007.
104. Sinding, K.; Searfass, C.; Malarich, N.; Reinhardt, B.; Tittmann, B.R. High Temperature Ultrasonic Transducers for the Generation of Guided Waves for Non-Destructive Evaluation of Pipes. *AIP Conf. Proc.* **2014**, *1581*, 302–307.
105. Hirao, M.; Ogi, H. *EMATs for Science and Industry: Noncontacting Ultrasonic Measurements*; Springer Science & Business Media: Berlin/Heidelberg, Germany, 2003; ISBN 1-4020-7494-8.
106. Ribichini, R. *Modelling of Electromagnetic Acoustic Transducers, Department of Mechanical Engineering*; Imperial College: London, UK, 2011.
107. Dixon, S.; Petcher, P.A.; Fan, Y.; Maisey, D.; Nickolds, P. Ultrasonic Metal Sheet Thickness Measurement without Prior Wave Speed Calibration. *J. Phys. D Appl. Phys.* **2013**, *46*, 445502. [CrossRef]
108. Mirkhani, K. *Characterization of Interfacial Degradation in Adhesive Joints Using EMAT's*; Library and Archives Canada= Bibliothèque et Archives: Ottawa, Canada, 2005; ISBN 0-612-94432-8.
109. Thompson, R.B. Physical Principles of Measurements with EMAT Transducers. *Phys. Acoust.* **1990**, *19*, 157–200.
110. Na, W.-B.; Kundu, T. A Combination of PZT and EMAT Transducers for Interface Inspection. *J. Acoust. Soc. Am.* **2002**, *111*, 2128–2139. [CrossRef]
111. Mirkhani, K.; Chaggares, C.; Masterson, C.; Jastrzebski, M.; Dusatko, T.; Sinclair, A.; Shapoorabadi, R.J.; Konrad, A.; Papini, M. Optimal Design of EMAT Transmitters. *NDT E Int.* **2004**, *37*, 181–193. [CrossRef]
112. Alers, G.A. Electromagnetic Induction of Ultrasonic Waves: EMAT, EMUS, EMAR. In Proceedings of the 16th World Conference on NDT, Montreal, QC, Canada, 30 August–3 September 2004.
113. Satyarnarayan, L.; Chandrasekaran, J.; Maxfield, B.; Balasubramaniam, K. Circumferential Higher Order Guided Wave Modes for the Detection and Sizing of Cracks and Pinholes in Pipe Support Regions. *NDT E Int.* **2008**, *41*, 32–43. [CrossRef]
114. Mu, J.; Zhang, L.; Rose, J.L. Defect Circumferential Sizing by Using Long Range Ultrasonic Guided Wave Focusing Techniques in Pipe. *Nondestruct. Test. Eval.* **2007**, *22*, 239–253. [CrossRef]
115. Lamb, H. On Waves in an Elastic Plate. *Proc. R. Soc. Lond. A* **1917**, *93*, 114–128. [CrossRef]
116. Tian, Z.; Yu, L. Lamb Wave Frequency–Wavenumber Analysis and Decomposition. *J. Intell. Mater. Syst. Struct.* **2014**, *25*, 1107–1123. [CrossRef]
117. Yang, Z.; Yang, H.; Tian, T.; Deng, D.; Hu, M.; Ma, J.; Gao, D.; Zhang, J.; Ma, S.; Yang, L.; et al. A Review on Guided-Ultrasonic-Wave-Based Structural Health Monitoring: From Fundamental Theory to Machine Learning Techniques. *Ultrasonics* **2023**, *133*, 107014. [CrossRef] [PubMed]
118. Yu, L.; Giurgiutiu, V. Advanced Signal Processing for Enhanced Damage Detection with Piezoelectric Wafer Active Sensors. *Smart Struct. Syst.* **2005**, *1*, 185–215. [CrossRef]
119. Auyeung, Y.; Balaguru, P.; Chung, L. Bond Behavior of Corroded Reinforcement Bars. *Mater. J.* **2000**, *97*, 214–220.
120. Lee, H.-S.; Noguchi, T.; Tomosawa, F. Evaluation of the Bond Properties between Concrete and Reinforcement as a Function of the Degree of Reinforcement Corrosion. *Cem. Concr. Res.* **2002**, *32*, 1313–1318. [CrossRef]
121. Khalili, P.; Cawley, P. Excitation of Single-Mode Lamb Waves at High-Frequency-Thickness Products. *IEEE Trans. Ultrason. Ferroelectr. Freq. Control* **2015**, *63*, 303–312. [CrossRef]
122. Khalili, P.; Cegla, F. Excitation of Single-Mode Shear-Horizontal Guided Waves and Evaluation of Their Sensitivity to Very Shallow Crack-like Defects. *IEEE Trans. Ultrason. Ferroelectr. Freq. Control* **2020**, *68*, 818–828. [CrossRef]
123. Balasubramaniam, K.; Annamalai, S.; Venkataraman, K.S. Higher Order Modes Cluster (HOMC) Guided Waves Technique for Corrosion Detection. In Proceedings of the 18th World World Conference on Non-Destructive Testing, Durban, South Africa, 16–20 April 2012.
124. Jayaraman, C.; Krishnamurthy, C.V.; Balasubramaniam, K. Higher Order Modes Cluster (Homc) Guided Waves—A New Technique for Ndt Inspection. *AIP Conf. Proc.* **2014**, *1096*, 121–128.
125. Howard, R.; Cegla, F. On the Probability of Detecting Wall Thinning Defects with Dispersive Circumferential Guided Waves. *NDT E Int.* **2017**, *86*, 73–82. [CrossRef]
126. Jankauskas, A.; Mazeika, L. Ultrasonic Guided Wave Propagation through Welded Lap Joints. *Metals* **2016**, *6*, 315. [CrossRef]
127. Mažeika, L.; Raišutis, R.; Maciulevičius, A.; Žukauskas, E.; Kažys, R.; Seniūnas, G.; Vladišauskas, A. Comparison of Several Techniques of Ultrasonic Lamb Waves Velocities Measurements. *Ultragarsas/Ultrasound* **2009**, *64*, 11–17.
128. Raišutis, R.; Tumšys, O.; Žukauskas, E.; Samaitis, V.; Draudvilienė, L.; Jankauskas, A. An Inspection Technique for Steel Pipes Wall Condition Using Ultrasonic Guided Helical Waves and a Limited Number of Transducers. *Materials* **2023**, *16*, 5410. [CrossRef] [PubMed]
129. Lenkov, S.V.; Zverev, N.N.; Muravieva, O.V.; Myshkin, Y.V. Hardware and Software System of the Guided Wave Pipe Testing Using Electromagnetic-Acoustic Transformation. In Proceedings of the 2016 International Siberian Conference on Control and Communications (SIBCON), Moscow, Russia, 12–14 May 2016; pp. 1–4.
130. Qing, X.P.; Beard, S.; Shen, S.B.; Banerjee, S.; Bradley, I.; Salama, M.M.; Chang, F.-K. Development of a Real-Time Active Pipeline Integrity Detection System. *Smart Mater. Struct.* **2009**, *18*, 115010. [CrossRef]
131. Farrar, C.R.; Worden, K. An Introduction to Structural Health Monitoring. *Philos. Trans. R. Soc. A Math. Phys. Eng. Sci.* **2007**, *365*, 303–315. [CrossRef]

132. Ostachowicz, W.; Soman, R.; Malinowski, P. Optimization of Sensor Placement for Structural Health Monitoring: A Review. *Struct. Health Monit.* **2019**, *18*, 963–988. [CrossRef]
133. Sohn, H. Effects of Environmental and Operational Variability on Structural Health Monitoring. *Philos. Trans. R. Soc. A Math. Phys. Eng. Sci.* **2007**, *365*, 539–560. [CrossRef] [PubMed]
134. Gao, H.; Guers, M.J.; Rose, J.L. Flexible Ultrasonic Guided Wave Sensor Development for Structural Health Monitoring. In Proceedings of the Nondestructive Evaluation and Health Monitoring of Aerospace Materials, Composites, and Civil Infrastructure V, San Diego, CA, USA, 28 February–2 March 2006; Volume 6176, pp. 325–336.
135. Gresil, M.; Poohsai, A.; Chandarana, N. Guided Wave Propagation and Damage Detection in Composite Pipes Using Piezoelectric Sensors. *Procedia Eng.* **2017**, *188*, 148–155. [CrossRef]
136. Leong, W.H.; Staszewski, W.J.; Lee, B.C.; Scarpa, F. Structural Health Monitoring Using Scanning Laser Vibrometry: III. Lamb Waves for Fatigue Crack Detection. *Smart Mater. Struct.* **2005**, *14*, 1387. [CrossRef]
137. Dürager, C.; Heinzelmann, A.; Riederer, D. A Wireless Sensor System for Structural Health Monitoring with Guided Ultrasonic Waves and Piezoelectric Transducers. *Struct. Infrastruct. Eng.* **2013**, *9*, 1177–1186. [CrossRef]
138. Pallarés, F.J.; Betti, M.; Bartoli, G.; Pallarés, L. Structural Health Monitoring (SHM) and Nondestructive Testing (NDT) of Slender Masonry Structures: A Practical Review. *Constr. Build. Mater.* **2021**, *297*, 123768. [CrossRef]
139. Zhang, P.; Venketeswaran, A.; Wright, R.F.; Lalam, N.; Sarcinelli, E.; Ohodnicki, P.R. Quasi-Distributed Fiber Sensor-Based Approach for Pipeline Health Monitoring: Generating and Analyzing Physics-Based Simulation Datasets for Classification. *Sensors* **2023**, *23*, 5410. [CrossRef]
140. Li, X.; Wang, Y.; Wan, X.; Tang, B.; Qin, Y.; Xu, C. Physics-Informed Deep Filtering of Ultrasonic Guided Waves for Incipient Defect Inspection of Large-Scale Square Tube Structures. *J. Sound Vib.* **2023**, *567*, 118066. [CrossRef]

coatings

MDPI

Article

Experimental Study on Bond Behavior between Steel Rebar and PVA Fiber-Reinforced Concrete

Xuhui Zhang, Qi Lu * and Yang Wang

College of Civil Engineering, Xiangtan University, Xiangtan 411105, China
* Correspondence: luqi.one@outlook.com

Abstract: An experimental test was performed to study the bond behavior between steel rebar and concrete reinforced by polyvinyl alcohol (PVA) fibers. Twenty specimens were prepared and subjected to the pull-out test. Four different fiber volume contents (0%, 0.2%, 0.4%, and 0.6%) were considered and the concrete with the strength grade of C35 was designed in the present study. The effects of PVA fibers, rebar diameter, and cover depth on bond behavior were clarified. The effects of PVA fibers on the mechanical property of concrete were also studied. The results show that PVA fibers decreased the compressive strength of concrete, but increased splitting tensile, flexural, and direct tensile strength. PVA fibers negatively affected bonding in the ascending branch both for the pull-out and the splitting failure cases, but improved the bonding in the descending branch after peak stress for the splitting failure case. In the present test, the maximum decrement of bond strength was about 16.2% for specimens with less than 0.6% PVA fibers. PVA fibers restricted both the macro-cracking and micro-cracking for the splitting cases, the former were much more significant than the latter. The effects of rebar diameter and cover depth on bonding became slight and significant with the increasing content of PVA fibers, respectively.

Keywords: polyvinyl alcohol fiber; bond behavior; mechanical performance; experimental study

1. Introduction

Polyvinyl alcohol (PVA) fibers are a kind of organic synthetic material, which has many advantages compared with other fiber materials. PVA fibers have high strength and ductility with a relatively high elastic modulus, outstanding resistance to corrosion without toxicity, high hydrophilicity in alkaline environments in concrete, strong bonding with cement matrix, low cost, and effective restriction of cracking of concrete over the long term [1–4]. PVA fibers have been widely used in many applications in civil engineering such as the fiber-cement used as an asbestos replacement, Engineered Cementitious Composites (ECC), strengthening of enlarged sections for concrete structures, and various kinds of shotcrete, and play a promising role in fire resistance, toughening, and anti-cracking in concrete [5].

Extensive work has been carried out to investigate the effects of PVA fibers on the mechanical properties of concrete. It has been reported that PVA fibers increased the splitting tensile strength and flexure strength of concrete compared to plain concrete [6–9]. The addition of PVA fibers slightly affects the pre-cracking behavior of concrete but substantially improved the post-cracking response, enhancing the ductility and toughness [10–12]. Nuruddin et al. [13,14] studied and compared the effects of PVA fibers on the static and dynamic behavior of concrete and structures. They found that the static behaviors of concrete were improved while the dynamic properties were not affected by a low content of PVA fibers [13]. PVA fibers, however, improved the ductility and damping ratio of the concrete beams under four-point loading cases [14]. Some works have also been performed to study the durability of PVA fiber-reinforced concrete. It was reported that PVA fibers enhanced the frost resistance, carbonization resistance, and fatigue life of concrete [15–18].

Zhou et al.'s study showed that the addition of PVA fibers can enhance the strain capacity and crack resistance of ECC under uniaxial tensile [19]. In addition, Ayub et al. added PVA fibers to FRC to increase the optimal compressive strength strain from 9.08% to 19.62% [20]. Nowadays, however, very few works have been performed to clarify the effects of PVA fibers on bond behavior between concrete and steel rebars.

The bonding between concrete and steel rebars is a key factor for concrete structures, which relates to the design, construction, and behavior of structures [21–25]. In the past few years, many works have been performed to study the bond performance between steel rebar and fiber-reinforced concrete [26–31]. Most of these works, however, focused on cases with steel fibers and steel–PVA hybrid fibers. Some of these reported that the addition of steel fibers resulted in more ductile bond behavior and the bond strength increased with increasing content of steel fibers [32]. While others hold that the addition of fibers showed no measurable effect or even a passive effect on bond strength [33]. There is still a disagreement about the effects of steel fibers on bond behavior. For the case of steel–PVA hybrid fibers, most of the works showed an obvious positive effect on bond strength due to the synergetic influences in inhibiting the cracking at multiple scales and stages [34,35]. The material properties and the mechanism of single PVA fibers (only PVA fibers) are different from that of steel fibers or hybrid fibers, resulting in a difference in bonding mechanism. The effects of single PVA fibers on the bond behavior between concrete and steel rebar are still not well understood.

In this paper, an experimental test is proposed to study the effects of PVA fibers on the bond behavior between rebar and concrete. Twenty pull-out specimens were prepared and subjected to a bonding test. The effects of PVA fibers, rebar diameter, and cover depth on bond behavior were clarified. Several conclusions are drawn based on the study.

2. Experimental Program

2.1. Pull-Out Specimens

Twenty pull-out specimens were prepared for the test. The specimens were designed as a steel rebar embedded inside a 150 mm × 150 mm × 150 mm concrete cube. The steel rebars were 360 mm long with an equal embedment length in the concrete for all the specimens. The embedment length was 60 mm, which was achieved using two pieces of PVC tubing installed adjacent to the embedment region to avoid the contact between the concrete and the bar, as shown in Figure 1.

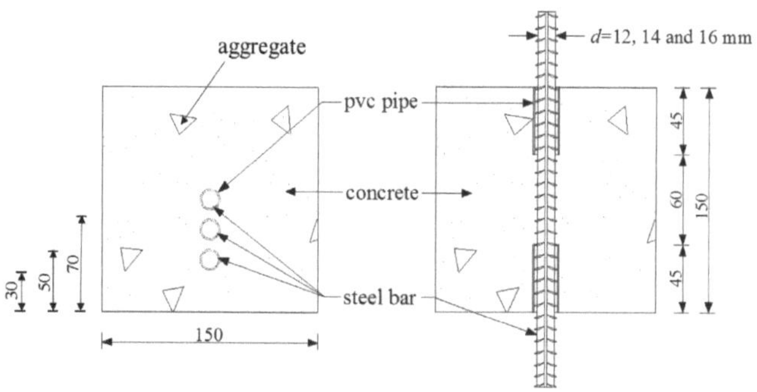

Figure 1. Detail of the specimen (unit: mm).

The factors considered in the test included four fiber volume contents (0%, 0.2%, 0.4%, and 0.6%), three rebar diameters (12 mm, 14 mm, and 16 mm), and three concrete cover thicknesses (30 mm, 50 mm, and 70 mm). It should be noted that the concrete cover depths were 69 mm, 68 mm, and 67 mm for specimens with 12 mm, 14 mm, and 16 mm rebars in the center region, respectively. The three cases were defined as the specimens with concrete cover depths of 70 mm.

The specimens designations consist of three parts. The first part indicates the PVA fiber volume ratio where 0, 0.2, 0.4, and 0.6 indicated the fiber volume contents of 0%, 0.2% 0.4%, and 0.6%, respectively. The second part of the specimen label references the rebar diameters. The third part marks the concrete cover depths. For example, PVA0.2-d12-c70 indicates that the fiber volume content is 0.2%, the rebar diameter is 14 mm, and the concrete cover depth is 70 mm. The details of the specimens are listed in Table 1.

Table 1. Details of the specimens.

Pull-Out Test	Content of PVA Fiber (%)	Rebar Diameters (mm)	Cover Thicknesses (mm)
PVA0.0-d12-c30	0.0	12	30
PVA0.2-d12-c30	0.2	12	30
PVA0.4-d12-c30	0.4	12	30
PVA0.6-d12-c30	0.6	12	30
PVA0.0-d12-c50	0.0	12	50
PVA0.2-d12-c50	0.2	12	50
PVA0.4-d12-c50	0.4	12	50
PVA0.6-d12-c50	0.6	12	50
PVA0.0-d12-c70	0.0	12	70
PVA0.2-d12-c70	0.2	12	70
PVA0.4-d12-c70	0.4	12	70
PVA0.6-d12-c70	0.6	12	70
PVA0.0-d14-c70	0.0	14	70
PVA0.2-d14-c70	0.2	14	70
PVA0.4-d14-c70	0.4	14	70
PVA0.6-d14-c70	0.6	14	70
PVA0.0-d16-c70	0.0	16	70
PVA0.2-d16-c70	0.2	16	70
PVA0.4-d16-c70	0.4	16	70
PVA0.6-d16-c70	0.6	16	70

2.2. Materials and Mixture Design

Ordinary Portland cement, coarse aggregate, local river sand, water, and polycarboxylate superplasticizer were used in the present study. The cement employed in the test was 42.5 ordinary Portland cement. The coarse aggregate was continuously graded crushed stone with a maximum aggregate size of 20 mm. The fine aggregate was natural siliceous sand from the Xiangjiang River in China. The fineness modulus of the sand was 1.82. PVA fibers produced by Kuraray in Japan were used in the present test, as shown in Figure 2. The length and diameter of the fiber were 12 mm and 39 μm, respectively. The tensile strength and the elastic modulus were 1.6 GPa and 42 GPa, respectively. Grade 400 strength ribbed rebars were employed with different diameters. In addition, the tested yield strength (f_y) and ultimate tensile strength (f_u) were 454 and 521 MPa for the 12 mm rebar, 439 and 585 MPa for the 14 mm rebar, and 494 and 615 MPa for the 16 mm rebar.

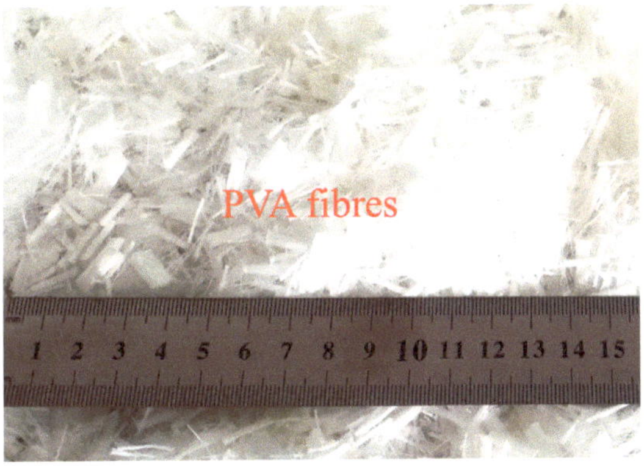

Figure 2. PVA fibers.

All the specimens had a similar mix proportion of cement, water, water-reducing agent, coarse aggregate, and fine aggregate with the different contents of PVA fibers. The content of cement was 485 kg/m^3. The water/cement ratio was 0.42. The contents of coarse aggregate and fine aggregate were 1092 kg/m^3 and 619 kg/m^3, respectively. The ratio of the coarse aggregate to the fine aggregate was 1.76. The content of the water-reducing agent was 0.8% of the cement. The water-reduction efficiency was 26%.

The concrete mixtures were prepared in a forced mixer. The cement, fine aggregate, and coarse aggregate were first mixed thoroughly for 2 min. Then, 70% of the water and water-reducing agents were added to the mixture and then the remaining 30%. After each addition, the mixtures were mixed for 1 min. After that, the PVA fibers would be gradually dispersed into the mixture and mixed for another 2 min. In the present test, all the specimens with the same content of PVA fibers were cast using the concrete mixture from the same batch, i.e., a total of four batches of the concrete mixture were prepared. Additionally, the PVA fibers were found to be uniformly distributed in the concrete mixtures in the present test by using the mixing program.

The concrete slump was employed to measure the fluidity of fresh concrete mixtures. The tested slumps of the four batches of concrete mixtures were 180 mm, 102 mm, 45 mm, and 12 mm, respectively. The content of PVA fiber in the four batches was 0%, 0.2%, 0.4%, and 0.6%, respectively. The slump decreased by about 93% when the content of PVA fibers increased from 0% to 0.6%. This indicated that the content of PVA fibers significantly affected the fluidity of the fresh concrete. The same phenomenon had also been reported by many other researchers [36]. The reasons could be attributed to the rough surface of PVA fibers, leading to significant friction between the fibers and cement mortar.

The pull-out specimens with the same content of PVA fibers were cast with the concrete mixture from the same batch. Meanwhile, some samples used for mechanical properties tests were also prepared by using the concrete mixture from the same batch (see details in the following section). All these specimens were demolded after 24 h and cured at a temperature of 20 ± 2 °C and relative humidity (RH) of 95%. After 28 days of curing, these specimens would be subjected to the mechanical property tests and pull-out test.

2.3. Mechanical Property of PVA Fiber-Reinforced Concrete

The compressive strength, splitting tensile strength, flexural strength, and direct tensile strength were tested for the four batches of concrete mixtures with different contents of PVA fibers. The compressive strength, splitting tensile strength, and flexural strength tests were tested based on the standard of GB/T 50081-2019 while the direct tensile strength was tested based on the method suggested by Dashti J [37]. Cubes with 150 mm length were used to test the compressive strength and splitting tensile strength of concrete. Prisms with 100 mm × 100 mm cross-section and 400 mm length were used to test the flexural strength. Dog-bone-shaped specimens were used to determine the tensile strength under monotonic axial tension. The middle part of the specimens is 100 mm wide, 100 mm thick, and 150 mm in length. Three samples were prepared and tested, and the mean value was reported. The details of the samples and the test set-up are shown in Figure 3.

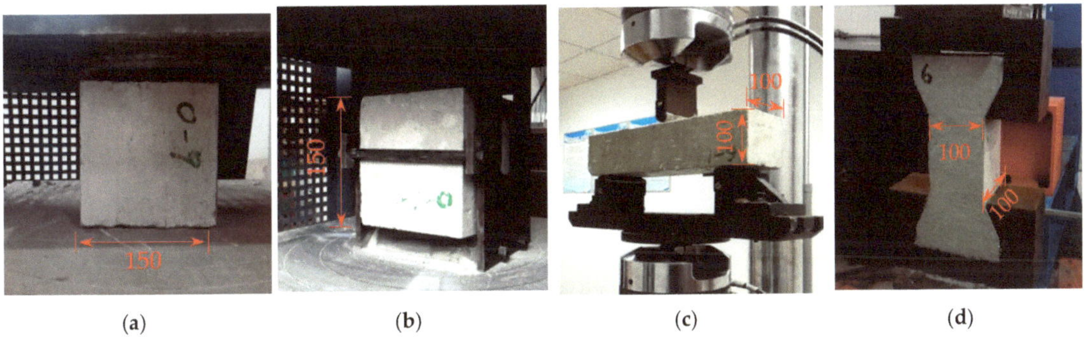

Figure 3. Details of mechanical property tests: (**a**) compressive strength; (**b**) splitting tensile strength; (**c**) flexural strength; (**d**) direct tensile strength.

Figure 4 shows the compressive strength, splitting tensile strength, flexural strength, and direct tensile strength of specimens with different contents of PVA fibers. PVA fibers showed different influences on these ultimate strengths. The compressive strength of the specimens gradually decreased with the increasing content of PVA fibers. For instance, the compressive strength decreased about 4.4%, 12.3%, and 6.9% for specimens with 0.2%, 0.4%, and 0.6% PVA fibers compared with the specimens without PVA fibers, respectively. The addition of the PVA fibers, however, improved the splitting tensile strength, flexural strength, and direct tensile strength of the specimens, as shown in Figure 4b. The splitting tensile strength, for example, increased about 21.38%, 25.72%, and 33.70% for specimens with 0.2%, 0.4%, and 0.6% PVA fibers compared with the specimens without PVA fibers, respectively.

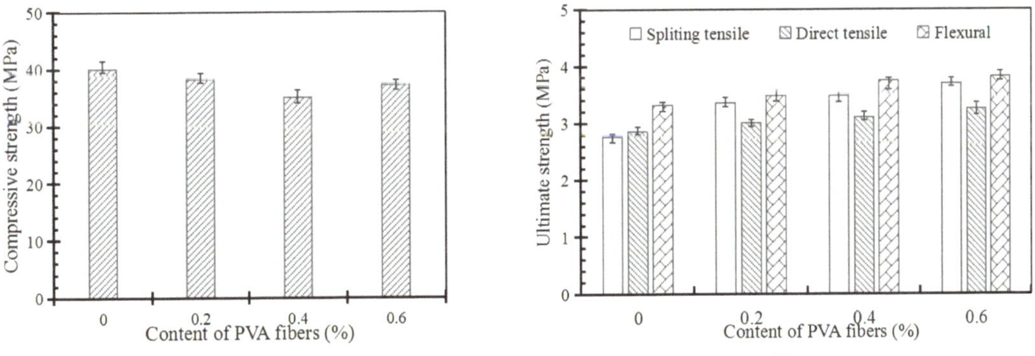

Figure 4. The effect of PVA fibers on the mechanical properties of concrete: (**a**) compressive strength; (**b**) splitting tensile strength, direct tensile strength, and flexural strength.

It was interesting to note that the increase in splitting tensile strength was much greater than that of the flexural strength and direct tensile strength. The increase in direct tensile strength were about 5.26%, 8.77%, and 14.03% for specimens with 0.2%, 0.4%, and 0.6% PVA fibers, respectively. The corresponding increase in flexural strength were 4.82%, 12.65%, and 14.76%. The increases were small compared with the increase in splitting tensile strength. This could be attributed to the reliable anchorage of PVA fibers under the compressive force for the case of the splitting tensile test. It was difficult to pull out the PVA fibers which improved the splitting tensile strength.

2.4. Pull-Out Test of Specimens

The pull-out test was carried out by using an MTS machine with a maximum axial force capacity of 100 kN. A special loading frame was designed for the present test, as shown in Figure 5. The loading end of the specimens and the loading frame were fixed by the upper and lower clamps of the MTS, respectively. The position of the upper plate of the loading frame can be adjusted to contact the specimen surface, which was used to minimize the effect of uneven stress and to prevent unexpected lateral movement of the specimens during the loading process. The pull-out force was automatically measured by the test machine. The relative slip between the steel bar and concrete was measured by an extensometer clamped at the free end of the specimens. Additionally, a dial indicator was set transversely on the specimen surface to measure the development of the crack during the loading test.

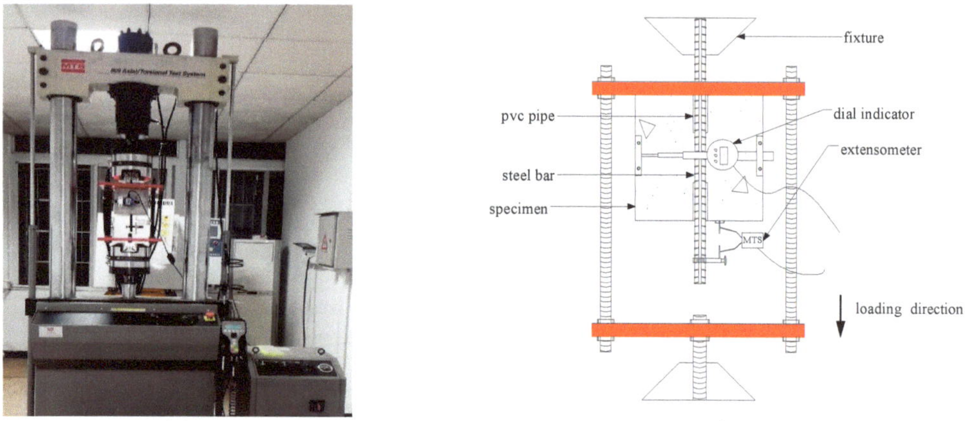

(a) (b)

Figure 5. Set-up of the pull-out test. (**a**) loading instrument; (**b**) detail device.

The pull-out tests were monotonically loaded using a displacement-control mode at a speed of 1 mm/min. The experimental data, including the pull-out force, relative slip, and crack width, were synchronously collected with a frequency of 1 s. The tests would be stopped as the specimens split or the residual pull-out force became constant. It should be noted that, in this study, the slip measured at the free end was used to represent the slip characteristics of the specimen. Additionally, the average bond stress along the embedded length of rebar was used to represent the stress characteristics, which can be expressed as

$$\tau = \frac{F}{\pi d L} \quad (1)$$

where F is the measured pull-out force; d is the nominal diameter of the steel bar; and L is the embedded length of the steel bar in the concrete.

3. Bond Behavior between Rebar and PVA Fiber-Reinforced Concrete

The test results of the pull-out specimens, including the failure mode, the bond stress at onset slip of free end τ_{ons}, residual bond stress τ_r, maximum bond stress τ_{max}, and slip corresponding to the maximum stress S_0 are summarized in Table 2. It should be noted that τ_r is the bond stress corresponding to the slip of 0.45 times of the spacing between steel ribs, which was proposed by Zhao et al. [38]. The bond stress–slip curves for all the specimens are shown in Figure 6.

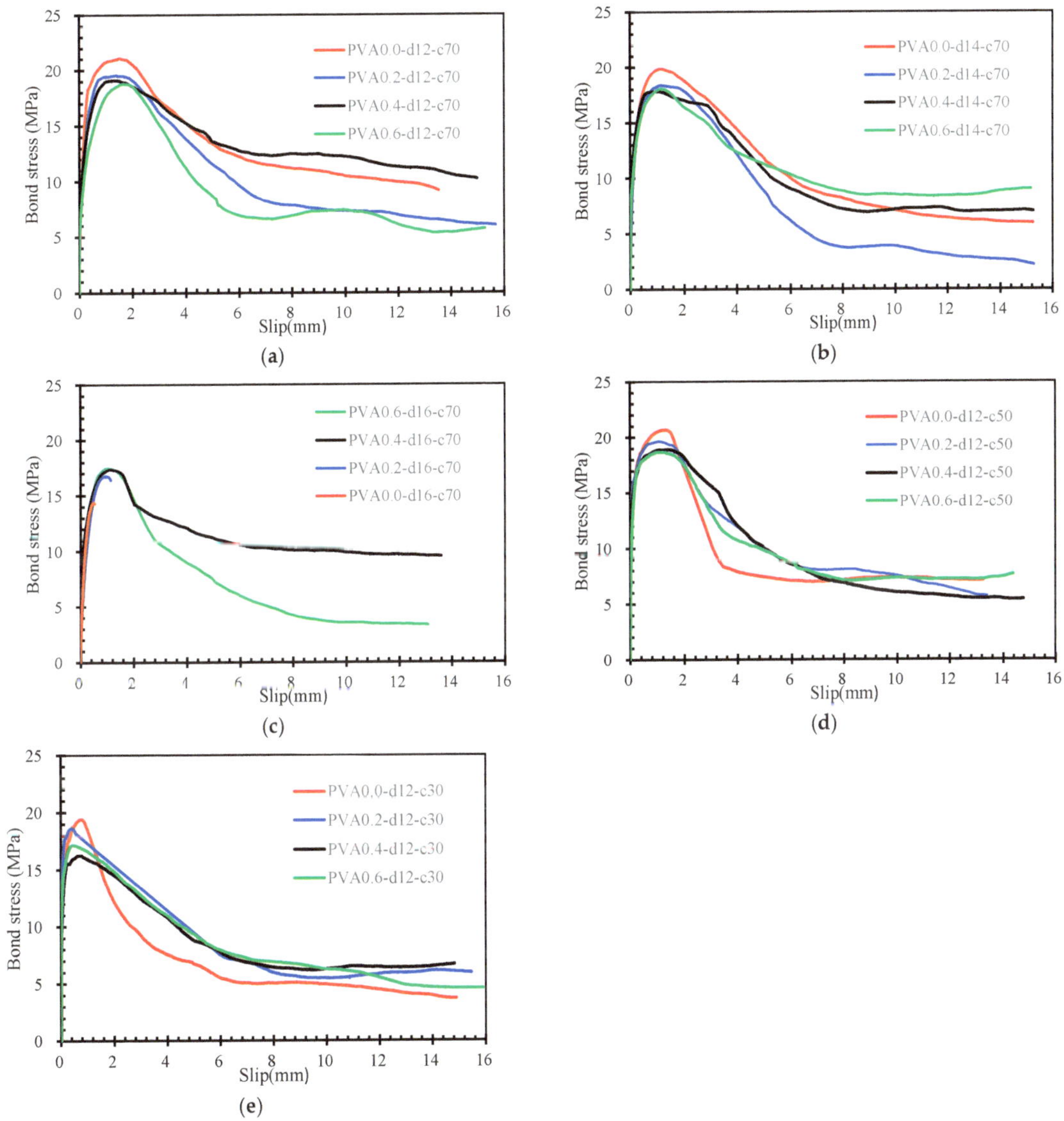

Figure 6. Bond stress–slip curves for the specimens. (**a**) d12-c70 series (c/d = 5.75); (**b**) d14-c70 series (c/d = 4.86); (**c**) d16-c70 series (c/d = 4.19); (**d**) d12-c50 series (c/d = 4.17); (**e**) d12-c30 series (c/d = 2.50).

Table 2. Experimental parameters and results of pull-out tests.

Test	Content of Fiber (%)	f_{cu} (MPa)	f_t (MPa)	f_{ts} (MPa)	f_{tf} (MPa)	Failure Mode	τ_{ons} (MPa)	τ_{max} (MPa)	τ_r (MPa)	S_0 (mm)	K_0 (MPa/mm)
PVA0.0-d12-c30	0	39.98	2.85	2.76	3.32	S-P	9.26	19.41	8.02	0.72	511
PVA0.2-d12-c30	0.2	38.21	3.00	3.35	3.48	S-P	10.87	18.61	12.20	0.39	931
PVA0.4-d12-c30	0.4	35.06	3.10	3.47	3.74	S-P	8.72	16.26	11.45	0.68	739
PVA0.6-d12-c30	0.6	37.23	3.25	3.69	3.81	S-P	9.83	17.17	11.59	0.47	973
PVA0.0-d12-c50	0	39.98	2.85	2.76	3.32	S-P	9.79	20.69	8.25	1.30	470
PVA0.2-d12-c50	0.2	38.21	3.00	3.35	3.48	P	10.17	19.66	12.58	1.04	756
PVA0.4-d12-c50	0.4	35.06	3.10	3.47	3.74	P	11.11	18.93	13.31	1.40	315
PVA0.6-d12-c50	0.6	37.23	3.25	3.69	3.81	P	8.56	18.70	11.29	1.13	283
PVA0.0-d12-c70	0	39.98	2.85	2.76	3.32	P	4.52	21.08	16.24	1.53	105
PVA0.2-d12-c70	0.2	38.21	3.00	3.35	3.48	P	5.79	19.57	14.94	1.42	109
PVA0.4-d12-c70	0.4	35.06	3.10	3.47	3.74	P	5.89	19.10	15.93	1.34	115
PVA0.6-d12-c70	0.6	37.23	3.25	3.69	3.81	P	5.65	18.91	12.79	1.69	65
PVA0.0-d14-c70	0	39.98	2.85	2.76	3.32	P	5.38	19.90	14.32	1.11	95
PVA0.2-d14-c70	0.2	38.21	3.00	3.35	3.48	P	5.26	18.38	12.06	1.19	97
PVA0.4-d14-c70	0.4	35.06	3.10	3.47	3.74	P	6.01	17.83	13.29	0.92	108
PVA0.6-d14-c70	0.6	37.23	3.25	3.69	3.81	P	5.81	18.06	12.33	1.18	106
PVA0.0-d16-c70	0	39.98	2.85	2.76	3.32	S	2.73	14.40	0	0.5	73
PVA0.2-d16-c70	0.2	38.21	3.00	3.35	3.48	S	2.90	16.74	0	0.97	51
PVA0.4-d16-c70	0.4	35.06	3.10	3.47	3.74	S-P	2.47	17.37	11.51	1.11	68
PVA0.6-d16-c70	0.6	37.23	3.25	3.69	3.81	S-P	3.72	17.47	8.30	1.05	64

Notes: P means pull-out failure of specimens; S means splitting failure of specimens; S-P means splitting–pull-out failure of specimens.

3.1. Failure of Pull-Out Specimens

Three different failure modes were observed in the present test: pull-out failure, splitting failure, and splitting–pull-out failure. For the first failure mode, the specimens failed gradually as the steel rebar was pulled out without concrete cracking. In this case, the bond stress–slip curves contained typical three-branch characteristics including ascending, descending, and residual branches. For the second failure mode, the specimens failed suddenly as the splitting of the concrete was accompanied by a loud crash. Afterwards, the concrete members were completely broken apart. The bond stress–slip curves of the contents only contained the ascending branch. For the third mode, the specimens maintained integrity and the bond stress was not completely lost after the splitting of the concrete. In this case, the bond stress–slip curve contained the typical three-branch characteristics and showed a quasilinear degeneration in the descending branch. The final failure of these three modes are shown in Figure 7.

(a) (b) (c)

Figure 7. Typical failure of specimens: (**a**) pull-out failure, (**b**) splitting failure, and (**c**) splitting-pull-out failure.

The failure modes of all specimens are summarized in Table 1. It can be found that the ratio of cover depth to steel rebar diameter (*c/d*) significantly affected the failure mode of the specimens. The d12-c70 series specimens (with a rebar diameter of 12 mm and cover depth of 70 mm) and the d14-c70 series specimens (with a rebar diameter of 14 mm and cover depth of 70 mm) failed by the pull-out mode. The corresponding ratios of cover depth to steel rebar diameter were 5.75 and 4.86, respectively. For the other specimens, the ratios of cover depth to steel rebar diameter were less than 4.19. Most of these specimens failed by splitting mode or splitting–pull-out mode except for some specimens with additional PVA fibers. This indicates that specimens are more prone to pull-out failures if the ratio of cover depth to rebar diameter is greater than 4.86. This is consistent with the results reported in previous research studies [39]. It was concluded that the specimen were more prone to pull-out failure rather than splitting failure when the ratio of cover depth to rebar diameter was greater than 4.5~5.

The addition of PVA fibers showed a positive effect on the failure of specimens with a small ratio of cover depth to rebar diameter. As summarized in Table 1, the d12-c50 series specimens and d16-c70 series specimens have similar ratios of cover depth to rebar diameter, which are 4.17 and 4.19, respectively. The failure mode of these specimens changed with the increase in the addition of PVA fibers. For the d12-c50 series, the plain specimen failed by the splitting–pull-out mode while the other fiber-reinforced specimens failed by the pull-out mode. For the d16-c70 series, specimens with 0% and 0.2% PVA fibers failed by the splitting mode. The failure changed to the splitting–pull-out mode as the PVA fibers increased to 0.4% and 0.6%. The bridging effect of PVA fibers delays the appearance and propagation of cracks, which improves the damage tolerance of the concrete and changes the specimens' failure in a more ductile way.

In addition, the rebar diameter also showed a significant effect on the failure of specimens with ratios of cover depth to rebar diameter less than 4.5~5. All the d12-c30 series specimens failed by the splitting-pull-out mode. The corresponding ratio of cover depth to rebar diameter is 2.5. However, some of the d16-c70 series specimens with a high ratio of cover depth to rebar diameter (c/d = 4.19) failed by the more brittle splitting mode. This indicates that for specimens with a low ratio of cover depth to rebar diameter, the rebar diameter, i.e., the height of the rib, plays an important role in its failure modes. Large diameter rebars with a high rib lead to greater expansion stress during the pull-out procedure, easily resulting in the splitting of the specimens.

After the pull-out test, the specimens were split apart to visually inspect the damage of the rebar and concrete. The split parts were similar for all the specimens. Typical split parts are shown in Figure 8, which come from splitting failure specimen PVA0.2-d16-c70. The concrete between the ribs of steel rebar was smashed while no damage was found on the rebars for specimens that failed by pull-out and splitting. This indicates that the failure of all the specimens is started by the trapping of concrete particles between the rebar ribs. For the pull-out case, the concrete within the ribs would be totally polished by the rebars. For the splitting case, the further slip would cause the splitting of the concrete cover, which is restricted by PVA fibers. In the present test, PVA fibers were found to be pulled out rather than pulled off (see details in Figure 8c). In addition, it can be seen from Figure 8c that the obvious PVA fiber bridging effect limited the development of cracks, thus improving the performance of the concrete.

Figure 8. Pull-out failure mode details: (**a**) PVA0.2-d16-c70 series; (**b**) bonding section; (**c**) external crack.

3.2. Effects of PVA Fibers

Figure 6 shows the bond stress–slip curves for specimens with different contents of PVA fibers. As mentioned before, the curves are significantly affected by the failure modes, i.e., the ratio of cover depth to rebar diameter. Moreover, the addition of PVA fibers also showed a significant effect on the bond stress–slip curves. The slope of the ascending branch and the bond strength decreased with the increase in PVA fibers. The slope of the ascending branch reflects the bond stiffness of the sample, which is mainly determined by the shear action between the concrete and the rib. According to Section 2.2, the addition of polyvinyl alcohol fiber would lead to a decrease of the flow performance of newly mixed concrete, which would affect the compactness of concrete, thus reducing the shear effect between the concrete and reinforcement, and ultimately leading to the decline of the bond strength between the concrete and reinforcement. The same phenomenon has been reported by many other researchers [40,41]. The details of the bond strengths of the specimens are shown in Figure 9a. All the specimens showed a descending trend in bond strength with the increase in PVA fibers except the d16-c70 series specimens. In this experiment, the maximum bond strength decrease of the d12-c30 series samples, d12-c50 series samples, and d12-c70 series samples with the increase in PVA fiber content was about 16.2%, 9.6%, 10.3%, and 10.4%, respectively. In the d16-c70 series, the additional of PVA fibers changed the failure of the specimens from the splitting mode to splitting–pull-out mode, which increased correspondingly the bond strength with a maximum increase of about 21.3%. In general, the addition of PVA fibers showed a negative effect on bond stiffness and bond strength between the rebar and concrete for both pull-out and splitting cases. The details of the residual bond strength of the samples are shown in Figure 9b. Similarly, except for the d12-c50 series of samples, the residual bond strength of all samples decreased with the increase in PVA fibers, and the maximum decrease was about 27.9%. This indicates that the content of PVA fibers has a more adverse effect on the residual binding strength of samples. It can be clearly seen from Figure 6 that for the PVA0.0-d16-c70 and PVA0.2-d16-c70 samples, due to the lack of effective restraint in the concrete, splitting failure occurred immediately after reaching the peak load, and then the load rapidly dropped to 0. Therefore, the bond stress–slip curve did not decline and the residual bond strength was 0.

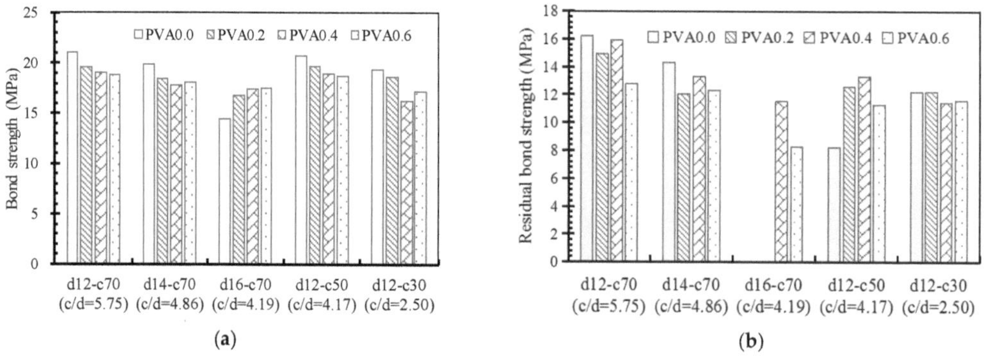

Figure 9. Bond strength and residual bond strength of specimens with different contents of PVA fibers. (**a**) Bond strength; (**b**) residual bond strength.

The slip corresponding to maximum bond stress and the bond stress at the onset of free end slip for specimens with different contents of PVA fibers are shown in Figures 10 and 11, respectively. It can be found that the PVA fibers had negligible effects on the corresponding slip to maximum bond stress and the bond stress at the onset of slip. For each series of specimens with the same cover depth and rebar diameter, the changes in the slip and the onset bond stress did not correspond with the increase in PVA fibers. The slip corresponding to maximum bond stress, however, was more significantly affected by the ratio of cover depth to rebar diameter, i.e., the failure mode of specimens. The specimens that failed by pull out mode usually had a large corresponding slip. The smaller the ratio of cover depth to rebar diameter, the smaller the corresponding slip. The bond stress at the onset of slip seemed to be further affected by the diameter of the rebars. The pull-out failure specimens had similar onset bond stresses. The splitting failure specimens had high onset bond stresses for cases with small rebar diameters. The details are discussed in the following sections.

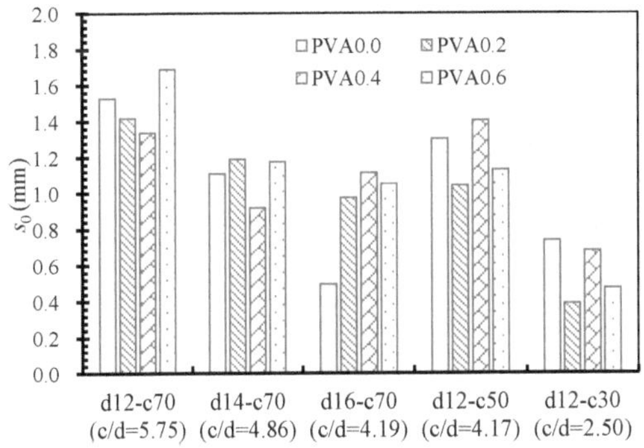

Figure 10. Slip corresponding to maximum bond stress of specimens.

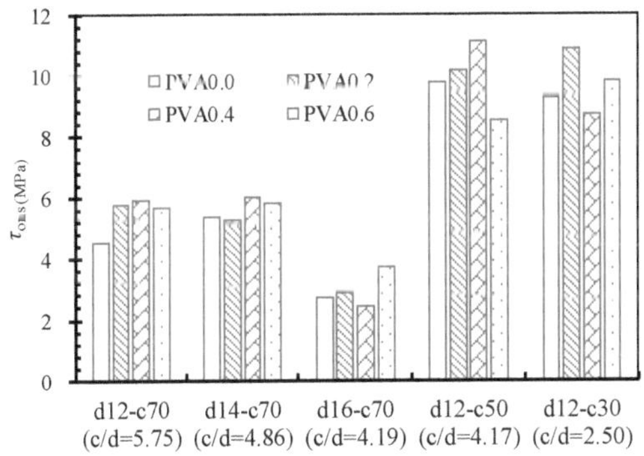

Figure 11. Bond stress at onset of free end slip of specimens.

The additional PVA fibers, however, had a positive effect on the descending branches for specimens that failed by the splitting–pull-out mode. Figure 6e shows the bond stress–slip curves for the d12-c30 series specimens. The bond stress decreased very rapidly in the descending branch for the plain specimen without additional PVA fibers. The decrease in bond stress in the other fiber-reinforced specimens, however, became slow. Additionally, the decrease in bond stress seemed to be linear with the increase in slip. This indicates that PVA fibers can restrict the degradation of bond stress in a linear manner for splitting–pull-out failure specimens. This could be attributed to the bridging effects of PVA fibers after the splitting of the specimens, which effectively restrained the cracking and slowed down the decrease in bond stress.

Figure 12 shows the propagation of splitting cracks for the d12-c30 series specimens. For the plain specimens, the splitting crack appeared at low load stress and gradually propagated with the increase in applied load stress. There was a sudden widening in cracking and the width rapidly increased to a greater value after the peak load stress. Afterwards, the splitting crack was almost unchanged during the further load procedure. The propagation of splitting cracks in the specimens with 0.2% PVA fibers was similar to that of the plain specimen, but with a narrow crack at the end. For specimens with 0.4% and 0.6% PVA fibers, the load stresses corresponding to the appearance of splitting cracks significantly increased. Additionally, there was no sudden widening of the cracks after the peak load stress. The final crack width became much narrower compared to that on the plain specimen. PVA fibers play a positive effect in restricting the splitting cracks.

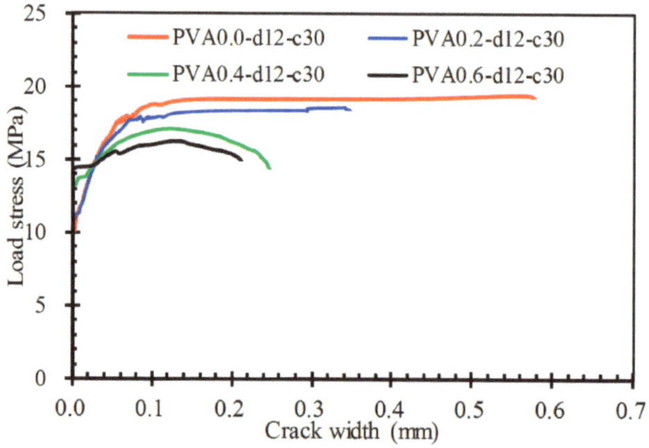

Figure 12. Propagation of splitting crack in d12-c30 series specimens.

The details about the comparison of cracking load stress and maximum crack width for specimens with different contents of PVA fibers are shown in Figure 13. Specimens with 0.2%, 0.4%, and 0.6% PVA fibers had increased cracking load stress by about 8.37%, 31.44%, and 42.43% and decreased the final crack width by about 36.23%, 59.02%, and 65.08%, respectively. The changes in the final crack width were more obvious. This indicates that the PVA fibers restrict both the micro-cracking and macro-cracking of specimens that failed by splitting. The restriction effect on macro-cracking seemed much more significant than that on micro-cracking. In the process of steel bar pull out, with the increase in stress, the PVA fibers effectively bridged the cracks generated by the steel bars in the process of concrete pull out, and the bridge effect diagram is shown in Figure 14.

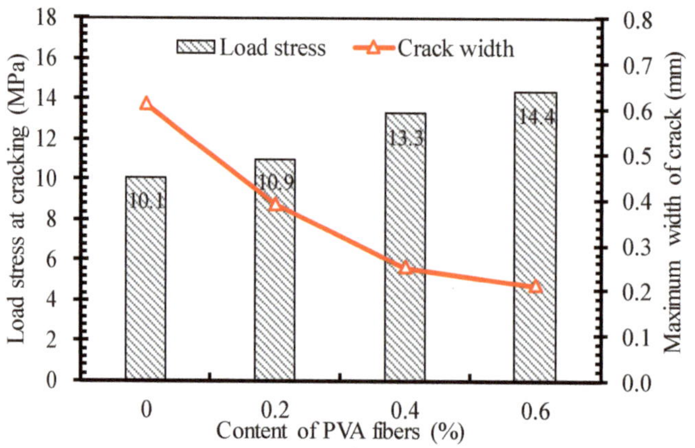

Figure 13. Effects of PVA fibers on splitting cracks.

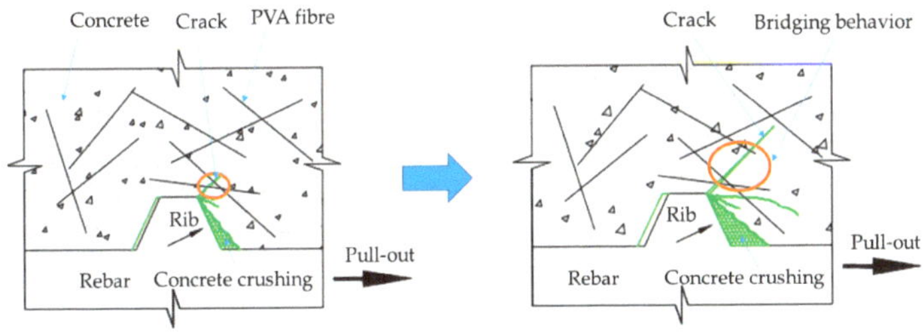

Figure 14. Bridging action of PVA fibers.

3.3. Effects of Rebar Diameter

Figure 15 shows the bond stress–slip curves for the specimens with different rebar diameters. Rebar diameter affected the failure mode of the specimens. The ratio of cover depth to steel rebar diameter is inversely linked to rebar diameter. In the present test, the failure mode of the specimens was changed from the pull-out mode to splitting mode as the diameter of the rebar exceeded 16 mm. The bond stress–slip curves degraded from three branches to one branch, as shown in Figure 15a,b. Moreover, rebar diameter also affected the initial bonding stiffness and bond strength of the specimens. The initial bonding stiffness is measured as the slope of the bond stress–slip curve at the initial slip. The initial bond stiffness K_0 is the slope of the bond stress–slip curve at the time of the initial slip. In this paper, the secant slope corresponding to one-half of the maximum bond stress was used to measure the initial bond stiffness of different specimens, and the results are shown in Table 2. Compared with large reinforcement specimens, the initial bond stiffness of the bond stress–slip curve of small reinforcement specimens during the initial slip stage was significantly larger. This indicates that specimens with small-diameter rebars have great initial bonding stiffness and bond strength compared to specimens with large-diameter rebars. The difference of bond stress–slip curves in specimens with 12 mm, 14 mm, and 16 mm rebars became smaller with the increase in PVA fibers.

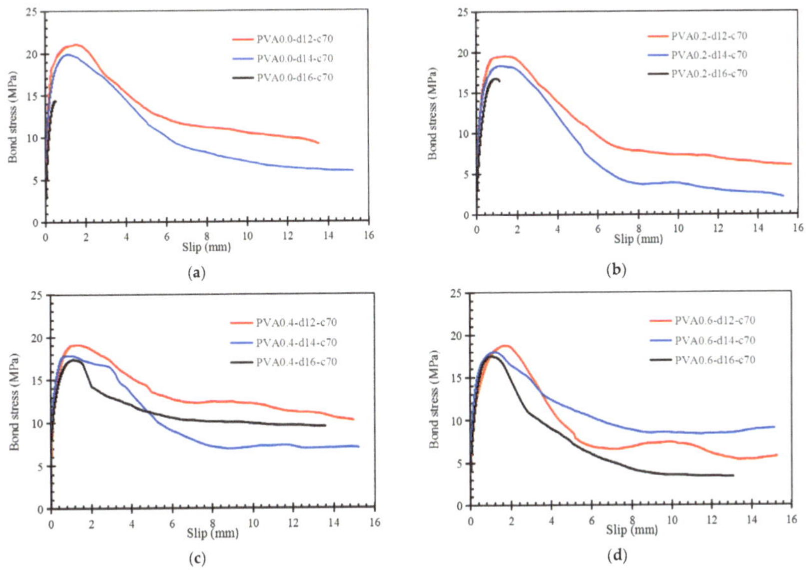

Figure 15. Bond stress–slip curves for the specimens with different rebar diameters. (**a**) PVA0.0-c70 series; (**b**) PVA0.2-c70 series; (**c**) PVA0.4-c70 series; (**d**) PVA0.6-c70 series.

The details about the comparison of bond strength, slip corresponding to bond strength, and the bond stress at the onset of free end slip for specimens with different rebars are summarized in Figure 16. Almost all of the bond strength, slip corresponding to bond strength, and the bond stress at the onset of free end slip decreased with the increase in rebar diameter. For example, the increase of rebar diameter from 12 mm to 14 mm and 16 mm in the PVA0.2-c70 series specimens decreased the bond strength by about 6.1% and 14.4%, decreased the corresponding slip by about 16.0% and 31.3%, decrease the bond stress at the onset of free end slip by about 9.1% and 49.9%, respectively. The larger the diameter of the steel rebar, the smaller the bond strength, the corresponding slip, and the bond stress at the onset of free end slip. These results are consistent with previous studies [42–44]. This could be attributed firstly to the more uneven distribution of bond stress along the rebars with large diameters than that of rebars with small diameters. Secondly, rebars with a large diameter tend to more easily trap bleeding water, resulting in more voids around the rebars, reducing the concrete and rib shear between the steel bars, thus reducing the interaction between the steel and concrete. Finally, the shear effect between the reinforced concrete and rib is the main part of the bonding force of reinforcement. According to the measured steel contour data, with the increase in the diameter of the reinforcement, the relative spacing of the reinforcement ribs decreases, and the relative height of transverse ribs decreases, leading to the reduction of the shear effect between the reinforcement and concrete, thus reducing the bonding force between the reinforcement and concrete.

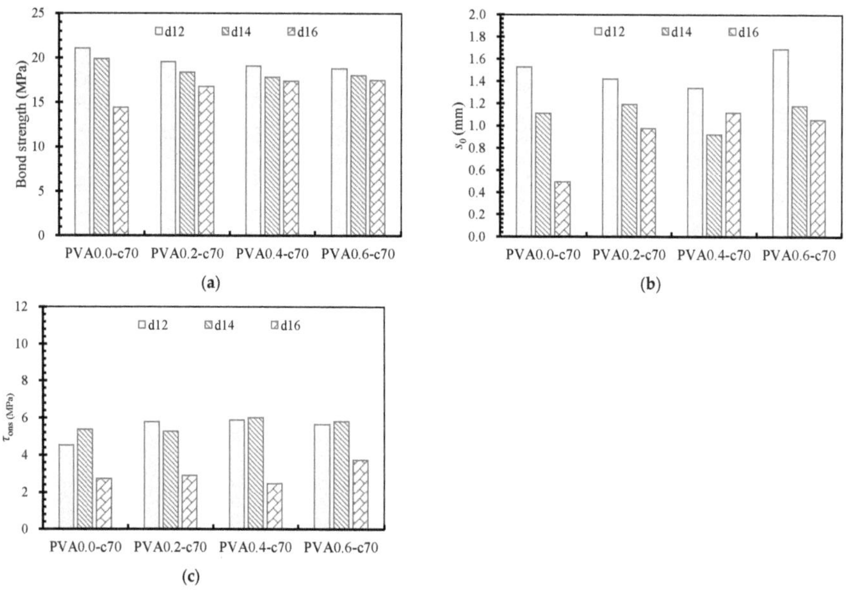

Figure 16. Effects of rebar diameter on bond behaviors. (**a**) Bond strength; (**b**) slip corresponding to bond strength; (**c**) bond stress at onset of free end slip.

The negative effects of a large rebar on bond behavior between concrete and rebar were slightly reduced with the increase in PVA fibers. For instance, the decreases in bond strength were 5.6% and 31.7% in the plain specimens as the rebar diameter increased from 12 mm to 14 mm and 16 mm, respectively. For specimens with 0.6% PVA fibers, the decrease in bond strength were only 4.0% and 7.1%, respectively. Similar effects were also been investigated in the slip corresponding to bond strength and the bond stress at the onset of free end slip. This indicates that PVA fibers have relatively positive effects on the bond behavior of specimens with large-diameter rebars.

3.4. Effects of Cover Depth

Figure 17 shows the bond stress–slip curves for specimens with different cover depths. The samples in each subgraph have the same content of PVA fibers and rebar diameter, but the cover depth is different. The curves match well with the failure modes for specimens with different cover depths. The curves show a rapid descent for specimens with thin covers after the maximum bond stress, while there was an obvious flat segment for specimens with thick covers. The bond strength of the specimens decreased with the decrease in cover depth. Moreover, the cover depth also affected the initial bond stiffness of the specimen. Specimens with thin coves had high initial stiffnesses. The more PVA fibers, the greater the difference in initial bond stiffness caused by cover depth.

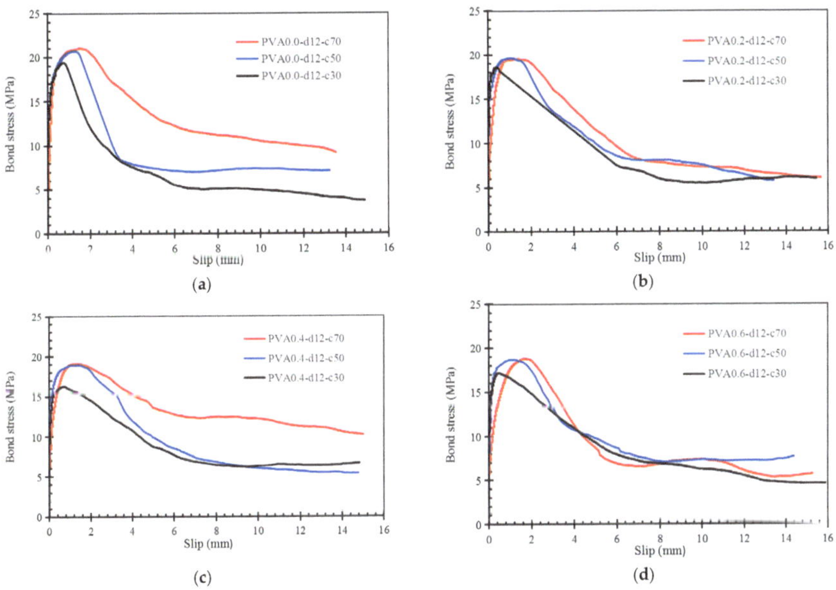

Figure 17. Bond stress–slip curves for the specimens with different cover depth. (**a**) PVA0.0-d12 series; (**b**) PVA0.2-d12 series; (**c**) PVA0.4-d12 series; (**d**) PVA0.6-d12 series.

These results could be attributed to the different methods of energy consumption during the loading procedure. For the specimens with a thick cover, the energy is consumed mainly through the shear crushing of local concrete between the ribs. Meanwhile, the energy consumption in the thin case is controlled by the cracking of the cover. The covers usually have large areas compared with the local concrete between the ribs. The splitting cracking of the cover would consume a large amount of energy without many slips, resulting in a corresponding high bond stiffness. The addition of PVA fibers improved the toughness and tensile strength of the concrete, increasing the energy dissipation of the concrete. This can overcome the adverse effects of additional voids and maintain the initial bond stiffness with the increase in PVA fibers. For specimens with thick covers, more additional voids appear in the concrete with the increase in fibers, causing corresponding degradation of energy consumption. Therefore, specimens with thick covers have lower initial bond stiffness than the specimen with thin covers, especially for the case with more PVA fibers.

The details about the comparison of bond strength, slip corresponding to bond strength, and the bond stress at the onset of free end slip for specimens with different cover depths are summarized in Figure 18. The bond strength and the corresponding slip of specimens decreased with the decrease in cover depth, while the bond stress at the onset of free end slip increase significantly with the decrease in cover depth. For instance, the decrease in cover depth from 70 mm to 50 mm and 30 mm in the PVA0.6-d12 series specimens decreased the bond strength by about 0.6% and 8.7%, decreased the corresponding slip about 33.1% and 72.0%, but increased the bond stress at the onset of free end slip by about 51.5% and 74.0%, respectively. The reason for the positive effect of a thin cover of the bond stress at the onset of free end slip could be similar to that for the initial bond stiffness, as discussed previously.

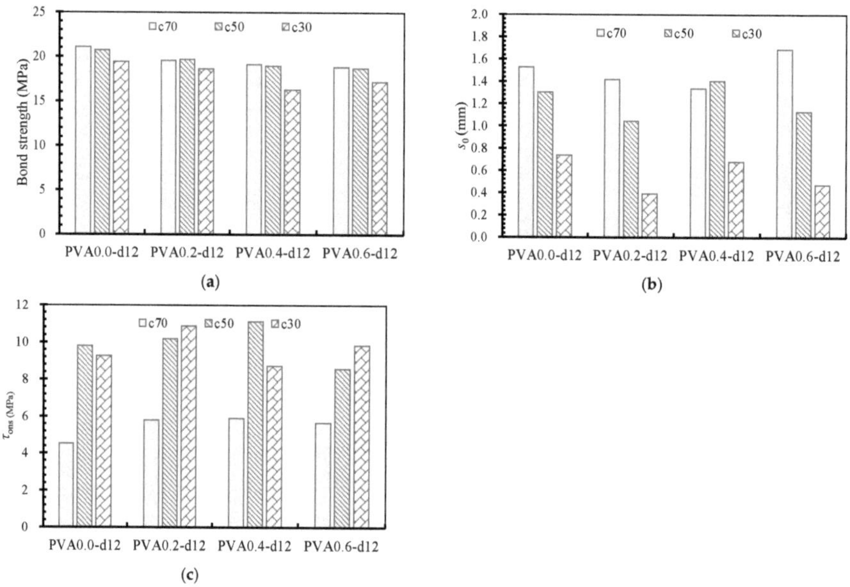

Figure 18. Effects of cover depth on bond behaviors. (**a**) Bond strength (**b**) Slip corresponding to bond strength (**c**) Bond stress at onset of free end slip.

4. Conclusions

This study experimentally investigated the effects of PVA fibers on bond behavior between rebar and concrete. The effects of PVA fibers, rebar diameter, and cover depth on bond behavior were clarified. The following preliminary conclusions can be drawn based on the present study:

(1) PVA fibers show a negative effect on the compressive strength of concrete, but a positive effect on splitting tensile, flexural, and direct tensile strength. In the present test, the maximum decrease in compressive strength was 12.3%, and the maximum increase in splitting tensile, flexural, and direct tensile strength were 33.7%, 12.9%, and 14.0%, respectively, for concrete with PVA fiber contents less than 0.6%. The increase in splitting tensile strength was more significant than that in flexural and direct tensile strength.

(2) The failure mode of bonding specimens depends mainly on the ratio of cover depth to steel rebar diameter. Specimen were more prone to pull-out failure when the ratio of cover depth to rebar diameter was greater than 4.5–5.0. For specimens with a small ratio of cover depth to rebar diameter, the splitting of concrete was more significantly affected by rebar diameter rather than cover depth. PVA fibers can restrict the splitting cracks and improve the failure of the specimen in a more ductile way.

(3) PVA fibers generally show a negative effect on bond behavior, but a positive effect on the descending branch in the case of splitting failure. In the present test, PVA fibers decreased the initial bond stiffness and bond strength for the cases with both pull-out and splitting failures. The maximum decrease in bond strength was about 16.2%

with PVA fiber contents less than 0.6%. PVA fibers restricted both the macro-cracking and micro-cracking for splitting cases, with the former effects seeming much more significant than latter.

(4) Specimens with small rebars had greater initial bonding stiffness and bond strength than specimens with large rebars. The difference in bond behavior caused by rebar diameter became smaller with the increase in PVA fibers. A thin cover decreased the bond strength of the specimens, but increased the initial bond stiffness of specimens. The more PVA fibers, the greater the difference in initial bond stiffness caused by cover depth.

It should be noted that only one sample was tested in the pull-out test rather than repeated samples, which would lead to some discrepancies. Further studies with more repeated samples are needed in the future.

Author Contributions: Writing—original draft, X.Z., Q.L. and Y.W.; Writing—review & editing, X.Z. All authors have read and agreed to the published version of the manuscript.

Funding: This work reported here was conducted with the financial supports from the Education Department Research Foundation of Hunan Province (Grant No. 21B0170), the Natural Science Foundation of Hunan Province of China (Grant No. 2021JJ40540), and the Special Funds for the Construction of Innovative Provinces in Hunan, China (Nos. 2019RS1059).

Informed Consent Statement: Informed consent was obtained from all subjects involved in the study.

Data Availability Statement: All data generated or analyzed during this study are included in this article.

Acknowledgments: This work was conducted with financial support from the Education Department Research Foundation of Hunan Province (Grant No. 21B0170), the Natural Science Foundation of Hunan Province of China (Grant No. 2021JJ40540), and the Special Funds for the Construction of Innovative Provinces in Hunan, China (No. 2019RS1059). The supports are gratefully acknowledged.

Conflicts of Interest: The authors declare no conflict of interest.

References

1. Pakravan, H.R.; Jamshidi, M.; Latifi, M. Study on fiber hybridization effect of engineered cementitious composites with low- and high-modulus polymeric fibers. *Constr. Build. Mater.* **2016**, *112*, 739–746. [CrossRef]
2. Juarez, C.A.; Fajardo, G.; Monroy, S.; Duran-Herrera, A.; Valdez, P.; Magniont, C. Comparative study between natural and PVA fibers to reduce plastic shrinkage cracking in cement-based composite. *Constr. Build. Mater.* **2015**, *91*, 164–170. [CrossRef]
3. Zanotti, C.; Borges, P.H.R.; Bhutta, A.; Banthia, N. Bond strength between concrete substrate and metakaolin geopolymer repair mortar: Effect of curing regime and PVA fiber reinforcement. *Cem. Concr. Compos.* **2017**, *80*, 307–316. [CrossRef]
4. Passuello, A.; Moriconi, G.; Shah, S.P. Cracking behavior of concrete with shrinkage reducing admixtures and PVA fibers. *Cem. Concr. Compos.* **2009**, *31*, 699–704. [CrossRef]
5. Said, M.; Abd El-Azim, A.A.; Ali, M.M.; El-Ghazaly, H.; Shaaban, I. Effect of elevated temperature on axially and eccentrically loaded columns containing Polyvinyl Alcohol (PVA) fibers. *Eng. Struct.* **2020**, *204*, 110065. [CrossRef]
6. Wang, J.; Dai, Q.; Si, R.; Guo, S. Investigation of properties and performances of Polyvinyl Alcohol (PVA) fiber-reinforced rubber concrete. *Constr. Build. Mater.* **2018**, *193*, 631–642. [CrossRef]
7. Abushawashi, N.; Vimonsatit, V. Material Classification and Composite Elastic Modulus of Hybrid PVA Fiber Ferrocement. *J. Mater. Civ. Eng.* **2016**, *28*, 04016073. [CrossRef]
8. Liu, F.; Ding, W.; Qiao, Y. Experimental investigation on the flexural behavior of hybrid steel-PVA fiber reinforced concrete containing fly ash and slag powder. *Constr. Build. Mater.* **2019**, *228*, 116706. [CrossRef]
9. Nuruddin, M.F.; Ullah Khan, S.; Shafiq, N.; Ayub, T. Strength Prediction Models for PVA Fiber-Reinforced High-Strength Concrete. *J. Mater. Civ. Eng.* **2015**, *27*, 04015034. [CrossRef]
10. Pan, J.; Cai, J.; Ma, H.; Leung, C.K.Y. Development of Multiscale Fiber-Reinforced Engineered Cementitious Composites with PVA Fiber and $CaCO_3$ Whisker. *J. Mater. Civ. Eng.* **2018**, *30*, 04018106. [CrossRef]
11. Sun, L.; Hao, Q.; Zhao, J.; Wu, D.; Yang, F. Stress strain behavior of hybrid steel-PVA fiber reinforced cementitious composites under uniaxial compression. *Constr. Build. Mater.* **2018**, *188*, 349–360. [CrossRef]
12. Shafiq, N.; Ayub, T.; Khan, S.U. Investigating the performance of PVA and basalt fibre reinforced beams subjected to flexural action. *Compos. Struct.* **2016**, *153*, 30–41. [CrossRef]
13. Noushini, A.; Samali, B.; Vessalas, K. Effect of polyvinyl alcohol (PVA) fibre on dynamic and material properties of fibre reinforced concrete. *Constr. Build. Mater.* **2013**, *49*, 374–383. [CrossRef]

14. Noushini, A.; Samali, B.; Vessalas, K. Ductility and Damping Characteristics of PVA-FRC Beam Elements. *Adv. Struct. Eng.* **2016**, *18*, 1763–1787. [CrossRef]
15. Nam, J.; Kim, G.; Lee, B.; Hasegawa, R.; Hama, Y. Frost resistance of polyvinyl alcohol fiber and polypropylene fiber reinforced cementitious composites under freeze thaw cycling. *Compos. Part B Eng.* **2016**, *90*, 241–250. [CrossRef]
16. Feng, J.; Su, Y.; Qian, C. Coupled effect of PP fiber, PVA fiber and bacteria on self-healing efficiency of early-age cracks in concrete. *Constr. Build. Mater.* **2019**, *228*, 116–118. [CrossRef]
17. Jang, J.G.; Kim, H.K.; Kim, T.S.; Min, B.J.; Lee, H.K. Improved flexural fatigue resistance of PVA fiber-reinforced concrete subjected to freezing and thawing cycles. *Constr. Build. Mater.* **2014**, *59*, 129–135. [CrossRef]
18. Şahmaran, M.; Özbay, E.; Yücel, H.E.; Lachemi, M.; Li, V.C. Frost resistance and microstructure of Engineered Cementitious Composites: Influence of fly ash and micro poly-vinyl-alcohol fiber. *Cem. Concr. Compos.* **2012**, *34*, 156–165. [CrossRef]
19. Zhou, J.; Pan, J.; Leung, C.K. Mechanical behavior of fiber-reinforced engineered cementitious composites in uniaxial compression. *J. Mater. Civ. Eng.* **2014**, *27*, 04014111. [CrossRef]
20. Ayub, T.; Khan, S.U.; Ayub, A. Nalytical model for the compressive stress–strain behavior of PVA-FRC. *Constr. Build. Mater.* **2019**, *214*, 581–593. [CrossRef]
21. Zhang, X.; Zhang, W.; Luo, Y.; Wang, L.; Peng, J.; Zhang, J. Interface shear strength between self-compacting concrete and carbonated concrete. *J. Mater. Civ. Eng.* **2020**, *32*, 04020113. [CrossRef]
22. Nguyen, H.-A.; Chang, T.-P.; Shih, J.-Y. Engineering Properties and Bonding Behavior of Self-Compacting Concrete Made with No-Cement Binder. *J. Mater. Civ. Eng.* **2018**, *30*, 04017294. [CrossRef]
23. Wardeh, G.; Ghorbel, E.; Gomart, H.; Fiorio, B. Experimental and analytical study of bond behavior betweenrecycled aggregate concrete and steel bars using a pullout test. *Struct. Concr.* **2017**, *18*, 811–825. [CrossRef]
24. Alabduljabbar, H.; West, J. Bond behavior of cleaned corroded rebar repaired by partial depth repair. *Structures* **2020**, *27*, 813–823. [CrossRef]
25. Zhang, X.; He, F.; Chen, J.; Yang, C.; Xu, F. Orientation of steel fibers in concrete attracted by magnetized rebar and its effects on bond behavior. *Cem. Concr. Compos.* **2023**, *138*, 104977. [CrossRef]
26. Shen, D.; Li, C.; Feng, Z.; Wen, C.; Qjha, B. Influence of strain rate on bond behavior of concrete members reinforced with basalt fiber-reinforced polymer rebars. *Constr. Build. Mater.* **2019**, *228*, 116755. [CrossRef]
27. Hou, L.; Zhou, B.; Guo, S.; Zhuang, N.; Chen, D. Bond-slip behavior between pre-corroded rebar and steel fiber reinforced concrete. *Constr. Build. Mater.* **2018**, *182*, 637–645. [CrossRef]
28. Alyousef, R.; Topper, T.; Al-Mayah, A. Effect of FRP Wrapping on Fatigue Bond Behavior of Spliced Concrete Beams. *J. Compos. Constr.* **2016**, *20*, 04015030. [CrossRef]
29. Wu, K.; Chen, F.; Chen, C.; Zheng, H.; Xu, J. Load-Transfer Mechanism and Bond-Stress Components in Steel and Steel Fiber–Reinforced Concrete Structure. *J. Struct. Eng.* **2019**, *145*, 04019160. [CrossRef]
30. Hou, L.; Ye, Z.; Zhou, B.; Shen, C.; Aslani, F.; Chen, D. Bond behavior of reinforcement embedded in steel fiber reinforced concrete under chloride attack. *Struct. Concr.* **2019**, *20*, 2242–2255. [CrossRef]
31. Akbarzadeh Bengar, H.; Ahmadi Zarrinkolaei, F. Effect of steel fibers and concrete cover on bond behavior between steel deformed bar and concrete under high temperature. *Structures* **2021**, *32*, 1507–1521. [CrossRef]
32. Aslani, F.; Nejadi, S. Bond characteristics of steel fiber and deformed reinforcing steel bar embedded in steel fiber reinforced self-compacting concrete (SFRSCC). *Cent. Eur. J. Eng.* **2012**, *2*, 445–470. [CrossRef]
33. Dancygier, A.N.; Katz, A.; Wexler, U. Bond between deformed reinforcement and normal and high-strength concrete with and without fibers. *Mater. Struct.* **2010**, *43*, 839–856. [CrossRef]
34. Deshpande, A.A.; Kumar, D.; Ranade, R. Temperature effects on the bond behavior between deformed steel reinforcing bars and hybrid fiber-reinforced strain-hardening cementitious composite. *Constr. Build. Mater.* **2020**, *233*, 117337. [CrossRef]
35. Ganesan, N.; Indira, P.V.; Sabeena, M.V. Bond stress slip response of bars embedded in hybrid fibre reinforced high performance concrete. *Constr. Build. Mater.* **2014**, *50*, 108–115. [CrossRef]
36. Tabatabaeian, M.; Khaloo, A.; Joshaghani, A. Experimental investigation on effects of hybrid fibers on rheological, mechanical, and durability properties of high-strength SCC. *Constr. Build. Mater.* **2017**, *147*, 497–509. [CrossRef]
37. Dashti, J.; Nematzadeh, M. Compressive and direct tensile behavior of concrete containing Forta-Ferro fiber and calcium aluminate cement subjected to sulfuric acid attack with optimized design. *Constr. Build. Mater.* **2020**, *253*, 118999. [CrossRef]
38. Zhao, W.; Xiao, J. On bond-slip constitutive model between ribbed steel bars and concrete, 04. *Eng. Mech.* **2011**, *28*, 164–171. (In Chinese)
39. Lin, H.; Zhao, Y.; Ozbolt, J.; Feng, P.; Jiang, C. Analytical model for the bond stress-slip relationship of deformed bars in normal strength concrete. *Constr. Build. Mater.* **2019**, *198*, 570–586. [CrossRef]
40. Alberti, M.G.; Enfedaque, A.; Gálvez, J.C. The effect of fibres in the rheology of self-compacting concrete. *Construction and Building Materials* **2019**, *219*, 144–153. [CrossRef]
41. Huang, L.; Xu, L.; Chi, Y.; Deng, F.; Zhang, A. Bond strength of deformed bar embedded in steel-polypropylene hybrid fiber reinforced concrete. *Constr. Build. Mater.* **2019**, *218*, 176–192. [CrossRef]

42. Hung, C.-C.; Lee, H.-S.; Chan, S.N. Tension-stiffening effect in steel-reinforced UHPC composites: Constitutive model and effects of steel fibers, loading patterns, and rebar sizes. *Compos. Part B Eng.* **2019**, *158*, 269–278. [CrossRef]
43. Yoo, D.-Y.; Kwon, K.-Y.; Park, J.-J.; Yoon, Y.-S. Local bond-slip response of GFRP rebar in ultra-high-performance fiber-reinforced concrete. *Compos. Struct.* **2015**, *120*, 53–64. [CrossRef]
44. Paswan, R.; Rahman, M.R.; Singh, S.K.; Singh, B. Bond Behavior of Reinforcing Steel Bar and Geopolymer Concrete. *J. Mater. Civ. Eng.* **2020**, *32*, 04020167. [CrossRef]

coatings

Article

Principle and Method for Determining the Calendar Safety Life of Aircraft Structural Protection Systems

Teng Zhang [1], Tianyu Zhang [1,*], Yuting He [1], Bo Hou [2] and Changfan Li [2]

1 Aeronautical Engineering College, Air Force Engineering University, Xi'an 710038, China; zt_gm@126.com (T.Z.)
2 Research Institute of Army Aviation, Beijing 101100, China
* Correspondence: zz664191970@163.com

Abstract: The calendar safety life of the surface protection system in aircraft structures is the time limit for it to be used without functional failure at a certain level of reliability and confidence. The reliability of such protection systems and the operational safety and economy of the structure are closely related. This paper firstly establishes two methods for determining the calendar safety life of aircraft structural protection systems under a single service environment and in multiple service environments. A method for determining the reliability of the calendar safety life of the aircraft structural protection system was proposed, and an expression of the relationship between the maintenance costs for the aircraft fleet and the reliability of the calendar safety life of the aircraft structural protection system based on the relationship between the amount of corrosion damage to the structural substrate and the corrosion time and the expression of the calendar safety life of the protection system was established. Finally, taking a hypothetical aircraft fuselage wall plate connection structure as an example, an alternating corrosion fatigue test with protection system specimens was carried out. The process for determining the calendar safety life of the structural protection system and its reliability are given. This method is important to ensure the safety of aircraft structures, improve the efficiency of use, and reduce maintenance costs. Generally speaking, the reliability of the calendar safety life of the structure is 99.9%, and after the analysis in this paper, the reliability of the structural protection system is about 70%.

Keywords: principal; method; protection system; calendar safety life; aircraft structure

1. Introduction

Protective systems, such as coatings applied on the surface of aircraft structures, can effectively protect the structure from corrosion and ensure the safe flight of the aircraft [1]. Surface coating systems play a very important protective role for aircraft performing cruise missions in severe weather areas (e.g., high temperature, high humidity, high salt, high sand, strong UV, etc.) [2,3]. The complete protective system for aircraft structures consists of an anodic oxide film, primer, intermediate and topcoat [2]. However, protective coatings are used in many cases as single or separate coating systems, which are related to the specific service environment and operating conditions. Moreover, the correct combination for the pre-treatment method of the substrate and the protective coating system determines the corrosion resistance of the whole protection system. Aircraft that are often in service in coastal areas, island areas or industrial areas are subject to harsh environmental corrosion [4–7]. Moreover, because the aircraft is a special closed structure, water often accumulates inside it and causes corrosion, such as in the seam area between the fuselage and the wing skin, the connection area between the bolts or rivets and the plates, the landing gear of the aircraft, the area where the battery is placed, the recess area between the flaps and the hinges where water easily accumulates, the area affected by the engine exhaust, and other places where water easily accumulates [8]. The main substrate materials studied are metallic materials, such as aluminum alloys, and the main coating materials studied are organic coatings, such as epoxy primers. Because the main object in this paper is the aircraft structure, most of the fuselage surfaces, landing gear, and various connectors in the aircraft are aluminum alloy materials; moreover, the aluminum alloy materials used in current civil aircraft structures in China account for about four-fifths of all materials, so aluminum-based materials are used in this paper.

The calendar safety life of the surface protection system in an aircraft structure is the time limit for its use without functional failure at a certain level of reliability and confidence [1]. Our research is focused on the force-bearing components of aircraft structures, and the main object of our research is the metallic structures and coating systems for aircraft surfaces. For critical structures affecting flight safety, and for structures with particularly poor corrosion resistance regarding the substrate materials (e.g., landing gear), the calendar safety life of protection systems should be selected at a high level of reliability and confidence based on the safety of their use [2]. For other aircraft structures, since it still takes a certain amount of time for the corrosion failure of the structural substrate to occur after the failure of the protection system, structural safety, overhaul time, and repair costs, among other factors, should be considered comprehensively [3]. Repair costs and other factors should be considered from an economical point of view to determine the reliability and confidence of the protection system [4]. Therefore, the calendar safety life of the protection system for these structures is mainly related to the use of economics [5].

There are many types of surface protection systems for aircraft structures, such as topcoats, primers, corrosion inhibitors, anodic oxidation layers, galvanized layers, and aluminum cladding layers [6]. The various types of protection systems can be divided into many different grades [7]. When the surface protection system for the structure is damaged after the aircraft has been in service for a period, some types of protection systems can be repaired during a structural overhaul, such as the topcoat and primer, while others cannot, such as the anodic oxide layer and the aluminum cladding layer [8]. If the surface protection system has corroded to the inner layer before the first overhaul (first flip), causing irreparable damage to the protection layer, the state of the protection system after the first flip is worse than when the aircraft leaves the factory, and its calendar safety life is shorter than the calendar safety life of the protection system before the first flip [9].

Therefore, the calendar safety life of the protection system is divided into two categories: the calendar safety life of the structural protection system before the first overturn, and that after the first overturn. The former is the calendar life limit with a specific probability of corrosion failure, corresponding to the state of the structural protection system after the aircraft leaves the factory. This is mainly related to the determination of the first overturn period of the structure. The calendar safety life of the structural protection system after the first overturn is the calendar life limit with a specific probability of corrosion failure corresponding to the state of the structural protection system after the aircraft is overhauled. The calendar life limit for the probability of failure of a structural protection system after the first overhaul is mainly related to the determination of the structural overhaul interval. It should be noted here that the difference between the calendar safety life for the protection system before the first flip and after the first flip is whether the inner protection system is damaged. If the inner protection system is intact at the time of the first flip, the protection system can be repaired to factory conditions at the time of the first flip. Then, the calendar safety life of the structure after the first flip is still managed based on the calendar safety life before the first flip.

This paper presents a principle and method for determining the calendar safety life of an aircraft structural protection system, and proposes a method for determining its reliability and, finally, validates and analyzes this method by means of an example. Failure of a protective coating on an aircraft structure does not necessarily mean that the structure itself has failed. The structure will be selected with a certain reliability in the life determination, and current common practice is to select a reliability level of 50%. This means that the maintenance cost for the aircraft is too high. The research in this paper found that the reliability of the aircraft structural protection system can meet the requirements at about 70%, which means that this can greatly reduce the maintenance costs and be more economical, while meeting the safety requirements.

2. Principles and Methods for Determining the Calendar Safety Life

2.1. Single Service Environment

The calendar safety life of an aircraft structural protection system is based on the number of years it has been in service [1]. If an aircraft is in a similar environment concerning the service area during its lifetime, it can be considered that the aircraft is in service in a single environment [2–4]. As such, an annual corrosion spectrum for the service area, where the aircraft is located, is the only thing needed to prepare [3,4]. Then, the calendar safety life of the aircraft structural protection system can be obtained through testing and analysis. If the aircraft is in service in different areas with large differences in the environment during its lifetime, it is considered that the aircraft is in service in multiple environments. In this case, an annual corrosion spectrum for the different areas for testing and analysis to obtain the calendar safety life of the structural protection system is needed to prepare [6,8].

The calendar safety life of a surface protection system is related to the materials and processes selected for the protection system, as well as the environment and service history of the aircraft, the location of the structure, and the reliability and confidence level selected according to the service requirements [1,3–5]. The first three parameters determine the real failure time for the protection system. The selected reliability and confidence levels determine the safety level for the calendar safety life of the protection system [6–8]. The reliability and confidence levels for the calendar safety life of the protection system should be determined according to the importance of the structure, maintenance costs, and other comprehensive considerations [8,9].

For the critical load-bearing parts in an aircraft structure or when the structural base material corrosion resistance is particularly poor, a higher reliability level should be selected to ensure that the probability of the protection system failing before the structural overhaul is very low [3,10]. This ensures that the pre-overhaul base structure does not fail due to corrosion [10]. For general structures, the reliability level should be determined mainly from the perspective of the economy of the structural use and maintenance [7]. The level should not be too high, such that the structure is frequently repaired, affecting the equipment integrity rate and increasing the maintenance costs. Nor should it be too low, such that the substrate corrosion becomes overly serious, leading to major repairs to the structure or even replacement parts, which also increases the maintenance costs. In addition, the selection of the calendar safety life reliability for the structural protection system must ensure that the structural matrix for this reliability corresponds to the calendar cycle without fracture failures due to the effects of corrosion [7–9,11].

Since determining the calendar life of a structure using the actual service conditions requires a long test cycle, corrosion problems are generally investigated in engineering by conducting equivalent tests under accelerated laboratory conditions [12,13]. The technical approach to determine the calendar safety life of a surface protection system is shown in Figure 1. Several major tasks involved in this process are described as follows [4].

(1) Designing and manufacturing simulated specimens for structural protection systems

The object of the study of the calendar safety life of a protection system should be selected according to the purpose of the study. For example, to determine the calendar safety life of the aircraft fuselage, the main components most vulnerable to corrosion are usually selected as the object of the study [14].

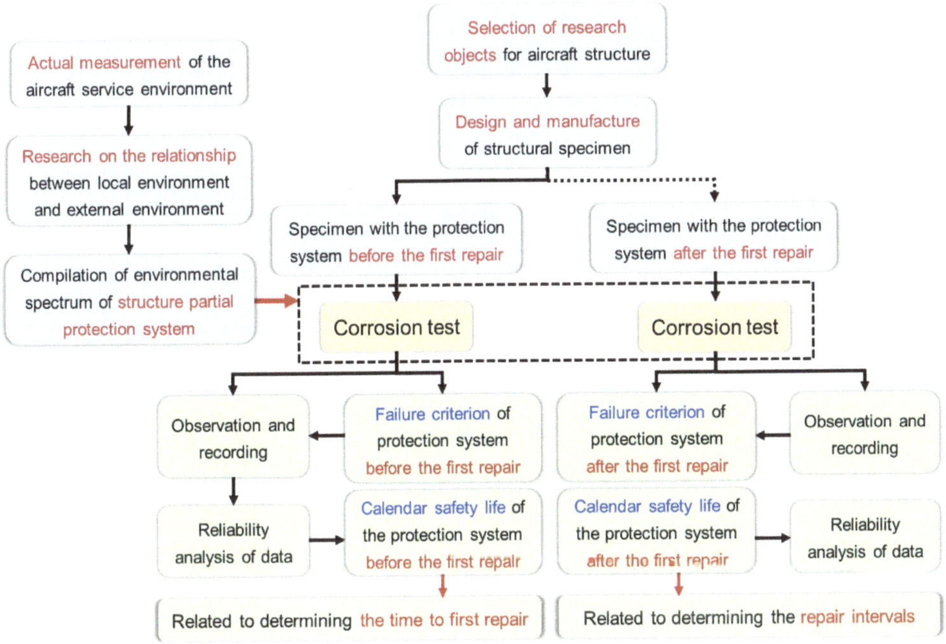

Figure 1. Technical ways to determine the safety life of protective systems.

The design of the protection system simulation specimens follows three principles [1–4]: (1) The specimens should use the same surface protection system materials as the actual structure of the aircraft, including the same base material, the same type of protection system and material grades, the same structure processing technology, and the same thickness of the protection layer. (2). The specimens should have the same local characteristics as the actual structure, such as the study object and the other structures connected through rivets. The specimens should be considered for the design of the contact parts, such as the actual structure of the existence of tabs, round holes, and other features affecting the corrosion patterns of the protection system. (3). The production of a fleet of aircraft requires a long time, so the actual fleet of protection system materials used may not be batch produced. For specimen processing, different batches of the same grade of protection system materials should be chosen where possible to reflect the dispersion of the material.

When producing mockups, the first mockup of the pre-rollover structure should be produced based on the state of the aircraft structure at the factory [15]. When determining the failure criterion for the protection system before the first overturning of the structure (i.e., the possible state of the surface protection system at the first overturning) based on this simulated part, the post-first overturning structural simulated part can be produced based on the criterion [16]. For example, if corrosion of the structural cladding occurs as the failure criterion for the structural mockup before the first overturning, the actual structure of the aircraft managed according to this calendar safety life will have some areas where cladding may fail during the overhaul. Since the cladding and the anodized layer outside the cladding are not repaired during the overhaul of the aircraft structure, generally only the primer and topcoat are repainted [17]. Therefore, in order to more reasonably simulate the state of the post-first overturn protection system, it is necessary to produce a mockup of the surface protection system. To more reasonably simulate the state of the post-first flip protection system and from the perspective of partial safety, the designed post-first flip structure's simulation parts should be sprayed with primer and topcoat only, without aluminum cladding and anodizing treatment.

(2) Preparation of the environmental spectrum for local structural protection systems

To conduct accelerated test research under laboratory conditions, it is necessary to ensure that the accelerated environmental spectrum used in the test is equivalent to the corrosion damage effect of the real service environment for the aircraft structure. This requires the preparation of a local accelerated environmental spectrum for the part where the structure is located.

First, the actual service environment of the aircraft is considered in order to carry out real measurements to determine the external environmental characteristics of the aircraft. Second, the local environment of the structure to be studied and the relationship between the external environment model are established. Again, based on the determined local environmental characteristics of the aircraft structure, the structural accelerated environment spectrum is prepared using existing methods [6–8].

(3) Judging the first flip before and the first flip after the failure of the protection system

The number of load cycles corresponding to the fracture of the structure during the fatigue test is the fatigue life. However, the study of the calendar life differs from this. The criteria for determining the calendar life and the life of the protection system are often not visually expressed. Therefore, to evaluate the calendar safety life of the protection system, it is necessary to give the criteria for the calendar life of the protection system. The failure of the protection system will cause corrosion to the structural substrate, so the failure of the protection system before the first overturning can be obtained by observing whether corrosion damage occurs to the structural substrate. Similarly, the failure of the protection system after the first overturning can be obtained by observing whether irreparable damage occurs to the inner protection system.

To obtain the specific process for the failure of the protection system, different cycles of corrosion tests to record the damage characteristics of the protective layer and the specific location of the damage must be carried out first. Then, through microscopic observation, energy spectrum analysis is conducted, and other means are used to determine whether the structural substrate (or the irreparable protective layer) is corroded. The specific location of the corrosion is, thus, recorded. Finally, through the location of the damage to the protective layer and the substrate (or the irreparable protective layer), the corrosion position of the corresponding relationship is used to find the corrosion of the substrate based on the corresponding damage characteristics of the protective layer. This is conducted to obtain the criteria for the pre-flip (or post-flip) failure of the protective system.

(4) Carry out corrosion tests on the simulated parts of different protection systems

In the preparation of the accelerated environmental spectrum to carry out the first turn over before and after the first turn over, the protection system simulation of the corrosion test takes place. In the test process, the equivalent corrosion spectrum is applied after a cycle (e.g., equivalent to a year) to a specimen for surface observation and recording. When the damage characteristics of the specimen of the surface protection system corresponds to the failure criterion, the specimen of the surface protection system reaches the calendar life limit in equivalent corrosion years.

From the perspective of saving costs, the corrosion test for the first post-turn over structural simulator can be omitted. As the protection system for the pre-flip structure mockup is better than that of the post-flip structure mockup, the failure criterion of the post-flip structure mockup usually appears before the failure criterion of the pre-flip structure mockup. (For example, damage to the paint layer often appears before damage to the aluminum cladding layer.) Therefore, only the corrosion test for the structural simulator before the first flip can be carried out. Each specimen is recorded twice when the above two judgments appear, and the calendar time corresponding to the failure judgment for the structural simulator after the first flip is used as the calendar life of the structural protection system after the first flip.

(5) Reliability analysis of the test data and the determination for the calendar safety life of the protection system

From the perspective of saving test funds, test parts are usually only made into simulated parts that reflect the characteristics of the actual structure, rather than the actual structural parts. Therefore, the surface area of the structure of the specimen is generally only a fraction of the surface area of the actual structure. Assuming the actual structure in the key parts of the surface area for the specimen of the key parts of the surface area is k times, the actual structure in the service environment can be equivalently seen as k specimens, while accepting corrosion. Any one of the surface protection system failures is to be regarded as the entire actual structure of the surface protection system failure. Therefore, the reliability of the specimen selected, and the reliability of the actual structure are different. When studying the calendar safety life of the surface protection system of an actual structure, it is necessary to take into account the difference in surface area between the specimen and the actual structural piece in critical areas.

The calendar safety life reliability for the surface protection system of an actual structure is chosen to be α. Then, for a sample of the actual structure with sample capacity n, there are approximately $n(1-\alpha)$ structures, with the surface protection system failing when the calendar safety life of the protection system is reached. From a the point of view of safety, for a sample of a specimen with sample capacity nk (the surface area of the actual structure is k times the surface area of the specimen; then, a sample of the specimen with sample capacity nk is equivalent to a sample of the actual structure with sample capacity n) it is still required that $n(1-\alpha)$ specimens of the surface protection system of the specimen fail when the calendar safety life of the protection system is reached. Assuming that the calendar safety life reliability should be selected as α' for the calendar safety life analysis of the surface protection system of the actual structure with the test data, the relation $nk(1-\alpha') = n(1-\alpha)$ can be obtained [1] as follows:

$$\alpha' = (k + \alpha - 1)/k \tag{1}$$

The analysis results obtained from the test data according to reliability α' can be used as the calendar safety life of the protection system for the actual structure at reliability α. The above issues are further elaborated below with the help of Figure 2.

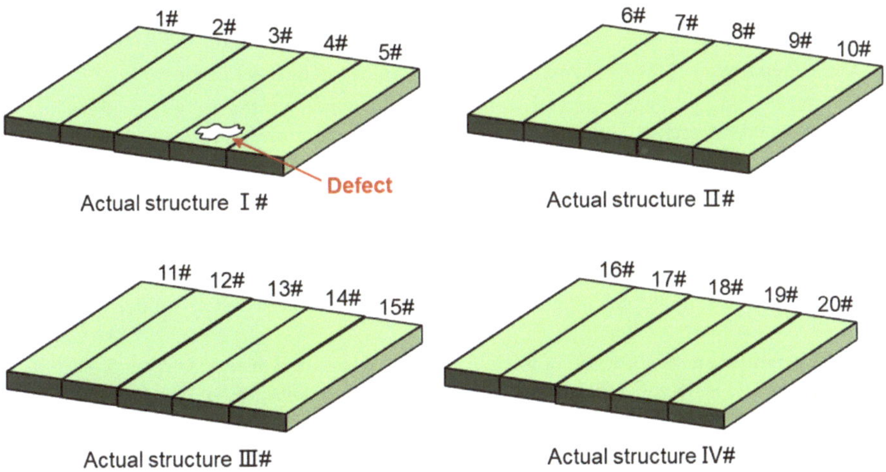

Figure 2. Schematic diagram of the proportional relationship between the specimen and the actual structure.

Figure 2 features four pieces of actual structure, with serial numbers from I# to IV#. Assuming that the surface area of the actual structure is five times that of the specimens, the sum of the surface area of the four pieces of the actual structure is equivalent to the surface area of 20 specimens. The serial numbers for the specimens are 1# to 20#. Then, it is assumed that the calendar safety life of the actual structure surface protection system is 10 years, and the reliability of the calendar safety life is 75%. This means that when four structures are in service/use for 10 years, about 4 × (1%–75%) = 1 structure surface protection system failure occurs, such as the defect in the I# structure. From a safety perspective, for a sample capacity of 4 × 5 = 20 specimen samples, when the calendar safety life of the protection system is 10 years, at most one specimen of the surface protection system is still required to fail, such as the defect in the 4# specimen. Then, the specimen surface protection system calendar safety life reliability is $(20 - 1)/20 = 95\%$. It can be seen that, because the surface area of the specimen is smaller than the actual structure, the reliability of the specimen surface protection system calendar safety life is obviously greater than the actual structure. The analysis results obtained from the test data, based on the above reliability, can be used as the calendar safety life of the protection system for the actual structure at a reliability of α. With the obtained test results and the given reliability, the calendar safety life of the structural protection system should be determined by first determining the distribution of the calendar safety life of the protection system. Then, the calendar safety life value that satisfies the given reliability and confidence levels according to the probability density function of the determined distribution were obtained. Since there is a general method for determining the form of the distribution and calculating the reliability value, it is not repeated here.

2.2. *Multiple Service Environments*

Aircraft are usually in service in multiple regions during their full life service [9]. Civil aircraft usually switch between multiple cities during the routes flown. Military aircraft also often switch between regions [10]. Due to the differences in the corrosive environments in different regions, the calendar safety life of the structural protection system determined by the equivalent environmental spectrum in one region alone cannot be used to manage the calendar life of the protection system for aircraft in service in multiple regions [11].

Existing studies have shown that the damage effect of corrosion on materials can be considered to obey a linear law [10–12]. On this basis, the determination for the calendar safety life of aircraft structural protection systems in a multiservice environment can be carried out by calculating the linear cumulative damage, as follows:

(1) Preparation of the local environmental spectrum for different service areas

According to the climatic environment parameters for different service areas and the selected research objects, the local environmental spectrum for the protection system corresponding to the aircraft in service in different areas is prepared [13]. For example, the aircraft will be in service in four areas, A, B, C, and D. Among them, the service environments in A and B are similar, so a total of three environmental spectra need to be compiled, as follows: (1) the environmental spectrum corresponding to A and B, (2) the environmental spectrum corresponding to C, and (3) the environmental spectrum corresponding to D.

(2) Determining the calendar safety life of the protection system under different service environments

According to the process shown in Figure 1, the tests and analyses on the protection system under different environmental spectra are carried out separately to determine the calendar safety life of the protection system under different environmental spectra [14,15]. For example, through testing and analysis, the calendar safety life of the protection system before and after the first overturning under the environmental spectrum: (1) is 10 years and 9 years, respectively; the calendar safety life before and after the first overturning under environmental spectrum (2) is 5 years and 4 years, respectively; and the calendar safety life before and after the first overturning under environmental spectrum (3) is 7 years and 6 years, respectively.

(3) Calendar safety life prediction based on the usage plan

According to the calculation method for linear cumulative damage, if the calendar safety life of the protection system in a certain environmental spectrum is T years, and the aircraft is in service/use in its corresponding area for t years ($t < T$), then the degree of damage to the protection system in these t years is t/T. When the cumulative degree of damage to the protection system reaches 1, the protection system is considered to have reached the expected safe use cycle limit, and the sum of the corresponding calendar life at this time is the multi-service area protection system calendar safety life value.

According to the above example, a new aircraft is scheduled to be in service for 5 years at site A, 3 years at site B, 5 years at site C, and 7 years at site D. According to the cumulative damage calculation, the cumulative damage before the first rollover of the protection system for 5 years of service at A and 3 years of service at B is 0.8. In other words, the aircraft protection system will need to be repaired after 1 year of service, at most, in site C. After the first flip, the protection system will continue to serve in site C for 4 years, and the accumulated damage will reach 1 after the first flip. Then, it needs to be repaired for the second time. The aircraft will be transferred to service in site D after the overhaul, and the protection system will need to be repaired for the third time after 6 years of service. After the aircraft is overhauled and transferred to site D, a third repair is required after 6 years of service, and the repaired protection system can be safely used in the remaining 1 year of the service cycle.

It should be noted that the first turn over period and overhaul interval of the aircraft structure are usually determined based on the fatigue safety life of the critical load-bearing structure [16,17]. The calendar safety life of the protection system is only related to the process for developing the first turn over period and overhaul interval, rather than playing a decisive role. When the aircraft structure reaches the overhaul period and the protection system has not yet reached its service limit, the protection system is usually also repaired. By contrast, when the protection system reaches its service limit and the aircraft has not yet reached the overhaul period, a comprehensive analysis based on damage to the protection system and the calendar safety life of the structural substrate is required to decide whether to advance the structural overhaul.

3. Principles and Methods for Determining the Calendar Safety Life Reliability

The fatigue safety life at this stage of developing the service and life limits for aircraft structures does not refer to the life of the aircraft structure at the time of damage, but rather to a service/life cycle limit value for an aircraft with high reliability [1–3]. Since structural damage can be a direct threat to flight safety, its reliability level is generally high. For example, if the fatigue life of an aircraft structure obeys a lognormal distribution, it needs to satisfy a 99.9% reliability level with 90% confidence [13,16–18]. However, for the protection system in an aircraft structure, its failure does not immediately lead to structural damage, and a suitable reliability level can be found in determining its calendar safety life, which can ensure structural safety and improve the economy of use of the aircraft structure at the same time.

The calendar safety life of an aircraft structural protection system is not only related to the protection system itself and the environment in which the aircraft is used, but also to the selected reliability [16–19]. If the reliability chosen is too low, more structural protection systems may fail when the calendar safety life of the protection system is reached, resulting in the corrosion of more structural substrate materials [20–22]. As such, more parts will need to be repaired during the aircraft overhaul, and some parts may not be repaired in time, due to the failure of the protection system, resulting in the replacement of the structure [23,24]. This increases aircraft maintenance costs. If the reliability selected is too high, it will lead to a short calendar safety life of the protection system [12,25]. This will make structural overhauls more frequent, not only increasing the inspection and maintenance costs of the aircraft, but also affecting the combat readiness rate of the aircraft [3–5,26]. In addition, the selection of the calendar life reliability for the structural protection system must ensure that the structural substrate does not fail within the calendar period corresponding to this reliability [1–3,27,28].

Therefore, it is important to select a suitable reliability level for the calendar life of the aircraft structure protection system, to ensure the safety of the aircraft structure, improve the efficiency of use, and reduce maintenance costs [3,5,29].

The selection of the reliability level for the calendar life of the aircraft structural protection system is closely related to the corrosion damage law for the structural substrate [28–31]. The selected reliability level should not only ensure the safety of the structure, but also optimize the costs of aircraft maintenance.

The protection system's calendar safety life reliability is determined using the following methods [1,2].

(1) Establish the relationship between the amount of structural substrate corrosion damage and the corrosion time model

In a corrosive environment, damage to the aircraft structure occurs in a variety of ways: an unloaded structure may suffer from pitting, spalling, etc., or a loaded structure may suffer from corrosion fatigue, stress corrosion cracking, etc.

According to the type of corrosion damage to the structural substrate, accelerated corrosion tests are conducted based on a structural substrate simulator in laboratory conditions. The test process requires a specimen without a surface protection system and one that can reflect the local characteristics of the structural substrate. The test conditions (the corrosion environment and load environment) can reflect the actual service environment of the aircraft structure and have a clear equivalence relationship. The specimens are divided into several groups. Each group undergoes different corrosion cycles. The amount of structural corrosion damage after different corrosion cycles were determined and the average amount of corrosion damage to the structural substrate and corrosion time relationship model were established [1–3,5] as follows:

$$h = f(t) \tag{2}$$

In the above formula, h is the average amount of corrosion damage, that is, the reliability of 50% of the corrosion damage; t is the equivalent corrosion time; and $f(t)$ is the average amount of corrosion damage as a function of the corrosion time of the substrate.

When the amount of structural damage reaches h_C, it is considered that structural repair is no longer economical, or that the repaired structure cannot guarantee flight safety and structural replacement is required. When the amount of structural damage reaches h_B, it is considered that the structure can no longer be used, owing to, for example, the risk of a sudden fracture.

An example was provided to explain the replacement critical damage value h_C and the scrap critical damage value h_B of an aircraft structure. The structure of an aircraft is a critical part of damage tolerance. The critical crack length in the structure in service/use load is 200 mm, but when the crack length exceeds 150 mm, damage to the structure is very serious and can no longer ensure flight safety. The structure must be stopped from being used further. That is, a 150 mm crack length is the critical damage value for the structure for scrapping h_B. When the length of the structure does not exceed 50 mm, the structure can be reamed to stop holes and cracking, or reinforcements and other measures can be used for the repair. Once the crack length exceeds 50 mm, however, the repair cost of this structure exceeds the cost of the replacement parts, and repair is no longer economical. That is, there is a 50 mm crack length in the structure of the replacement at the critical damage value h_C. For h_B, when the structural damage is greater than h_C and less than h_B, the structure can continue to maintain safe service, but the repair is uneconomical.

(2) Carry out a corrosion test on the protective system specimens to establish the expression of the calendar safety life of the protective system

A corrosion test on the specimen of the protection system is carried out in an equivalent accelerated corrosion environment, according to the protection system calendar life of the failure judgment. This is conducted to derive the equivalent corrosion time when different specimens reach the failure judgment, that is, the calendar life of different specimens.

According to the test results, the distribution law on the calendar life of the protection system was determined and the expression of the calendar safety life of the protection system was derived as follows [4–6]:

$$N_\alpha = \overline{N} - k_\alpha \cdot S \tag{3}$$

In the above equation, N_α is the calendar safety life of the protection system under reliability α; $\overline{N X_T X_T}$ is the average value of the calendar life obtained from the test, which can be obtained from Equation (4); k_α is a one-sided tolerance coefficient to meet the reliability α with a given confidence level, which can be obtained by consulting the relevant standard [14], for the value under the normal distribution function, which can be approximated from Equation (5); and S is the standard deviation for the calendar life obtained from the test, which can be obtained from Equation (6) [1–3,5,6]:

$$\overline{N} = \frac{1}{n} \cdot \sum_{i=1}^{n} N_i \tag{4}$$

$$k_\alpha = \frac{\mu_\alpha + \mu_\gamma \sqrt{\frac{1}{n}\left[1 - \frac{\mu_\gamma^2}{2(n-1)}\right] + \frac{\mu_\gamma^2}{2(n-1)}}}{1 - \frac{\mu_\gamma^2}{2(n-1)}} \tag{5}$$

$$S = \sqrt{\frac{1}{n-1} \cdot \sum_{i=1}^{n} \left(N_i - \overline{N}\right)^2} \tag{6}$$

In the above equations, n is the number of data, N_i is the calendar life of the i-th specimen, $\mu\alpha$ is the standard normal bias associated with the reliability α, and $\mu\gamma$ is the standard normal bias associated with the confidence γ.

When the number of structures in the actual fleet is m, there is a confidence level of γ, whereby approximately $m(1-\alpha)$ structures have failed in the protection system when the calendar safety life $N\alpha$ is reached.

Similarly, for an arbitrary reliability α', when the calendar safety life $N\alpha'$ determined by this reliability is reached, there is a confidence level of γ that the protection system for approximately $m(1-\alpha')$ structures fail.

(3) Establish a model of the relationship between the maintenance cost and reliability, and find the reliability value that minimizes the maintenance cost of the fleet

Let the cost for the economic repair of the structure be C_0, the cost for replacing the structure be C_1, the total cost for the repair of the entire fleet of m structures be C, and other costs for the fleet repair (such as disassembly, transportation, etc.) be C_2.

The damage threshold for a structure to reach the replacement requirement is h_C, and the average corrosion damage to the structural substrate versus the corrosion time is modeled as $h = f(t)$. This means that approximately half of the structural substrates with failed protection systems require replacement after a time of $t_C = f-1(h_C)$. Assuming a reliability of α' ($\alpha' > \alpha$), among the $m(1-\alpha)$ pieces of protection system failure structures that reach a calendar safety life of $N\alpha$, there are $m(1-\alpha')$ pieces of structural substrates that experience at least t_C time in a corrosive environment, i.e., $N\alpha - N\alpha' = t_C$, where $N\alpha'$ is the calendar safety life determined by the reliability α'. At a calendar safety life of $N\alpha$, the number of structural pieces that reach the damage threshold required for replacement is $0.5 \times m(1-\alpha')$. Therefore, the following equation can be obtained [1,3,6]:

$$N_{\alpha} - N_{\alpha'} = \overline{N} - k_{\alpha} \cdot S - \left(\overline{N} - k_{\alpha'} \cdot S\right) = \left(k_{\alpha'} - k_{\alpha}\right) \cdot S = f^{-1}(h_C) \tag{7}$$

That is,

$$k_{\alpha'} = \frac{f^{-1}(h_C)}{S} + k_{\alpha} \tag{8}$$

The relationship between the reliability α and α' can be expressed by the following equation [1,3,6]:

$$\alpha' = g(\alpha) \tag{9}$$

Therefore, when a reliability of α is selected to determine the calendar safety life of the protection system, the total fleet maintenance cost to reach the calendar safety life of the protection system is [1–3]:

$$C = C_0 \cdot \{m(1-\alpha) - 0.5m[1-g(\alpha)]\} + C_1 \cdot 0.5m[1-g(\alpha)] + C_2 \tag{10}$$

The total maintenance cost for the fleet at a single overhaul is not yet a measure of the economy of overhauling the fleet over its entire life. If the selected protection system calendar safety life reliability is very high, it makes the fleet's single overhaul costs very low, but increases the number of repairs. In terms of all the overhauls in the whole life of the fleet, the sum of the cost may be very high. Therefore, the overhaul economy should be measured by the overhaul cost $C\alpha$ apportioned per year of service of the fleet [1,3]:

$$C_{\alpha} = \frac{C}{N_{\alpha}} = \frac{C}{\overline{N} - k_{\alpha} \cdot S} \tag{11}$$

In Equations (10) and (11), m, C_0, C_1, C_2, $\overline{N}$, and S are known quantities. Therefore, the overhaul cost for the fleet is a function of the reliability α of the calendar safety life of the protection system. By taking the derivative of Equation (11) to 0, the value of the reliability α that minimizes the maintenance cost of the fleet is obtained.

(4) Determining the reliability constraints

The reliability α that minimizes the maintenance cost of the fleet can be determined by the above steps, but it is not yet certain that this reliability is the calendar safety life reliability to be selected for the protection system. From the point of view of structural safety, the selected calendar safety life reliability for the protection system must also ensure that the aircraft structure has high reliability without failure fractures. That is, the reliability of the calendar safety life of the protection system is limited by the safety of the structure.

The damage threshold for a structure to reach scrapping requirements is h_B. That is, approximately half of the structural matrix for a protection system failure requires scrapping after a time of $tB = f - 1(h_B)$. Assuming a reliability of β, among the $m(1-\alpha)$ pieces in protective system failure structures that reach the calendar safety life $N\alpha$, there are $m(1-\beta)$ structural substrates that experience at least $t\beta$ time in the corrosive environment. Then, $N_{\alpha} - N_{\beta} = t\beta$, where $N\beta$ is the calendar safety life determined by the reliability β; at the calendar safety life $N\alpha$, the number of structural pieces that reach the critical damage value for the scrapping requirement is $0.5 \times m(1-\beta)$. Therefore, the following equation can be obtained [1,3,6]:

$$N_{\alpha} - N_{\beta} = \overline{N} - k_{\alpha} \cdot S - \left(\overline{N} - k_{\beta} \cdot S\right) = \left(k_{\beta} - k_{\alpha}\right) \cdot S = f^{-1}(h_B) \tag{12}$$

That is,

$$k_{\beta} = \frac{f^{-1}(h_B)}{S} + k_{\alpha} \tag{13}$$

The functional relationship between the reliability α and β is as follows [1,3,6]:

$$\beta = l(\alpha) \tag{14}$$

Then, the reliability of the structural substrate is [1,3,6]:

$$\lambda = 1 - \frac{0.5m(1-\beta)}{m} = 0.5[1 + l(\alpha)] \tag{15}$$

If the actual reliability of the structural substrate determined by the value of α meets the structural reliability requirements, the value of α can be selected as the reliability of the protection system. If not, the minimum reliability requirement of the structural substrate $\underline{\lambda}$ (e.g., 99.9%) needs to be inferred from Equation (15), to determine the minimum required value for the calendar life reliability of the protection system [1,3,6]. That is,

$$\underline{\alpha} = l^{-1}(2\underline{\lambda} - 1) \tag{16}$$

4. Example Analysis

4.1. Protection System Failure Test

A schematic diagram of the connection parts in the aircraft structure is shown in the left half of Figure 3. The structural connection form is a single-sided rivet lap type. The upper and lower wall panels are connected by three rows of semi-circular head rivets, and the number of rivets used for the connection between the two adjacent partition frames is 54 × 3. The thickness of the wall plate is 2 mm, and the material is 7B04 aluminum alloy. The rivet diameter is 5 mm, and the material is 2024-T4 aluminum alloy. The load direction for the structure is vertical in Figure 3. According to the characteristics of the hypothetical structure, the designed structural simulation specimen adopts the form of a multi-rivet lap, and the specimen is a single-sided lap riveted by semi-circular head rivets. The lap plate is attached according to the rolling direction. The protective layer on the surface of the specimen is epoxy primer. The configuration dimensions for the specimen are shown in the right half of Figure 3.

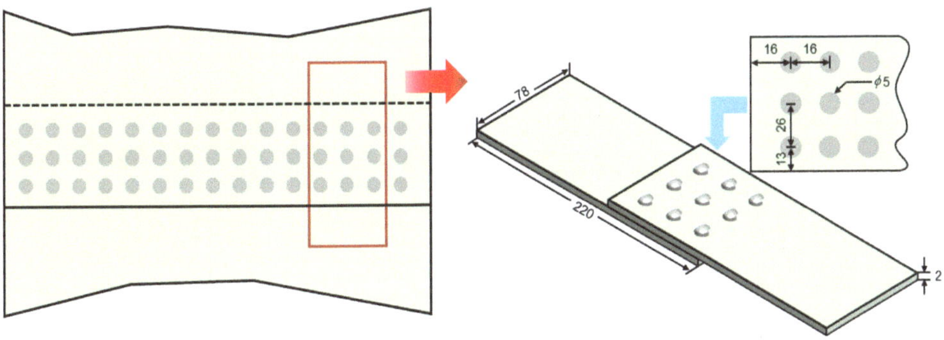

Figure 3. The specimen with the material and specific size.

The type of load spectrum used in this paper is the program block spectrum, the specific values are shown in Table 1, and the specific loading history is shown in Figure 4, with one spectrum block for every 1239 cycles [32]. The fatigue test loading equipment is an MTS-810-500 kN fatigue testing machine, the loading frequency is 15 Hz, and the test loading dynamic load error is less than 1% of the maximum load [32].

Table 1. Program block spectrum.

Number of Program Blocks	1	2	3	4	5	6	7	8	9	10	11
F_{max}/kN	13.88	18.50	22.63	22.63	24.75	22.63	22.63	20.63	26.75	16.50	15.25
F_{min}/kN	2.06	4.13	18.50	2.06	0.00	4.13	10.31	4.13	6.19	0.00	4.13
Cycles	305	28	12	31	3	10	202	95	3	123	427

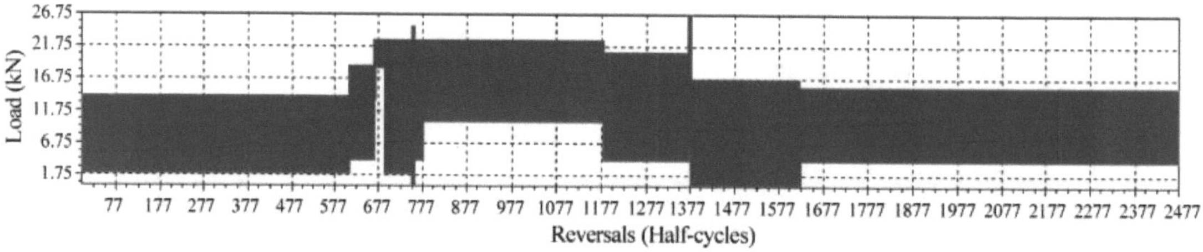

Figure 4. Fatigue test loading history.

Alternating tests on the specimens for corrosion and fatigue were carried out. Four specimens were subjected to a pure fatigue test. The corrosion fatigue alternating test was divided into five groups. The total corrosion time was 325 h, 975 h, 1300 h, 1625 h, and 1950 h, and there were four specimens in each group. The test sequence for the five groups in the corrosion fatigue alternating test is in the form of "corrosion–fatigue–corrosion–fatigue", that is, a twice alternating corrosion fatigue test. The corrosion time per corrosion and the number of cycles in the first fatigue loading are shown in Table 2. The specimen is loaded directly onto the fracture during the second fatigue load. Table 3 shows the failure time of the protection system for 16 specimens. Figure 5 shows the microscopic morphology for the failed part of the coating. As can be seen from the figure, the coating failure has produced many cracks and pits on the surface, allowing corrosive media to penetrate through the surface defects. From the sectional point of view, the coating has been spalled and the metal has been eroded by the corrosive medium, and severe intergranular corrosion (IGC) has occurred.

Table 2. Conditions for applying the corrosion fatigue alternating test.

Group	Corrosion Time (Hours)	Number of First Fatigue Loading Cycles
1	162.5	80,000
2	487.5	60,000
3	650.0	50,000
4	812.5	50,000
5	975.0	45,000

Table 3. Failure time of the specimen protection system.

Number of Alternating Times	8	9	10	11	12	13	14
Number of specimens where the protection system failed	1	2	4	4	3	1	1

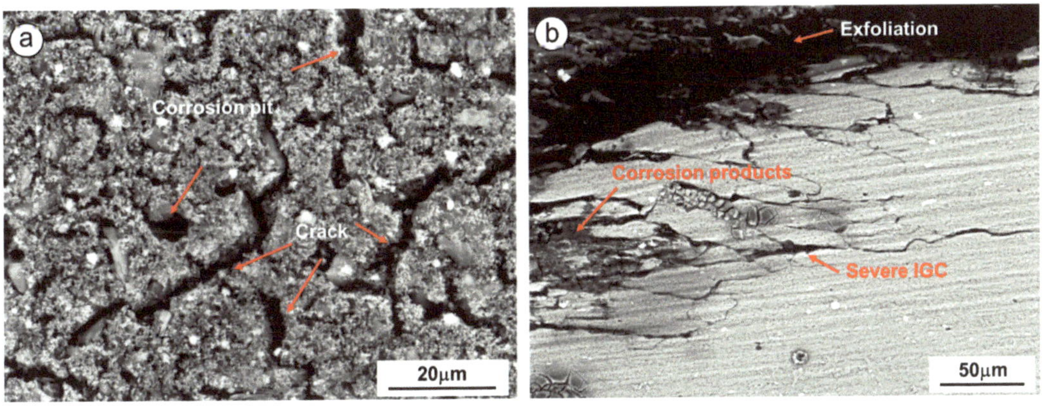

Figure 5. Microscopic morphology of the coating after failure: (**a**) surface morphology; (**b**) sectional morphology.

The failure of the protection system in the aircraft structure is the failure of the protection system near the rivet. The degradation of the fatigue life of the structure is such that the corrosion medium penetrates into the hole edge area along the rivet gap. During the overhaul, the initial protection state of the structure can basically be restored by replacing the rivets, repairing the hole edge, and repainting the protective layer, etc. As such, the test results can be used to analyze the calendar safety life before the first turn and the calendar safety life after the first turn of the protection system at the same time. That is, the calendar safety life before the first turn of the protection system and the calendar safety life after the first turn are equal. The pure fatigue test results for the specimen are shown in Figure 6. The results of the corrosion fatigue alternation test are shown in Figure 7.

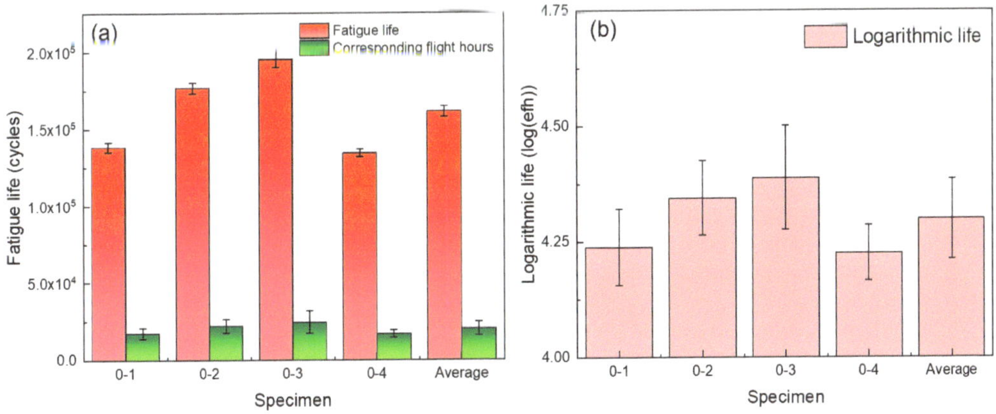

Figure 6. Results of the fatigue test: (**a**) fatigue life and corresponding flight hours; (**b**) logarithmic life.

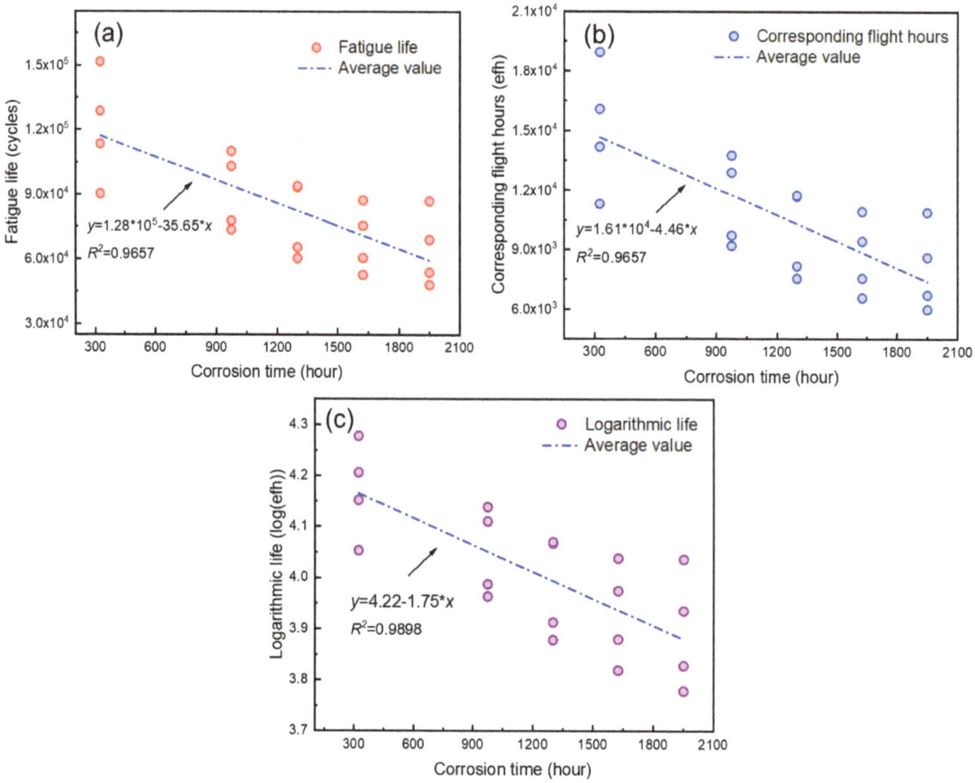

Figure 7. Results of the corrosion fatigue alternation test: (**a**) fatigue life; (**b**) corresponding flight hours; (**c**) logarithmic life.

4.2. Determining the Reliability of the Safety Life of the Protection System Calendar

The method proposed above is used to determine the reliability of the calendar safety life of the protection system in the research object. The reliability of the calendar safety life reflects the degree of safety of the protection system itself and has nothing to do with where the aircraft is in service. Thus, in order to simplify the calculation and unify the time measurement scale for an accelerated corrosion spectrum and an aging coating spectrum, the equivalent number of corrosion years of structural service in A is used as the time unit.

(1) Establish a model of the relationship between the amount of corrosion damage with the structural matrix and the corrosion time

The amount of matrix corrosion damage can be directly expressed by parameters, such as the corrosion depth, corrosion weight loss, resistance change, etc., or it can be indirectly reflected by the degradation in the life of the structure. That is, it is considered that the degradation in the life of the structure after a certain period of corrosion reaches the critical value for corrosion damage.

Therefore, the corrosion influence coefficient curve can be used as the relationship between the amount of corrosion damage and the corrosion time in the structural matrix [12,15]. The general logarithmic form of the corrosion effect coefficient is as follows [11]:

$$\lg[1-h] = b\lg T + \lg a \tag{17}$$

where h is the average corrosion damage in the matrix indirectly reflected by the corrosion influence coefficient, T is the equivalent corrosion years, and a and b are the fitting parameters.

According to the test corrosion time, as well as the test results in Tables 2 and 3, the data required for the corrosion effect coefficient curve fitting are listed in Table 4. Here, h_i is the ratio of the average structure life with accelerated corrosion for t hours in the laboratory to the average life of the pure fatigue test, that is, the corrosion influence coefficient of 50% of the data reliability, $h_i = N_{50}(T)/N_{50}(0)$.

Table 4. Data used for the corrosion impact factor curve fitting.

Equivalent Corrosion Years T of a-Site (a)	0	10	30	40	50	60
Mean life $N_{50}(t)$ (efh)	20,147.2	15,136.0	11,394.6	9794.3	8636.3	8064.4
Corrosion impact factor h_i	1.0000	0.7513	0.5656	0.4861	0.4287	0.4003

From Equation (17), least squares fitting is used to derive the amount of structural matrix corrosion damage and the corrosion time relationship model, as follows:

$$h = f(T) = 1 - 0.0784T^{0.5041} \tag{18}$$

The correlation coefficient is $R^2 = 0.997$. According to the requirements for the structural overhaul, when the corrosion impact factor is reduced to 0.8, the structure reaches the replacement damage threshold value h_C. That is, structural repair is no longer economical, or the repaired structure cannot guarantee the safety of flight and structural replacement is needed [10]. According to the requirements on the residual strength of the structure, when the corrosion impact factor is reduced to 0.6, the structure reaches the scrap damage threshold value h_B. That is, the structure can no longer continue to be used, owing to the risk of a sudden fracture [11].

(2) Establish the calendar safety life expression for the protection system

According to the data in Table 2, the mean value for the calendar life of the protection system is 10.81 and the standard deviation is 1.56, calculated from Equations (4) and (6), respectively. Assuming that the calendar life of the protection system follows a normal distribution, the expression for the calendar life of the protection system, according to Equations (3) and (5) is:

$$N_\alpha = 10.81 - 1.56 \times \frac{\mu_\alpha + \mu_\gamma \sqrt{\frac{1}{16}\left[1 - \frac{\mu_\gamma^2}{30}\right] + \frac{\mu_\alpha^2}{30}}}{1 - \frac{\mu_\gamma^2}{30}} \tag{19}$$

where μ_α is the standard normal bias associated with the reliability α, and μ_γ is the standard normal bias associated with confidence γ.

Assuming that the number of fuselage wall panel attachment structures in the actual fleet is 1000, there is γ confidence that at most $1000(1 - \alpha)$ structures will fail in the protection system when the calendar safety life $N\alpha$ corresponding to an arbitrary reliability α is reached.

(3) Establish the relationship between the maintenance costs and the reliability model, and find the reliability value that minimizes the maintenance costs for the fleet

Assume that repair cost for the fuselage wall plate connecting structure is C_0 = 20,000 CNY/piece, that the replacement cost is C_1 = 300,000 CNY/piece, and that the total costs for other repairs to the 1000 pieces in the fuselage wall plate connecting structure in the fleet (including transportation and disassembly costs) is C_2 = CNY 30 million.

According to Equations (10) and (11), the annual apportioned overhaul costs for the cluster when the calendar safety life of the protection system is reached are:

$$C_\alpha = \frac{19,000 - 14,000g(\alpha) - 2000\alpha}{N_\alpha} \quad (20)$$

where $g(\alpha)$ is the expression of a particular reliability value α'. According to Equations (8) and (18), the relationship between the reliability α and α' of the sought protection system calendar safety life is:

$$k_{\alpha'} = k_\alpha + 4.11 \quad (21)$$

The expression for k_α is shown in Equation (5).

Therefore, the expression for $g(\alpha)$ can be obtained according to Equation (21), which leads to the expression for C_α. Then, the expression for C_α can be derived to obtain the value of the reliability α of the protection system that minimizes the overhaul costs of the fleet. With a confidence level of 0.90, the minimum value of C_a, the annual apportioned maintenance cost of the fleet before overhaul, is 3,876,600 CNY/year, corresponding to a reliability α of 0.763.

(4) Reliability verification

According to Equations (13)–(15), the calendar safety life reliability for the protection system α = 0.763 corresponds to a structural reliability of 1. That is, from the perspective of the calendar safety life, it is basically impossible for the structure to fail before reaching overhaul. Through the above analysis, the calendar safety life reliability for the protection system in this model of the fuselage wall plate connection structure is 0.763. When the calendar safety life reliability of the protection system is taken as 0.999 according to the existing fatigue life reliability, the corresponding calendar safety life for the protection system is 4.41 years, and the annual maintenance cost *Ca* shared by the fleet is 6.8073 million CNY/year. This not only increases the cost of maintenance, but also affects the combat readiness of the aircraft.

4.3. Determining the Calendar Safety Life of the Protection System

According to the protection system aging test basic spectrum block for A, B, C, and D, four regions of equivalent corrosion years for conversion, and the data in Figure 7, the data used to determine the calendar safety life of the protection system in the four regions are listed in Table 5.

Table 5. Equivalent failure time of the protection system under four regions, A, B, C, and D.

Number of Specimens with Protection System Failure	**1**	**2**	**4**	**4**	**3**	**1**	**1**
Corrosion (aging) basic block	8	9	10	11	12	13	14
Equivalent corrosion years in A	8.00	9.00	10.00	11.00	12.00	13.00	14.00
Equivalent corrosion years in B	5.33	6.00	6.67	7.33	8.00	8.67	9.33
Equivalent corrosion years in C	4.00	4.50	5.00	5.50	6.00	6.50	7.00
Equivalent corrosion years in D	3.20	3.60	4.00	4.40	4.80	5.20	5.60

According to the data in Table 1, the average values for the calendar life of the protection system in A, B, C, and D, can be calculated from Equations (4) and (6) as 10.81, 7.21, 5.41, and 4.33 years, respectively, with standard deviations of 1.56, 1.04, 0.78, and 0.62 years, respectively. Since the surface area of the specimen at the test part (rivet lap) is 1/18 of the actual structure, the reliability of the selected calendar safety life is 76.3%. According to Equation (1), the reliability α' that should be selected for the analysis of the test data obtained from the specimen is (18 + 0.763 − 1)/18 = 98.68%.

Using the calculation from Equation (5), the reliability 98.68% and confidence 90% corresponds to the normal distribution one-sided tolerance coefficient k_α = 2.9897. According to Equation (3), the calendar safety life of the fuselage wall plate connection structure in four places, A, B, C, and D, with 76.3% reliability and 90% confidence is 6.14 years, 4.10 years, 3.08 years, and 2.48 years, respectively.

Since this paper presents a new principle and method, there are no results from other literature to compare. The study in this paper is about reliability, which can be found in the literature [7,33]. The methods used are based on the average value, which means that the reliability is chosen to be 50%. The method in this paper finally determines the reliability to be around 70%, which can achieve both the safety requirements and makes the process more economical.

5. Conclusions

In this paper, the principles and methods for determining the calendar safety life of aircraft structural protection systems were studied from the perspective of reliability.

(1) The methods and specific technical approaches to determine the calendar safety life for two types of aircraft structural protection systems in a single service environment and in multiple service environments were established. The method for determining the calendar safety life of the protection system can lay the foundation for determining the overall calendar safety life of the structure;

(2) The method for determining the reliability of the calendar safety life of the aircraft structure protection system was studied. Based on the relationship between the corrosion damage of the structural substrate and the corrosion time and the expression of the calendar safety life of the protection system, the expression for the relationship between the maintenance cost of the fleet and the reliability of the calendar safety life of the aircraft structure protection system was established. The reliability value for the lowest maintenance cost of the fleet can be obtained by taking 0 for the derivative of the formula. The reliability limit is established from the perspective of structural safety in service. The reliability of the calendar safety life of a reasonably selected protection system is important to ensure the safety of the aircraft structures, improve the efficiency of use, and reduce the maintenance costs;

(3) Taking a hypothetical aircraft fuselage wall plate connection structure as an example, an alternating corrosion fatigue test using protection system specimens was carried out. Based on the test results, an example for determining the life of a structural protection system under hypothetical service conditions and service environments was given. This included determining the calendar safety life reliability of the protection system and determining the calendar safety life of the protection system. In general, the reliability of the calendar safe life of a structure is 99.9%, and after the analysis in this paper, the reliability of the structural protection system is about 70%.

Author Contributions: Conceptualization, T.Z. (Tianyu Zhang) and T.Z. (Teng Zhang); validation, Y.H., T.Z. (Tianyu Zhang) and T.Z. (Teng Zhang); formal analysis, T.Z. (Tianyu Zhang); investigation, B.H. and C.L.; writing—original draft preparation, T.Z. (Tianyu Zhang); writing—review and editing, T.Z. (Teng Zhang) and B.H.; supervision, Y.H.; funding acquisition, Y.H. All authors have read and agreed to the published version of the manuscript.

Funding: The authors gratefully acknowledge the financial support from the National Science and Technology Major Special Funding (No.: J2019-I-0016-0015), the National Natural Science Foundation of China (52175155) and the National Natural Science Foundation of China (52005507).

Institutional Review Board Statement: Not applicable.

Informed Consent Statement: Not applicable.

Data Availability Statement: All data included in this study are available upon request by contacting the corresponding author.

Conflicts of Interest: The authors declare no conflict of interest.

References

1. He, Y.T. *Aircraft Structural Life Control Principle and Technology*; Defense Industry Press: Beijing, China, 2017.
2. Zhang, H.W.; He, Y.T.; Wu, L.M.; Shao, Q.; Du, J.Q. Experimental Determination of Aircraft Structural Material Life Envelope. *Adv. Mater. Res.* **2011**, *299–300*, 1198–1202.
3. Zhang, T.; Zhang, T.; He, Y.; Wang, Y.; Wang, C. Probabilistic model of the pre-corrosion fatigue life of epoxy-coated aluminum alloys based on the single point-group model of the maximum likelihood method. *Int. J. Fatigue* **2023**, *171*, 107580.
4. Tianyu, Z.; Teng, Z.; Yuting, H.; Xianghong, F.; Yaping, B. Probabilistic model of the fatigue life of epoxy-coated aluminum alloys considering atmospheric exposure. *Int. J. Fatigue* **2022**, *162*, 106899.
5. He, Y.; Li, C.; Zhang, T.; Liu, J.; Gao, C.; Hou, B.; Wu, L. Service fatigue life and service calendar life limits of aircraft structure: Aircraft structural life envelope. *Aeronaut. J.* **2016**, *120*, 1746–1762.
6. Zhang, T.; He, Y.; Zhang, H.W. Life Prediction Method for Aircraft Metal Structures Based on Life Envelope. *J. Nanjing Univ. Aeronaut. Astronaut.* **2014**, *46*, 6.
7. Shreepathi, S.; Guin, A.K.; Naik, S.M.; Vattipalli, M.R. Service life prediction of organic coatings: Electrochemical impedance spectroscopy vs. actual service life. *J. Coat. Technol. Res.* **2011**, *8*, 191–200.
8. Zhang, T.; Zhang, T.; He, Y.; Wang, Y.; Bi, Y. Corrosion and aging of organic aviation coatings: A review. *Chin. J. Aeronaut.* **2022**, *36*, 1–35.
9. Zhang, T.; Zhang, T.; He, Y.; Zhang, S.; Ma, B.; Gao, Z. Long-Term Atmospheric Aging and Corrosion of Epoxy Primer-Coated Aluminum Alloy in Coastal Environments. *Coatings* **2021**, *11*, 237.
10. Molent, L. Alternative methods for derivation of safe life limits for a 7050-T7451 aluminium alloy structure. *Int. J. Fatigue* **2015**, *74*, 55–64. [CrossRef]
11. Shekhter, A.; Crawford, B.R.; Loader, C.; Hu, W. The effect of pitting corrosion on the safe-life prediction of the Royal Australian Air Force P-3C Orion aircraft. *Eng. Fail. Anal.* **2015**, *55*, 193–207.
12. Santos, D.; Costa, M.R.; Santos, M.T. Performance of polyester and modified polyester coil coatings exposed in different environments with high UV radiation. *Prog. Org. Coat.* **2007**, *58*, 296–302. [CrossRef]
13. Deflorian, F.; Rossi, S.; Fedrizzi, L.; Zanella, C. Comparison of organic coating accelerated tests and natural weathering considering meteorological data. *Prog. Org. Coat.* **2007**, *59*, 244–250. [CrossRef]
14. Feng, Z.; Frankel, G.S. Evaluation of Coated Al Alloy Using the Breakpoint Frequency Method. *Electrochim. Acta* **2016**, *187*, 605–615. [CrossRef]
15. Huang, X.W.; Jiang, X.H.; Li, C.C.; Wan, Q.; Chen, Y.P.; Pan, L.; Jiang, C.G.; Jiang, Y. Research Progress of Coating Life Prediction and Reliability Evaluation. *Mater. Prot.* **2018**, *51*, 110–114,143.
16. Lan, C.; Bai, N.; Yang, H.; Liu, C.; Li, H.; Spencer, B.F., Jr. Weibull modeling of the fatigue life for steel rebar considering corrosion effects. *Int. J. Fatigue* **2018**, *111*, 134–143. [CrossRef]
17. Cui, T.; Liu, D.; Cai, J.; Zhang, X. Effect of pre-corrosion and corrosion/fatigue alternation frequency on the fatigue life of 7B04-T6 aluminum alloy. *J. Mater. Res.* **2016**, *31*, 3869–3879. [CrossRef]
18. Liu, W.T.; Li, Y.H. *Aircraft Structure Calendar Life System Evaluation Technology*; Aviation Industry Press: Beijing, China, 2004.
19. Jiang, Z.G.; Tian, D.S.; Zhou, Z.T. *Aircraft Structural Load/Environmental Spectrum*; Electronic Industry Press: Beijing, China, 2012.
20. Chen, Y.L. *Corrosion Control and Strength Assessment of Naval Aircraft Structures*; Defense Industry Press: Beijing, China, 2009.
21. Wang, B.W.; Chen, X.M.; Su, Y.L. Research progress and prospect of fatigue and structural integrity for aeronautical industry in China. *Acta Aeronaut. Astronaut. Sin.* **2021**, *42*, 6–44.
22. Chen, Q.Z.; Kang, X.H.; Liu, J.G.; Fang, Z.Q.; Zhang, L. Military aircraft corrosion protection and calendar life study. *China Surf. Eng.* **2010**, *23*, 1–6.
23. Liu, W.T.; Li, Y.H.; Chen, Q.Z.; He, X.F. Accelerated test environment spectrum study of coatings on corroded parts of aircraft structures. *J. Beijing Univ. Aeronaut. Astronaut.* **2002**, *28*, 109–112.
24. Wood, K. A natural-weathering approach to predict the color service life of fluoropolymer-based coatings with organic pigments. *Serv. Life Predict. Polym. Coat.* **2020**, *4*, 49–63.
25. Zhang, T.; Zhang, T.; He, Y.; Du, X.; Zhang, S. Aging and corrosion behavior of epoxy primer coated aluminum alloys in UVA, UVA-neutral and UVA-acidic alternating-immersion environments. *Eng. Fail. Anal.* **2021**, *130*, 105759. [CrossRef]
26. He, X.F. Research on Key Technologies for the Reliability of Aircraft Structure Life under Corrosive Conditions. Ph.D. Thesis, Beijing University of Aeronautics and Astronautics, Beijing, China, 2003.
27. Gao, Z.T. *Fatigue Reliability*; Beijing University of Aeronautics and Astronautics Press: Beijing, China, 2000.
28. Zhang, T.; He, Y.; Li, C.; Zhang, T.; Zhang, S. Effect of Alternate Corrosion and Fatigue on Fatigue Crack Growth Characterization of 2024-T4 Aluminum Alloy. *Math. Probl. Eng.* **2020**, *2020*, 7314241. [CrossRef]
29. Wang, Q.; Min, Z.; Wong, Y.D.; Shi, Z.; Huang, W. Aging degradation of anhydride-cured epoxy asphalt binder subjected to ultraviolet exposure. *Int. J. Pavement Eng.* **2023**, *24*, 2171037. [CrossRef]
30. *ISO 3534-2*; Statistics-Vocabulary and Symbols. International Organization for Standardization: Geneva, Switzerland, 2006.
31. Chen, Q.Z. Research on Aircraft Structure Calendar Life Technology System under Corrosive Environment. Ph.D. Thesis, Beijing University of Aeronautics and Astronautics, Beijing, China, 1999.

32. Teng, Z. Research on the Principle and Some Key Issues of Aircraft Structure Life Control under the Influence of Corrosive Environment. Ph.D. Thesis, Air Force Engineering University, Xi'an, China, 2015.
33. Li, J.; Wu, Y.Z.; Li, B.S.; Gao, Y.D. Estimate Method for Calendar Life of Helicopter's Coating Based on Electrochemical Impedance. *Equip. Environ. Eng.* **2017**, *14*, 79–82.

MDPI AG
Grosspeteranlage 5
4052 Basel
Switzerland
Tel.: +41 61 683 77 34

Coatings Editorial Office
E-mail: coatings@mdpi.com
www.mdpi.com/journal/coatings

www.ingramcontent.com/pod-product-compliance
Lightning Source LLC
Chambersburg PA
CBHW041408210926
51850CB00013B/3212
* 9 7 8 3 7 2 5 8 4 3 1 5 2 *